T0329125

Airworthiness: An Introduction to Aircraft Certification

Airworthiness: An Introduction to Aircraft Certification

Second edition

Filippo De Florio

AMSTERDAM • BOSTON • HEIDELBERG • LONDON
NEW YORK • OXFORD • PARIS • SAN DIEGO
SAN FRANCISCO • SINGAPORE • SYDNEY • TOKYO

Butterworth-Heinemann is an Imprint of Elsevier

Butterworth-Heinemann is an imprint of Elsevier
The Boulevard, Langford Lane, Oxford OX5 1GB, UK
30 Corporate Drive, Suite 400, Burlington, MA 01803, USA

First edition 2006
Second edition 2011

British Library Cataloguing-in-Publication Data
A catalogue record for this book is available from the British Library

Library of Congress Cataloging-in-Publication Data Control Number
A catalog record for this book is available from the Library of Congress

ISBN: 978-0-08-096802-5

For information on all Butterworth-Heinemann publications
visit our website at elsevierdirect.com

Typeset by TNQ Books and Journals
www.tnq.co.in

Printed and bound in Great Britain

11 12 13   10 9 8 7 6 5 4 3 2 1

Working together to grow
libraries in developing countries

www.elsevier.com | www.bookaid.org | www.sabre.org

ELSEVIER   BOOK AID
International   Sabre Foundation

Table of Contents

Preface

The First Edition of *Airworthiness*, published by Elsevier in 2006, was based on *Aeronavigabilità* (Airworthiness), written in Italian at the end of the 2002, and published by IBN Editore, Rome.

Here is what I wrote in 2002 in the Preface for *Aeronavigabilità*:

> *I wrote these notes in order to provide the book I wish I had when, 'in the last century', I began to be interested in aircraft certification.*
>
> *The book has an informative character; it is written to offer a panoramic view of airworthiness and it is not intended to be a 'certification manual'. I have tried to express the concepts of airworthiness from a general point of view, without going into the detail of procedures which are likely to evolve quickly with the substantial changes that are foreseen in the aviation certification authorities. Regardless of this, the basic philosophies of airworthiness are unlikely to change significantly and familiarity with the basic principles of the subject — either from the point of view of the regulating authority, or the aircraft owner or operator — will assist any engineer or other aviation professional in their work. This is a subject that depends not only on formalities and equations, but on a good deal of common sense and on the collective experience of engineers and professionals acquired over more than a century of aeronautical activity.*
>
> *I hope this book will contribute to the understanding and mastering of the regulations and procedures affecting the professional training and practical work that certification engineers have to undertake in both regulatory authorities and in aircraft engineering companies.*

AIRWORTHINESS, SECOND EDITION

The last few years have been very eventful and for this reason I have updated and significantly developed the text of the First Edition.

In the Preface of the First Edition was highlighted the establishment of the European Aviation Safety Agency (EASA) as the most important event for the European airworthiness regulation. Likewise, the approval of Light-Sport aircraft was considered a significant event in the United States, destined to have an impact on general aviation worldwide. Other key issues, such as the development of unmanned aircraft, were taken into consideration. The recent dissolution of the JAA represents a very important moment in the history of airworthiness regulations. This book however will keep considering what was **vii** the organisation that prepared Europe and the World to the current EASA, with its international relations.

Furthermore, the first suborbital flights in the USA and the publication by FAA of a set of requirements on 15th December 2006 titled 'Human Space Flight Requirements for Crew and Space Flight Participants', officially opened Space to civil traffic, just as happened with civil air transportation. This is also why I've added a new chapter to the book to explain that Airworthiness could probably evolve into *'Spaceworthiness'*.

Although JAR requirements and JAA regulations have been for the most part replaced by EASA requirements, they are still referred to in the text for the sake of continuity or where, in certain cases, the JAR requirements remain valid, awaiting to be replaced by the corresponding EASA requirements. This is also a choice of a historical nature that gives an idea of the evolution of such regulations in Europe.

This book is not a *certification manual*. When standards or official procedures are discussed, they are summarized and, in order to give an idea of their content, I quote the most noteworthy articles, often partially or referring only to the titles. This is done for practical reasons and for reference; however I am not suggesting that this could replace the good practice of reading the original texts in full.

A word of caution: there are variations between British English and American English usage for terms that describe the same things, for example, aeroplane − airplane; aerobatic − acrobatic; etc. Furthermore, JAA/EASA use a form of English spellings such as organisation, authorisation, etc. that for the FAA are spelled organization, authorization, etc. In this book these have been standardized as far as possible to the '-ize' variants throughout for consistency. Elsewhere, other differences of spelling have been standardized to the US usage. Although clearly these will differ in the actual JAA/EASA documents the basic meaning is unaffected.

NOTE. The correct denomination of the FAA regulations should be 14 CFR Part XX (Ex. 14 CFR Part 11). For the sake of practicality, and to clearly see the difference from JAA and EASA requirements, the current denomination "FAR XX (Ex. FAR 11)" is used.

Filippo De Florio

Acknowledgments

I thank Francesca De Florio, my daughter, who provided fundamental linguistic and editing support. I am also grateful to my wife Giovanna and my son Sergio for their encouragement.

A special thanks to Jonathan Simpson, Elsevier, for his invaluable contribution in assessing the content of my original manuscript of the First Edition and his advice on updating and expanding the scope and depth of its content.

I am also grateful to all the team at Elsevier for their professional assistance.

The EASA, FAA, ICAO, and JAA websites have been a fundamental source of information for the content relating to the history and organization of these institutions.

About the Author

Mr Filippo De Florio, an aeronautical engineer, was Director of the Italian RAI-ENAC Type Certification Division between 1992 and 2000. In the same period, he was a member of the JAA Certification Committee.

As a member of the JAR 22 and JAR-VLA Study Groups since the 1970s, he has contributed to the creation and development of such standards.

He performed flight activity as a sailplane and aeroplane pilot for over 25 years, and he is a member of the OSTIV Sailplane Development Panel and Honorary Member of UVS International. In June 2008, he was awarded the first UAS Pioneer Award instituted *'to honour and thank individuals for their exceptional and dedicated services to the international UAS community.'*

Mr De Florio presently lives in France with his wife Giovanna. They have two children, Sergio and Francesca.

Chapter one

Flight Safety

Safety is a concept generally ingrained in the human mind; we consider "absence of danger" as its principal definition. Safety is something related to all human activities and, therefore, every civil society is organized (or should be organized) to guarantee public safety in relation to one's own or others' activities. This is certainly a moral obligation, and it is also a practical demand because accidents, causing damage to persons and properties, have a social cost. This is also the reason why human activities that could cause damage to persons and properties are controlled by national states through regulations.

We specifically deal with safety related to aeronautical activities, starting by considering what we have defined as the main conventional **flight safety factors: man**, the **environment**, and the **machine**.

(1) Man is intended here as an active part of the flight operations; we then consider pilots, maintenance manpower, air traffic controllers, and others. Clearly, it is important to be able to rely on very skilled people to avoid errors that cause accidents or catastrophes in flight operations. It is then of paramount importance to place these people in a legislative and organized context to guarantee a suitable level of professional training, updating of techniques and procedures, and psychological and physical fitness. National states entrust special public institutions with the responsibility for such obligations.

(2) The **environment** covers all the external factors that can have an influence on the flying of an aircraft. This includes meteorological conditions, traffic situations, communications, aerodromes, and so on. It is equally important to avoid situations that could jeopardize the aircraft itself. Then, we should consider correct meteorological information, rules for the vertical and horizontal separation of the aircraft, suitable aerodromes, and so on.

(3) The **machine** does not need a definition, but it is easy to understand the importance of a good project, sound construction, and efficiency in relation to the operations to be carried out. Also, in this case, national states entrust special public bodies with the responsibility of assuring that the project, the construction, and the operating instructions comply with flight safety.

An important point regarding these safety factors is that they act in series and not in parallel. They can be seen as three links of a chain representing flight safety (Fig. 1.1).

1

Man

Environment

Machine

SAFETY

FIGURE 1.1 Flight safety represented as three links in a chain

The failure of a single link is sufficient for an accident to happen. A pilot's error can put the best aircraft in jeopardy, and the best pilot cannot compensate for a serious failure in an aircraft. Accident reports offer countless examples of this; however, accidents are often caused by a combination of factors that could involve all these safety factors. Nevertheless, the accident always begins with the failure of one of the above-mentioned links.

In this book, we deal particularly with one of these safety factors: the machine.

We discuss design rules, the people who make them, who formulates the verifications from design to construction, and who is responsible for the organization of manufacturers and operators.

We are going to deal with **airworthiness**.

Chapter two

Airworthiness

A definition of "airworthiness" could be found in the Italian RAI-ENAC Technical Regulations: "For an aircraft, or aircraft part (airworthiness), is the **possession of the necessary requirements** for flying in **safe conditions**, within **allowable limits**."

In this definition, three key elements deserve special consideration: safe conditions, possession of the necessary requirements, and allowable limits.

(1) We can take for granted the meaning of **safe conditions** relating to the normal course and satisfactory conclusion of the flight.

According to one definition, safety is the freedom from those conditions that can cause death, injury or illness, damage to/loss of equipment or property, or damage to the environment.

(2) Possession of the necessary requirements means that the aircraft, or any of its parts, is designed and built according to the studied and tested criteria to fly in safe conditions, as mentioned above.

Regulations are intended to promote safety by eliminating or mitigating conditions that can cause death, injury, or damage.

These regulations are established by the airworthiness authorities appointed by the states. These are obtained through the publication of airworthiness standards (see details in the following chapters) containing a series of design requirements: from the strength of the structures to the flight requirements (flight qualities and performance), criteria for good design practice, systems, fatigue and flutter, necessary tests, flight and maintenance manual content, and so on. These standards are different for different types of aircraft. Obviously, it is not possible to design a sailplane, a "Jumbo," or a helicopter using the same rules. An important peculiarity of these standards is their evolution as time passes. Generally, a standard does not precede aeronautical progress; it follows it and sometimes accompanies it. A "blocked" standard would prevent aeronautical progress. It follows that the rules have to continuously fit with technical aeronautical evolution. Moreover, very often accident analysis leads to additional rules that, had they been applied to the design, might have prevented the accident or at least limited its effects; this process could be regarded as "afterthoughts," but it is better to consider it as "experience." The changing of the standards (normally with the purpose of adding something new or different) makes the design compliance to the rules more and more expensive, but this is the price to pay to improve flight safety.

3

(3) Allowable limits. Aircraft are designed for operation within a certain "flight envelope," which depends mainly on speed and structural load factors. In addition, the maximum weight of the aircraft can be established differently for different types of operations. Operational conditions of the aircraft, such as day-visual flight rule, night flight, instrumental flight, in or out of icing conditions, and so on, are also established. Exceeding these conditions and limits can cause accidents. Overweight takeoff, aerobatic manoeuvres performed with aircraft designed with load factors for nonaerobatic operations, flights in icing conditions without suitable protection, and exceeding the speed limits are just a few examples of the importance of flying within the allowable limits. Pilots are made aware of these limits through the flight manual, the markings and placards displayed in the cockpit, and, of course, training.

Chapter three

The ICAO and the Civil Aviation Authorities

3.1. THE INTERNATIONAL CIVIL AVIATION ORGANIZATION

The first recorded flight by a heavier-than-air machine was by the Wright brothers on 17 December 1903 in North Carolina.

Since the earliest years of aviation, far-seeing people envisaged a new dimension of transport that would go beyond national boundaries. In 1910, the first conference on air navigation international law was hosted by France in Paris, with the attendance of 18 European states.

The First World War fostered considerable development of aeronautical techniques, also demonstrating the potential for transport of goods and people. After the war, it became increasingly evident that this advanced means of transport would require international attention.

These problems were debated at the Paris Conference of Peace in 1919, and the discussions led to the establishment of an Aeronautical Commission. To succeed in the purpose of making aviation an instrument of peace, an International Air Convention was written and ratified by 38 states. The Convention contemplated all aspects of civil aviation and also the establishment of an International Commission for Air Navigation to monitor the development of civil aviation and to propose measures for this development.

The years between the two World Wars marked a continuous development of civil aviation in both the technical and the commercial fields.

The Second World War, apart from the horrors also caused by the operations of progressively more sophisticated military aeroplanes, had a major effect on the technical development of the aeroplane, compressing a quarter of a century of normal peacetime development into 6 years.

The possibility of carrying a great number of people and a large quantity of goods over long distances became a reality. For these reasons, the Government of the United States conducted exploratory discussions with other allied nations from the early months of 1944. On the basis of these talks, invitations were sent to 55 allied and neutral states to meet in Chicago in November 1944. Of these 55 states, 52 attended the meeting. The outcome of 5 weeks of meetings was the Convention on International Civil Aviation, consisting of a preamble and 96 articles.

Airworthiness: An Introduction to Aircraft Certification.

The International Civil Aviation Organization (ICAO) officially came into existence on 4 April 1947. At the invitation of the Government of Canada, Montreal was chosen as the site for its headquarters. Presently, the Contracting States number more than 180.

The aims and objectives of the ICAO are to develop the principles and techniques of international air navigation and to foster the planning and development of international air transport so as to

(1) Ensure the safe and orderly growth of international civil aviation throughout the world.

(2) Encourage the arts of aircraft design and operation for peaceful purposes.

(3) Encourage the development of airways, airports, and air navigation facilities for international civil aviation.

(4) Meet the needs of the peoples of the world for safe, regular, efficient, and economical air transport.

(5) Prevent economic waste caused by unreasonable competition.

(6) Ensure that the rights of the Contracting States are fully respected and that every Contracting State has fair opportunity to operate international airlines.

(7) Avoid discrimination between Contracting States.

(8) Promote safety of flight in international air navigation.

(9) Promote generally the development of all aspects of international civil aeronautics.

3.1.1. The International Standards

Since the ICAO was created, a main technical task of the organization has been the achievement of standardization in the operation of a safe, regular, and efficient air service. This has resulted in high levels of reliability in many areas that collectively shape international civil aviation, particularly in relation to the aircraft, their crews, and the ground-based facilities and services.

Standardization has been achieved through the creation, adoption, and amendments of 18 Annexes to the Convention, identified as **International Standards** and **Recommended Practices**.

Standards are directives that ICAO members agree to follow. If a member has a standard different from an ICAO Standard, that member must notify the ICAO of the difference.

Recommended practices are desirable but not essential. The basic principle for deciding whether a particular issue should be a standard is an affirmative answer to the question: "Is uniform application by all Contracting States essential?"

On the basis of the Convention, the Contracting States are engaged to achieve the highest practical degree of worldwide uniformity in regulations, organizing procedures in relation to aircraft, personnel, airways, and auxiliary services, whenever this will facilitate and improve air safety, effectiveness, and regularity.

The 18 Annexes are described as follows:

- **Annex 1. Personnel Licensing**—provides information on licensing of flight crews, air traffic controllers, and aircraft maintenance personnel, including medical standards for flight crews and air traffic controllers.

- **Annex 2. Rules of the Air**—contains rules relating to visual- and instrument-aided flight.
- **Annex 3. Meteorological Service for International Air Navigation**—provides meteorological services for international air navigation and reporting of meteorological observations from aircraft.
- **Annex 4. Aeronautical Charts**—contains specifications for the aeronautical charts used in international aviation.
- **Annex 5. Units of Measurement To Be Used in Air and Ground Operations**—lists dimensional systems to be used in air and ground operations.
- **Annex 6. Operation of Aircraft**—enumerates specifications to ensure a level of safety above a prescribed minimum in similar operations throughout the world.
- **Annex 7. Aircraft Nationality and Registration Marks**—specifies requirements for registration and identification of aircraft.
- **Annex 8. Airworthiness of Aircraft**—specifies uniform procedures for certification and inspection of aircraft.
- **Annex 9. Facilitations**—provides for the standardization and simplification of border-crossing formalities.
- **Annex 10. Aeronautical Telecommunications**—Volume 1 standardizes communications equipment and systems, and Volume 2 standardizes communications procedures.
- **Annex 11. Air Traffic Services**—includes information on establishing and operating air traffic control (ATC), flight information, and alerting services.
- **Annex 12. Search and Rescue**—provides information on organization and operation of facilities and services necessary for search and rescue.
- **Annex 13. Aircraft Accident and Incident Investigation**—provides uniformity in notifying, investigating, and reporting on aircraft accidents.
- **Annex 14. Aerodromes**—contains specifications for the design and equipment of aerodromes.
- **Annex 15. Aeronautical Information Services**—includes methods for collecting and disseminating aeronautical information required for flight operations.
- **Annex 16. Environmental Protection**—Volume 1 contains specifications for aircraft noise certification, noise monitoring, and noise exposure units for land-use planning and Volume 2 contains specifications for aircraft engine emissions.
- **Annex 17. Security—Safeguarding International Civil Aviation against Acts of Unlawful Interference**—specifies methods for safeguarding international civil aviation against unlawful acts of interference.
- **Annex 18. The Safe Transport of Dangerous Goods by Air**—specifies requirements necessary to ensure that hazardous materials are safely transported in aircraft while providing a level of safety that protects the aircraft and its occupants from undue risk.

Because aeronautical technology is continuously developing, the Annexes are constantly reviewed and updated whenever necessary. The typical content of an Annex is based on the following.

(1) Standards intended as specifications when their application is considered as *necessary* for the safety and regularity of international air navigation.

(2) Recommended practices intended as specifications when their application is considered as *a recommendation* in the interest of safety, regularity, and efficiency of international air navigation.

(3) Appendices dealing with the preceding points.

(4) Definitions of the used terminology.

The Contracting States have issued norms not strictly copying the contents of the Annex, which essentially state some of the principles or objectives to attain. The norms contain the requirements used to reach the objectives. Furthermore, although the principles remain the same, the requirements are often influenced by the state-of-the-art (technical evolution, new technology, and acquired experience), and they are then likely to be improved and amended.

The applicable Joint Aviation Authorities (JAA)/Federal Aviation Administration (FAA)/European Aviation Safety Agency (EASA) airworthiness standards for the certification of aircraft to be internationally recognized are issued in accordance with the ICAO Annexes. Then, from a practical point of view, the certification process is based on these airworthiness standards rather than (directly) on the ICAO International Standards.

To remain within the scope and objectives of this book, we consider the content of four Annexes directly connected with airworthiness:

- **Annex 6. Operation of Aircraft.** This Annex contains the standards and recommendations relating to the operation of aircraft for international commercial air transport, including the regulation for the certification of the operators. It also contains the technical and operational regulations for international general aviation activities, including maintenance.

The essence of Annex 6 is that the operation of aircraft engaged in international air transport must be as standardized as possible to ensure the highest levels of safety and efficiency.

The purpose of Annex 6 is to contribute to the safety of international air transport by providing criteria for safe operating practices and to contribute to the efficiency and regularity of international air navigation by encouraging ICAO's Contracting States to facilitate the passage over their territories of commercial aircraft belonging to other countries that operate in conformity with these criteria.

To keep pace with a new and vital industry, the original provisions have been and are being constantly reviewed.

Part I. In 1948, the Council first adopted Standards and Recommended Practices for the operation of aircraft engaged in international commercial air transport. They are the basis of Part I of Annex 6.

This document specifies international Standards and Recommended Practices for aeroplanes used in international commercial air transport operation carrying passengers or freight. The Annex addresses flight operations; performance operating limitations; aeroplane instruments, equipment, and flight documents; aeroplane communication and navigation equipment; aeroplane maintenance; flight crew; flight operations officers/flight dispatchers; manuals, logs, and records; cabin crew; security; lights to be displayed in the air and on the ground during operations; contents of an operations manual; and flight time and flight duty period limitations.

Part II. A second part to Annex 6, dealing exclusively with international general aviation, became applicable in September 1969.

ICAO recognizes that international general aviation pilots and their passengers may not necessarily enjoy the same level of safety as the fare-paying passengers in commercial air transport because crews and equipments may not meet the same standard as in commercial transport aircraft. Part II, however, was designed specifically to ensure an acceptable level of safety to third parties (persons on the ground and persons in the air in other aircraft). Thus, operations involving commercial and general aviation aircraft in a common environment are required to adhere to these minimum safety standards.

Part III. Similarly, a third part to Annex 6, dealing with all international helicopter operations, became applicable in November 1986.

Part III originally addressed only helicopter flight recorders. However, an amendment completing the coverage of helicopter operations in the same comprehensive manner as aeroplane operations covered in Parts I and II was adopted for applicability in November 1990.

Hence, Part III covers international commercial in transport operations and general aviation operations in helicopters.

The human factor is an essential component for the safe and efficient conduct of aircraft operation. Annex 6 spells out the responsibilities of States in supervising their operators, particularly in respect of flight crew.

An important aspect covered in Annex 6 is the requirement for operators to establish rules limiting the flight time and flight duty shifts for flight crew members.

Critical to safe aircraft operations is the knowledge of the operating limits of each particular type of aircraft. The Annex sets out minimum performance operating limitations for aircraft currently in service.

The threat of hijacking civil aircraft has placed an additional burden on the pilot in command. The various safety measures that such acts necessitate, in addition to methods of a purely technical nature, have been studied by ICAO and made to cover as many emergency situations as possible.

- **Annex 8. Airworthiness of Aircraft.** This Annex contains the standards defining the minimum level of airworthiness for the development of the type certification requirements *as a basis for the international recognition of the certificates of airworthiness for aircraft* (according to Article 33 of the Convention) to fly to and land in the Contracting States. Each state is free to develop its own comprehensive and detailed code of airworthiness or to select, adopt, or accept a code established by another Contracting State. The level of airworthiness that must be maintained by a national code is indicated by the broad standards of Annex 8.

Part I of the Annex provides definitions.

Part II contains general airworthiness procedures applicable to all aircraft together with the standard format for the certificate of airworthiness.

Part IIIA contains the minimum airworthiness characteristics of aeroplanes over 5700 kg for which application for certification was submitted on or after 13 June 1960.

Part IIIB contains the minimum airworthiness characteristics of aeroplanes over 5700 kg for which application for certification will be submitted on or after 2 March 2004.

Part IVA contains the minimum airworthiness characteristics of helicopters for which application for certification was submitted on or after 22 March 1991.

Part IVB contains the minimum airworthiness characteristics of helicopters for which application for certification will be submitted on or after 13 December 2007.

Part V contains the minimum airworthiness characteristics of aeroplanes over 750 kg but not exceeding 5700 kg for which application for certification will be submitted on or after 13 December 2007.

The technical standards dealing with the certification of aeroplanes include requirements related to performance, flying qualities, structural design and construction, engine and propeller design and installation, systems and equipment design and installation, and operating limitations including procedures and general information to be provided in the aeroplane flight manual, crashworthiness of aircraft and cabin safety, operating environment, and human factors and security in aircraft design.

Special consideration is given to requirements for design features affecting the ability of the flight crew to maintain controlled flights. The layout of the flight crew compartment must be such as to minimize the possibility of the incorrect operation of controls due to confusion, fatigue, or interference. It should allow for a sufficiently clear, extensive, and undistorted field of vision for the safe operation of the aeroplane.

Aeroplane design features also provide for the safety, health, and well being of occupants by granting an adequate cabin environment during the foreseen flight and ground and water operating conditions, the means for rapid and safe evacuation in emergency landings and the equipment necessary for the survival of the occupants in the foreseen external environment within a reasonable time span.

Requirements for the certification of engines and accessories are designed to ensure that they function reliably under the foreseen operating conditions. Following the recent events of hijacking and terrorist acts on board of transport aircraft, special security features have been included in aircraft designs to improve the protection of the aircraft.

These include, for example, special features in aircraft systems, strengthening of the cockpit door, ceilings, and floors of the cabin crew compartment.

An annex that is not only directly linked to airworthiness but also capable of influencing the airworthiness requirements is Annex 13.

- **Annex 13. Aircraft Accident and Incident Investigation.** This Annex provides the international requirements for the investigation of aircraft accidents and incidents.[1]

The objective of the investigation of an accident or incident is its **prevention**. Subsequently, the causes of an aircraft accident or a serious incident must be identified to prevent repeated occurrences.

Under Annex 13, the particular state in which the accident or incident occurs will lead the investigation, but it may delegate all or part of the investigation to another state.

If the occurrence takes place outside the territory of any state, the State of Registry has the responsibility to conduct the investigation.

Representatives of the State of Registry, Operator, and Manufacturer are entitled to take part in the investigation.

The investigation process is aimed to the determination of the causes of the accident or incident and leads to the issue of a final report including appropriate safety recommendations to prevent similar occurrences.

The ICAO operates a computerized database known as the Accident/Incident Data Reporting system allowing the exchange of safety information in any Contracting State.

The safety recommendations are evaluated by the airworthiness authorities to issue, when deemed necessary, airworthiness directives (ADs) (for mandatory modifications, inspections, etc.), amendments of the relevant airworthiness requirements, useful information, and advisory material.

An important feature of the Annex 13 is the statement (in Chapter 3) that "*the sole objective of the investigation of an accident or incident shall be the prevention of accidents and incidents. It is not the purpose of this activity to apportion blame or liability.*"

In other words, the investigation is aimed at finding the causes but not the responsibilities of the accident or incident.

The judiciary of a state must usually carry out a judiciary inquiry to assess and punish possible penal responsibilities. Then, if the Contracting State has not

---

[1] **Accident.** Annex 13 defines accident as an occurrence associated with operation of an aircraft, which takes place between the time any person boards the aircraft with the intention of flight until such time as all persons have disembarked in which:

  a a person is fatally or seriously injured […]

  b the aircraft sustains damage or structural failure which adversely affects the structural strength, performance, or flight characteristics of the aircraft and would normally require major repair or replacement of the affected component […]

  c the aircraft is missing or is completely inaccessible.

  Incident. An occurrence, other than an accident, associated with the operation of an aircraft which affects or could affect the safety of operation.

developed provisions to avoid conflicts between the judiciary inquiry and the technical investigation, normally the first one prevails, sometimes making impossible a prompt development of the technical investigation.

Unfortunately, this is still happening despite the engagement of the Contracting States to follow the ICAO Standards.

Another important feature of the Annex 13 is (in Chapter 8) the institution of "*a mandatory incident reporting system to facilitate collection of information on actual or potential safety deficiency.*"

As a recommendation "*a State should establish a voluntary incident reporting system to facilitate the collection of information that may not be captured by a mandatory incident reporting system.*"

Of course, a **voluntary** incident reporting system should be "*not* punitive *and afford protection to the sources of the information.*"

- **Annex 16. Environmental Protection.** This Annex contains the standard applicable to the *aircraft noise* certification in relation to different noise levels proportionate to the type of aircraft (propeller-driven, jet-propelled, and helicopters). It states with accuracy the test procedures for an effective and unequivocal measurement. The standard contained in this Annex is normally used as proposed because it is directly applicable to all the technical requirements. The Annex contains the standard relating to the *aircraft engine emission* certification with reference to the toxicity of some chemical components, such as nitrogen oxides.

Annex 16 Volume I applies to aircraft noise and specifies the standards and recommended practices that apply to a wide range of aircraft.

Annex 16 Volume II applies to aircraft engine emissions applicable to specified aircraft engines.

These annexes influence the design of aircraft for reasons that are different from the typical compliance to the airworthiness standards. The aim of these annexes is not the safety of flight but the environment protection from the damage that can be caused by the aircraft operation.

Noise is the most evident environmental impact of aviation for people living in the proximity of airports. But this impact is also concerning millions of people living under the takeoff and landing paths.

The growing of aviation and the increasing of flight frequency make the problem more and more dramatic because the aircraft noise is likely to affect the quality of life of all the people concerned.

Therefore, together with operational rules to limit the damage, such as restrictions on certain category of aircraft at night, the Annex provides specific noise limitations for different types of aircraft.

Another important environmental impact of aviation is pollution, a cause of growing concern also because of the sharp increase in air traffic worldwide.

Emissions from aircraft affect climate change through greenhouse effect and depletion of the ozone layer.

The millions of civil and military flights per year have a significant negative effect on the atmosphere, already penalized by the emissions coming from industries and transport at ground level.

If we consider that the emissions at high cruising altitudes in the stratosphere have a multiple effect on the global warming than when they are released at ground level and that the traffic of aircraft is likely to double in the next 20 years, it is clear that there is a need to control such emissions.

The Annex 16, for the certification of aircraft engines, prescribes the control of emissions such as smoke, unburned hydrocarbons, carbon monoxide (CO), and nitrogen oxides.

3.2. THE CIVIL AVIATION AUTHORITIES

3.2.1. Origins

The national states of developed countries have established institutions and authorities to guarantee flight safety. In many cases, these organizations evolved from pre-existing institutions for the safety of marine and river navigation. It is of interest to point out that, historically, the mainspring for the improvement of the safety of navigation is not a social principle, but an economical choice made by insurance companies.

The word "register" was adopted by various navigational institutions and has a precise origin. In fact, it is derived from a register that a certain Edward Lloyd, owner of a tavern situated in the area of the river port of London at the end of the seventeenth century, filled with information on marine traffic gathered while talking to customers, such as ship owners and sailors. The collected information could be related to ships, traffic and, most importantly, to accidents resulting in the loss of men, goods, and ships. This was the origin of the highly esteemed newsletters, "Lloyd's News," that were first issued in 1696.

At the same time, marine insurance began to flourish and Lloyd's tavern rapidly became an important negotiation center. Lloyd was a practical man, well aware of the importance of the information he owned for the insurance business. Finally, Lloyd's, the incorporated society of underwriters in London, was born and was destined to become a world reference in the insurance field.

When Lloyd died in 1713, his heirs continued his work; "Lloyd's List," filled with lists, data, and marine news, highly appreciated in the circle of marine traffic, was first published in 1734; the List, originally handwritten, first appeared in printed form in 1760.

Meanwhile, other lists with various ship classification criteria were published by different ship owners, until all the publications were unified into the "Lloyd's Register" in 1833, the first register in the world, which acquired legal status in 1871. Other national registers were subsequently instituted in Europe.

Safety is obviously a matter of great importance for insurance companies: fewer accidents mean fewer indemnities to pay. It is also for this reason that the registers began to issue safety requirements for navigation.

Since the beginning of aviation, the operation of aircraft posed problems of an analogous nature to that of marine traffic, hence the necessity of the establishment of specific institutions, similar to the already existing institutions for marine traffic. In some cases, particular marine institutions took on the responsibilities of aviation regulations and control. Later, the growth of aviation led to the creation of autonomous registers and national authorities, dealing with aircraft and air navigation.

3.2.2. Tasks of airworthiness authorities[2]

From a general point of view, an airworthiness authority has the following tasks:

(1) **To prescribe** airworthiness requirements and procedures. In the following chapters, we deal with these prescriptions, ranging from aircraft type certification, construction, and operation to the relevant organizations.
(2) **To inform** the interested parties regarding the above-mentioned prescriptions. This is performed in different ways. The authority publishes technical regulations, technical standards, circulars, and so on, to be obtained on request or by other means. At present, much information can be found on the Internet.
(3) **To control** aeronautical material, design, manufacturing organizations, and aircraft operators. This is to ensure that all pertinent prescriptions are complied with. Control operations can be performed in different ways, with the appropriate involvement of the relevant authority.
(4) **To certify** aeronautical material and organizations. This is to declare in a legal form compliance with the applicable requirements of an aircraft or part of it, or a change to a type certificate, the capability of an organization, and so on.

3.3. THE JOINT AVIATION AUTHORITIES

The JAA was an associated body of the European Civil Aviation Conference (ECAC)[3] representing the civil aviation regulatory authorities of a number of European States who had agreed to cooperate in developing and implementing common safety regulatory standards and procedures. This cooperation was intended to provide high and consistent standards of safety and a "level playing field" for competition in Europe. Much emphasis was also placed on harmonizing the JAA regulations with those of the United States.

The JAA Membership was based on signing the "JAA Arrangements" document, originally signed by the then current Member States in Cyprus in 1990.

---

[2] These can be considered as a part of aviation authorities dealing with airworthiness.
[3] The ECAC was founded in 1955 as an intergovernmental organization. The ECAC's objective is to promote the continued development of a safe, efficient, and sustainable European air transport system. In so doing, the ECAC seeks to harmonize civil aviation policies and practices among its Member States, and promote understanding on policy matters between its Member States and other parts of the world. Close liaisons are maintained with the ICAO, EUROCONTROL (see Note 5), and the EASA.

Based on these Arrangements and related commitments, the objectives and functions of JAA may be summarized as follows.

3.3.1. Objectives

1. **Aviation Safety**. To ensure, through cooperation amongst Member States, that JAA members achieve a high, consistent level of aviation safety.
2. **Cooperation with EASA**. To cooperate with the EASA in performing its functions and tasks in accordance with an agreed program ensuring the involvement of the JAA non-EASA countries with the aim of maintaining the present unity in regulations on a pan-European dimension and the mutual acceptance/recognition of certificates/approvals and of implementing the Future of the JAA (FUJA)[4] decisions.
3. **Business Effectiveness**. To achieve a cost-effective safety system so as to contribute to an efficient civil aviation industry.
4. **Consolidation of Common Standards**. To contribute, through the uniform application of the highest possible common standards and through regular review of the existing regulatory situation, to fair and equal competition within Member States.
5. **International Cooperation**. To cooperate with other regional organizations or national authorities of States playing an important role in Civil Aviation to reach at least the JAA safety level and to foster the worldwide implementation of harmonized safety standards and requirements through the conclusion of international arrangements and through participation in technical assistance programs without affecting community competence.

3.3.2. Functions

The JAA's work began in 1970 (when it was known as the Joint Airworthiness Authorities). Originally, its objectives were only to produce common certification codes for large aeroplanes and engines. This was to meet the needs of European industries and particularly for products manufactured by international consortia (e.g., Airbus). Since 1987, its work has been extended to operations, maintenance, licensing, and certification/design standards for all classes of aircraft. With the adoption of the Regulation (EC) No. 1592/2002 by the European Parliament and the Council of the European Union (EU) and the subsequent setup of the EASA, a new regulatory framework was created in European aviation.

According to this Regulation, for EU Member States, national regulation in the airworthiness domain has been replaced by EU Regulation, and certification tasks have been transferred from National Authorities to EASA. Non-EU States maintain their responsibility in all fields.

---

[4] A working group was established in 2004 to develop a document ("roadmap") to define clear milestones for the **FUJA**.

Consequently, a "Roadmap" for the establishment of clear milestones for JAA's future was developed and adopted by the JAA Board (JAAB) and by the DGs of ECAC in August 2005 (FUJA Report) proposing a transformation from JAA to JAA T (T for "transition"), comprising a Liaison Office (LO) in Cologne (Germany) and a Training Office (TO) in Hoofddorp (the Netherlands).

In November 2005, the EU Commission began the legislative process to amend EASA Regulation (EC) 1592/2002 to extend the competences of EASA to the fields of operations and licensing.

In May 2006, minor amendments to the FUJA Report were agreed by the JAAB and the DGs of ECAC contemplating practical arrangements to take into account the revised anticipated dates for the extension of EASA competences.

Furthermore, EU Regulation 1899/2006, dated 12 December 2006 was published on 27 December 2006. This Regulation amends Council Regulation 3922/1991. The amendment contained a new Annex II dealing with commercial flight operations and is referred to as EU OPS. Following an implementation period of 18 months, EU OPS became directly applicable as of 16 July 2008.

3.3.3. JAA T Function

The JAA T existed and functioned with two offices: the LO and the TO:

- The *Liaison Office* "JAA LO" liaised between EASA and the Civil Aviation Authorities of the non-EASA JAA Member States to integrate the activities of these States with those of EASA. In addition, JAA LO ensured the general management of the rulemaking, including that in the fields of operations and licensing. The technical work was undertaken by EASA for all JAA members.
- The *Training Office* "JAA TO" provided relevant training to the aviation community to ensure that it was sufficiently familiar with the European aviation safety rules and regulations and to assist the non-EASA JAA Member States in their efforts to obtain EASA membership. As of 1 July 2009, after disbanding of JAA T, JAA-TO continued to provide training courses as a Dutch Foundation and associated body of ECAC.

3.3.4. Membership

a. Membership was open to members of the ECAC, which currently consists of 44 member countries. Membership took effect when the 1990 "Arrangements" were signed. There were 43 member countries in the JAA-T.

b. "Three-Phase" membership of the JAA.

The JAA T had a three-phase membership system. The procedure, consistent with the Arrangements, started with a familiarization visit by a "candidate" Authority to JAA T (Transition), leading to a report to the Chairman of the JAA Committee (JAAC) after a satisfactory conclusion. The Authority could then formally apply to the Chairman of the JAAB for membership, expressing its willingness to commit itself to the terms and commitments in the Arrangements.

The JAAC submitted its report to the JAAB and subject to a two-third majority positive vote, the applicant Authority could sign the Arrangements. At this stage, the Authority would become a "candidate member" and would have access to meetings, documentation, and so on, but would not have

- voting rights and
- the right or obligation to automatic recognition of the approvals issued by its own authority or those of other states.

In Phase 2, subsequent to the signing of the JAA Arrangements, JAA T would arrange a visit by a fact-finding team to the Authority. This team consisted of representatives from the JAAC and JAA T. A report was prepared and sent to the JAAC Chairman and when considered satisfactory, the JAAC recommended to the JAAB to grant full membership. At this stage, JAA's standardization team visits were arranged. This process could be very prolonged for some countries. It was felt, however, that such a process was essential to safeguard the high standards and credibility of the JAA T. The third phase was the one leading to full recognition for Member States.

The JAA T comprised 37 full Members and six candidate Members.

3.3.5. The Governing Bodies

JAA Board (JAAB) was formed by DGs (Director Generals) of the JAA Member States. It considered and reviewed the general policies and the long-term objectives of JAA. Among others it decided on the acceptance of a new member of JAA and on any changes to the Cyprus Arrangements.

JAA Committee (JAAC) was composed of one member from each authority (high-level safety expert). It was responsible for the administrative and technical implementation of the Cyprus Arrangements, especially for the adoption of JARs (Joint Aviation Requirements).

JAA Executive Board (EB) was formed by seven Members of the JAAC and one representative from EASA. It formed the management of JAAC responsibilities on a continuous basis, in between the regular meetings of the JAAC.

JAA Foundation Board (FB) was formed by the seven Members of the JAAC, which were members of the EB. It dealt mainly with the legal and financial aspects of JAA as Foundation established under Dutch law (Stitching JAA Beheer).

With the continuation of JAA-TO as a Dutch Foundation and associated body of ECAC, a JAA-TO Foundation Board was (re)established.

JAA T ensured the secretariat of all Governing Bodies.

3.3.6. General remarks

The activity of this worthy organization, which has led the way to the EASA, has very often been limited by its own nature. It is worth mentioning that we were talking about "authorities," not "authority." This means that the JAA did not have the legal status of an authority and therefore a legally recognized power.

JAA did not have the power, for example, to issue certificates; they could only "recommend" the release of such certificates to the national authorities under the relevant terms and conditions. For the same reasons, they could not impose rules and procedures—unless they became European directives—but only "recommend" their implementation. The shortcomings of such situations are clear, considering the variety of rules and laws that were in force in the Member States. This is why the institution of a true European authority was increasingly felt necessary. This is now a reality with the institution of EASA, which has benefited from the substantial and complex work carried out by the JAA.

3.3.7. Closing

Based on a decision of DGs of the ECACs in adopting the FUJA II Report, it was decided to disband the JAA system per 30 June 2009 and to keep the JAA Training Organization running.

3.4. THE EUROPEAN AVIATION SAFETY AGENCY

The EASA is an independent European Community body with a legal identity and autonomy in legal, administrative, and financial matters.

This single authority has been created by the adoption of a European Parliament and Council Regulation (EC) No. 1592/2002 of 15 July 2002 to put in place a Community system of air safety and environmental regulation.

Meanwhile, on 20 February 2008, the European Parliament and Council have adopted Regulation 216/2008 repealing Regulation 1592/2002, which extends the scope of EASA to operations, flight crew licensing, and third-country operators. EASA now has the mandate to work on Implementing Rules concerning the aforementioned areas.

The activity of the EASA started, as planned, on 28 September 2003 and, after a transitory period in Brussels, the Agency moved to Cologne (Germany).

3.4.1. Executive and regulatory tasks

The main tasks of the Agency currently include
1. Rulemaking: drafting aviation safety legislation and providing technical advice to the European Commission and to the Member States;
2. Inspections, training, and standardization programs to ensure uniform implementation of European aviation safety legislation in all Member States;
3. Safety and environmental type certification of aircraft, engines, and parts;
4. Approval of aircraft design organizations worldwide and of production and maintenance organizations outside the EU;
5. Authorization of third-country (non-EU) operators;
6. Coordination of the European Community program Safety Assessment of Foreign Aircraft regarding the safety of foreign aircraft using Community airports;
7. Data collection, analysis, and research to improve aviation safety.

In a few years, the Agency will also be responsible for safety regulations related to airports and Air Traffic Management (ATM) systems.

3.4.2. EASA partnerships

The EASA works closely with representatives of other organizations to ensure that it takes their views into account:

1. Interested parties in industry, which are subject to rules drafted by the EASA, are pivotal in ensuring the success of civil aviation safety standards by assisting in the drafting and correct application of European Community and EASA rules.
2. European aviation authorities perform a critical role in assisting the EASA with the performance of its core rulemaking, certification, and standardization functions.
3. International aviation organizations such as the JAA, EUROCONTROL,[5] and the ICAO work together with the EASA to promote international civil aviation standards.
4. EASA is developing close working relationships with counterpart organizations across the world including the FAA and the aviation authorities of Canada, Brazil, Israel, China, and Russia. Working arrangements between the Agency and these organizations are aimed at harmonizing standards and promoting best practice in aviation safety worldwide.
5. Accident investigation bodies issue safety recommendations and analysis that guide the Agency's safety strategy.

3.4.3. Structure of the EASA (Fig. 3.1)

The EASA Headquarters includes

1. Executive Director,
2. Rulemaking Directorate,
3. Certification Directorate,
4. Approval and Standardization Directorate, and
5. Administrative Directorate.

The **Executive Director** is appointed by the Agency's **Management Board**. This Board, which brings together representatives of the Member States' authorities and the Commission, is responsible for the definition of the Agency's priorities, the establishment of the budget, and for monitoring the Agency's operation.[6]

---

[5] EUROCONTROL has the role of coordinating the development of a uniform system of ATM throughout Europe (38 states), working with its partners in the air transport industry to provide a range of services: from air traffic controller training to managing air traffic flow and from regional control of airspace to development of innovative technologies and procedures.

[6] The **Advisory Body of Interested Parties** assists the Management Board in this work. It comprises organizations representing aviation personnel, manufacturers, commercial and aviation operators, the maintenance industry, training organizations, and air sport.

The **Rulemaking Directorate** contributes to the production of all EU legislation and implementation of material related to the regulation of civil aviation safety and environmental compatibility. It submits opinions to the European Commission and must be consulted by the Commission on any technical question in its field of competence. It is also in charge of the related international cooperation. Experts within the Rulemaking Directorate have direct contact with all relevant stakeholders and make use of the knowledge available within the industry and national administrations across the EU. The Agency's team of experts is comprised of people with a recognized background in aviation and Community regulations.

Currently, the Basic Regulation establishes Community competence only for the regulation of the airworthiness and environmental compatibility of aeronautical products, parts, and appliances. Work is underway to extend the scope of this regulation to embrace the regulation of pilot licensing, air operations, and third-country aircraft. It is also envisaged to extend the scope of the Basic Regulation to the safety regulation of airport operations and ATC services.

On 28 September 2003, the **Certification Directorate** took over responsibility for the airworthiness and environmental certification of all aeronautical products, parts, and appliances designed, manufactured, maintained, or used by persons under the regulatory oversight of EU Member States.

The Agency's certification work also includes all postcertification activities, such as the approval of changes to, and repairs of, aeronautical products and their components, as well as the issuing of ADs to correct any potentially unsafe situation. All type certificates are therefore now issued by the EASA and are valid throughout the EU.

On the same date, the Agency became the competent authority to approve and oversee the organizations involved in the design of aeronautical products, parts, and appliances. It also carries out the same role for foreign organizations involved in the manufacture or maintenance of such products.

To execute its tasks within the present period of building up its resources, the Agency relies on *national aviation authorities* who have historically filled this role and concludes contractual arrangements to this effect.

Where Community law is implemented at Member State level, the **Approval and Standardization Directorate** assists the Commission in overseeing its effective application and its uniform understanding.

The necessary standards are therefore being developed and maintained properly, uniformly, and consistently across the EU.

Accordingly, the Agency conducts inspections of undertakings as well as national authorities throughout the EU, both to monitor the application of EU rules on aviation safety and to assess the effectiveness of these rules. The Agency also provides technical training, which is essential to achieve overall consistency.

The **Administrative Directorate** supports the operational activities of the Agency. Its role is to help the Agency to plan and manage its resources within the limits set out in the regulatory framework. The Directorate's specialists deal with human resource issues, budgeting and finance, infrastructure, legal affairs, and procurement.

3.4.4. EASA certification

3.4.4.1. DESIGN APPROVAL

According to Regulation (EC) No. 1592, the EASA takes responsibility for the design approval of products, parts, and appliances designed, manufactured, or used by persons under the regulatory oversight of EU Member States, **except for those excluded by its Annex II**[7] or by its Article 1.2 (products engaged in military, customs, police, or similar services).

The European Commission then adopted *Regulation (EC) 1702/2003*, which specifies inter alia the requirements applicable to products, parts, and appliances, and also provides for the grandfathering of pre-existing certificates under conditions that aim at ensuring that they meet the level of safety required by the Basic Regulation (EC) No. 1592/2002 and its rules of implementation.

The Basic Regulation recognized the need for some transition to facilitate the transfer of responsibility from national administrations to the Agency. Therefore, Article of the Basic Regulation established the possibility for the Member States to continue to issue, during transition period, certificates and approvals by way of derogation of the provisions of the Basic Regulation under the conditions specified in its implementing rules, in particular Commission Regulation 1702/2003. This transition period ended on 28 March 2007.

As a consequence, the Agency's responsibilities for design-related activities (Type certificates, supplemental type certificates, approval of changes and repair design, and other post type-certification activities, including ADs) now include the following:

- Products with type certificates issued by EASA in accordance with Commission Regulation 1702/2003 as of 30 March 2007.
- Products with type certificates issued by the EU Member States that are deemed to have been issued in accordance with Commission Regulation (EC) No. 1702/2003.
- Products with specific airworthiness specifications issued by EASA in accordance with Regulation (EC) No. 1592/2002, to support restricted certificates of airworthiness.

In addition, EASA is responsible for the approval of the flight conditions on the basis of which a permit to fly[8] can be issued by the authority designated by the Member State of Registry.

Products that do not benefit from the grandfathering provisions will remain under the national administrations' oversight.

---

[7] **ANNEX II**. Lists the categories of aircraft to which the basic principle of Article 4(1) of the Regulation (EC) No. 1592/2002 (now 216/2008) does not apply, namely aircraft for which a type certificate or a certificate of airworthiness has not been issued on the basis of this Regulation and its implementing rules.

[8] See Section 8.4.3 of Chapter 8.

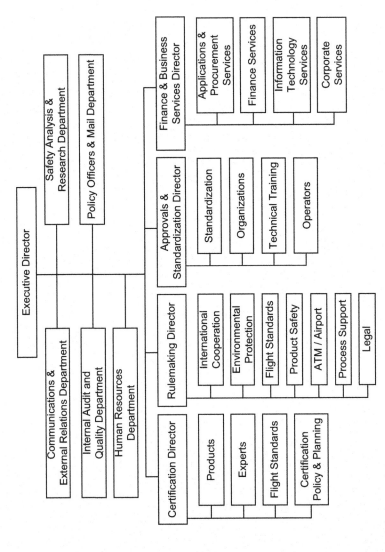

FIGURE 3.1 EASA organizational chart

In relation to the products already type certificated, the Agency has expedited, in cooperation with the concerned Member States of design, the review of the type certification bases of these products with the view to determine their EASA type certificate and thus take over responsibility for their continued airworthiness.[9]

Aircraft that were permitted to fly before 20 September 2003 and cannot be issued an EASA type certificate will remain under the responsibility of the Member State of Registry under applicable national regulations.

3.4.4.2. ORGANIZATION APPROVAL

The EASA Organizations Approval department is responsible for the following activities:

(1) Design Organizations
   (i) The management of all design organization applications;
   (ii) The issue of related Design Organization Approval[10] certificates and their continued surveillance;
   (iii) The issue of compliance statements for alternative procedures.

(2) Production Organizations
   (i) The management of all applications from non-EU countries (or from an EU country on request of the competent authority of that country) for Production Organization Approvals.[11]
   (ii) The issue of related certificate and their continued surveillance.

(3) Continuing Airworthiness Organizations[12]
   (i) The management of all applications from non-EU countries for maintenance (MOA), maintenance training organization approvals (MTOA), Part-M sub-part G continuing airworthiness management organization (CAMO), and Part-M Subpart F approvals (Subpart F).
   (ii) The issue of related certificates and their continued surveillance.

3.4.4.3. GENERAL REMARKS

At the end of 2004, the EASA was still in the organization phase. According to *Flight International* (October 2004):

> The EASA is currently engaged in extending its powers beyond its existing responsibility for airworthiness and maintenance into the operations arena. According to approved plans for centralizing all aviation safety rulemaking, the EASA is preparing to assume responsibility for operational issues, including air traffic management, airports, and pilots, mirroring the US Federal Aviation Administration.

> Mr. Goudou[13] used a speech to the European Parliament to address claims from several national aviation authorities that supplementary national requirements

---

[9] See Chapter 5, "Instructions for continued airworthiness."
[10] See Chapter 5, "Design Organization Approval."
[11] See Chapter 7, "Production Organization Approval."
[12] See Section 9.1.2 of Chapter 9.
[13] Patrick Goudou, Executive Director of the EASA.

licensing will continue to be enforced in the future. But, unlike the predecessor, the Joint Aviation Authorities, the EASA will not merely recommend regulations. As an agency of the EU's executive, the European Commission, it will have the power to enforce compliance.

Despite Mr. Goudou's goodwill, in the same article, *Flight International* mentioned that the EASA has had a setback in its recruitment of 95 certification staff by the end of 2004, mainly because of the Agency's move from Brussels to Cologne, which could not be considered attractive for experienced people living in other locations.

In any case, it is worth reading what Goudou wrote in an article (for a UVS International Publication):

During the set-up and transition phase, the keyword of the Agency's activities is 'continuity'. Indeed, it goes without saying that the Agency is not going to reinvent the wheel, as its initial tasks are based on the activities and existing procedures of the Joint Aviation Authorities (JAA), and on national know-how, which enables the Agency to provide continuity in terms of the certification work and the progressive resumption, without major upheaval, of the work carried out now by the JAA and national authorities. As such, no project has been delayed since the Agency has become operational.

The Agency already employs approximately 500 professionals coming from all Member States. It will continue to recruit highly qualified specialists and administrators during the next few years as it consolidates its position as Europe's centre of excellence in aviation safety.

Having completed the transition phase, the Agency's responsibilities are now growing to meet the challenges of the fast-developing aviation sector. In a few years, the Agency will also be responsible for safety regulations regarding airports and ATM systems.

During the last years, after rightly giving precedence to the regulation of commercial aviation, EASA has tried to put order into the regulation of **general aviation**.

The Advanced Notice of Proposed Amendment (A-NPA) 14-2006[14] issued in October 2006 was the object of thousands of comments, showing how deeply this issue was felt in Europe.

After the publication of a Comment Response Document, EASA issued the NPA No. 2008-07 on April 2008.

The intention is to create a lighter regulatory regime based around a new process for the **European Light Aircraft** and to introduce a concept of standard changes and repairs.

ELA is **not** a new category of aircraft defined by criteria such as stalling speed or certification code but is a substantially simpler new **process** for the regulation of aircraft and related products, parts, and appliances. The intention

---

[14] See Note 30 in Chapter 4.

is to issue type certificates for the type and certificates of airworthiness for the individual aircraft.

The ELA is sub-divided into two sub-processes: **ELA 1** and **ELA 2** related to aeroplanes, sailplanes or powered sailplanes, balloons, airships, engines, and propellers (ELA 2 also includes the Very Light Rotorcraft). The interested aircraft should not be classified as *complex-motor-powered aircraft*.[15] For aeroplanes, the maximum takeoff mass (MTOM) is 1000 kg for ELA 1 and 2000 kg for ELA 2.

We will not comment in detail the NPA in this book. However, there is some disappointment for the lack of adoption in Europe of a regulation bearing similarity with the FAA Light Sport Aircraft (LSA) (see Section 8.5.2.4 of Chapter 8) as it was required by several sector professionals. The introduction of this aircraft category, which has been very successful in the United States, could have satisfied the exigencies of basic aviation without the need of staying within the weight limits of the current rules for ultralights, which may be adequate for *true* ultralights, but are too low for aeroplanes. The FAA LSA also includes substantial simplifications for the aircraft certification without penalization of the overall safety as demonstrated after 3 years of operation.

It is interesting to note that most of the LSA-type aircraft sold in the United Sates are produced in Europe where they cannot fly.[16] The NPA tries to solve this problem, but the issue of a type certificate is nevertheless required.

It is true that the Article 5.2(a) of the Regulation 216/2008 requires a TC for the products, but the Point 4 of the same article presents a series of derogations: the LSA cat. could have been one of them.

In summary, it is not clear why after years of discussions on the harmonization with FAA, EASA have taken a different route for the "basic aviation," losing what could have been a real simplification for the ELAs and a great benefit for what is considered an important sector of aviation, including the simplification of the export and import of those products.

In terms of "basic aviation," as the above-mentioned Regulation 216/2008 does not apply to aircraft referred to in Annex II, aircraft generally defined as "ultralight" are the object of several different regulations depending on the European states in which they operate.

---

[15] According to Article 3(j) of Regulation (EC) No. 216/2008, *"complex motor-powered aircraft"* shall mean:
  (i) an aeroplane:
     − with a certificated MTOM exceeding 5700 kg or
     − certificated for a maximum passenger seating configuration of more than 19 or
     − certificated for operation with a minimum crew of at least two pilots or
     − equipped with (a) turbojet engine(s) or more than one turboprop engine or
  (ii) a helicopter certificated:
     − for an MTOM exceeding 3175 kg or
     − for a maximum passenger seating configuration of more than nine or
     − for operation with a minimum crew of at least two pilots or
  (iii) a tilt rotor aircraft.

[16] For aircraft developed according to the US "Light Sport Airplane," EASA can grant a Permit to Fly according to Part 21A.701(15).

During the last few years, these aircraft, at first considered and regulated as "leisure tools," have spread, often assuming for weight and technological complexity, the characteristics of superior class aircraft.

This is a big problem that sooner or later will need attention by EASA on the basis of the European principles in the field of civil aviation assuring a high and uniform level of protection of the European citizen by the adoption of common safety rules. This should also contribute to facilitating the free movement of these products in the internal market.

3.5. THE FEDERAL AVIATION ADMINISTRATION (FAA)

3.5.1. Origins

The Air Commerce Act of 20 May 1926 was the cornerstone of the Federal government's regulation of civil aviation. This landmark legislation was passed at the behest of the aviation industry, whose leaders believed that the aircraft could not reach its full commercial potential without Federal action to improve and maintain safety standards. The Act charged the Secretary of Commerce with fostering air commerce, issuing and enforcing air traffic rules, licensing pilots, certificating aircraft, establishing airways, and operating and maintaining aids to air navigation. A new Aeronautics Branch of the Department of Commerce assumed primary responsibility for aviation oversight.

3.5.2. Early responsibility

In fulfilling its civil aviation responsibilities, the Department of Commerce initially concentrated on functions such as safety rulemaking and the certification of pilots and aircraft.

In 1934, the Aeronautics Branch was renamed as the Bureau of Air Commerce to reflect its enhanced status within the Department. As commercial flying increased, the Bureau encouraged a group of airlines to establish the first three centers for providing ATC along the airways. In 1936, the Bureau itself took over the centers and began to expand the ATC system.

3.5.3. The Civil Aeronautics Act

In 1938, the Civil Aeronautics Act transferred the Federal civil aviation responsibilities from the Commerce Department to a new independent agency, the Civil Aeronautics Authority.

In 1940, President Franklin Roosevelt split the Authority into two agencies, the Civil Aeronautics Administration (CAA) and the Civil Aeronautics Board (CAB). The CAA was responsible for ATC, airman and aircraft certification, safety enforcement, and airway development. The CAB was entrusted with safety rulemaking, accident investigation, and economic regulation of the airlines. Both organizations were part of the Department of Commerce.

3.5.4. The birth of the FAA

The approaching introduction of jet airliners and a series of midair collisions spurred passage of the Federal Aviation Act of 1958. This legislation transferred the CAA's functions to a new independent body, the FAA, which had broader authority to combat aviation hazards. The act took safety rulemaking from the CAB and entrusted it to the new FAA. It also gave the FAA sole responsibility for developing and maintaining a common civil–military system of air navigation and ATC, a responsibility that the CAA previously shared with others.

3.5.5. From agency to administration

In 1966, Congress authorized the creation of a cabinet department that would combine major Federal transportation responsibilities. This new Department of Transportation (DOT) began full operations on 1 April 1967. On that day, the FAA became one of the several modal organizations within the DOT and was given a new name, the Federal Aviation Administration. At the same time, the CAB's accident investigation function was transferred to the new National Transportation Safety Board (NTSB).

3.5.6. Structural changes

The FAA's organizational structure has continued to evolve since its creation. The agency's first Administrator favored a management system under which officials in Washington exercised direct control over programs in the field. In 1961, however, his successor began a decentralization process that transferred much authority to regional organizations. This pattern generally endured until a 1988 "straight lining" again charged managers at national headquarters with more direction of field activities.

3.6. FAA ACTIVITIES

3.6.1. Safety regulations

The FAA issues and enforces regulations and minimum standards covering manufacturing, operating, and maintaining aircraft. It also certifies airmen and airports that serve air carriers.

3.6.2. Airspace and traffic management

The safe and efficient use of navigable airspace is one of the FAA's primary objectives. The FAA operates a network of airport towers, air route traffic control centers, and flight service stations. It also develops air traffic rules, assigns the use of airspace, and controls air traffic.

3.6.3. Air navigation facilities

The FAA builds or installs visual and electronic aids to air navigation. It also maintains, operates, and assures the quality of these facilities, and sustains other systems to support air navigation and ATC, including voice and data communications equipment, radar facilities, computer systems, and visual display equipment at flight service stations.

3.6.4. Civil aviation abroad

The FAA promotes aviation safety, encourages civil aviation abroad, and takes part in international conferences. Aeronautical information is exchanged with foreign authorities. The FAA certifies foreign repair shops, airmen, and mechanics; provides technical aid and training; and negotiates "Bilateral Aviation Safety Agreements" (BASA) with other authorities with the "Implementation Procedures for Airworthiness" to allow and facilitate the mutual certification of aeronautical products that are imported or exported between the United States and a signatory country, as well as promoting technical cooperation in matters of airworthiness, including maintenance, flight operations, and environmental certification.

The FAA deals with all the problems related to flight safety in the United States, but it has representatives on five continents committed to ensuring and promoting the safety, security, and efficiency of international civil aviation. The FAA engages in dialog with its counterparts in 188 countries and works closely with the ICAO. This effort includes providing technical assistance and training, ensuring that countries with airlines flying to the United States meet international standards, and harmonizing global standards so that passengers can benefit from a seamless air transportation network.

It is clear that all these international activities have the final and institutional purpose of guaranteeing flight safety in the United States. However, we cannot ignore the considerable drive given by the FAA for the growth of safety on a global scale.

3.6.5. Commercial space transportation

The FAA regulates and encourages the US commercial space transportation industry. It licenses commercial space launch facilities and private launches of space payloads on expendable launch vehicles.

3.6.6. Research, engineering, and development

The FAA conducts research on and develops the systems and procedures needed for a safe and efficient system of air navigation and ATC. It helps develop better aircraft, engines, and equipment, and it tests or evaluates aviation systems, devices, materials, and procedures. The FAA also carries out aeromedical research.

3.6.7. Other programs

The FAA registers aircraft and records documents reflecting title or interest in aircraft and their parts. It administers an aviation insurance program, develops specifications for aeronautical charts, and publishes information on airways, airport services, and other technical subjects in aeronautics.

3.6.8. Summary of FAA activities

The FAA is responsible for the safety of civil aviation (Fig. 3.2). Its main roles include

(1) Regulating civil aviation to promote safety.
(2) Encouraging and developing civil aeronautics, including new aviation technology.
(3) Developing and operating a system of ATC and navigation for both civil and military aircraft.
(4) Researching and developing the National Airspace System and civil aeronautics.
(5) Developing and carrying out programs to control aircraft noise and other environmental effects of civil aviation.
(6) Regulating US commercial space transportation.

3.7. FAA CERTIFICATION

The organization of the FAA is very complex; this is understandable considering the plurality of tasks, the size of the United States, and its relationship with the rest of the world.

From an airworthiness point of view, we will try to describe which structure deals with each relevant issue.

In the vast FAA organizational chart, we can find the **Aviation Safety** headquarters located in Washington which, among its many offices (such as the Office of Accident Investigation, Office of Aerospace Medicine, etc.), hosts the **Aircraft Certification Service**, structured as shown in Figs 3.3 and 3.4.

Figure 3.5 summarizes the main tasks of this Service.

3.7.1. The Aircraft Certification Service

The Aircraft Certification Service of the FAA is the office responsible for

(1) Administering safety standards governing the design, production, and airworthiness of civil aeronautical products;
(2) Overseeing design, production, and airworthiness certification programs to ensure compliance with prescribed safety standards;
(3) Providing a safety performance management system to ensure continued operational safety of aircraft; and
(4) Working with aviation authorities, manufacturers, and other stakeholders to help them successfully improve the safety of the international air transportation system.

Aircraft Certification is organized into the Office of the Director and three divisions located in Washington, DC Headquarters, and four geographic directorates. The Aircraft Certification Service headquarter's offices and the

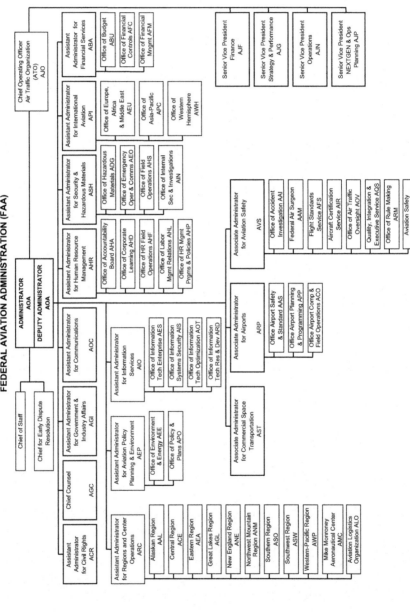

FIGURE 3.2 Organization of the Federal Aviation Administration (FAA)

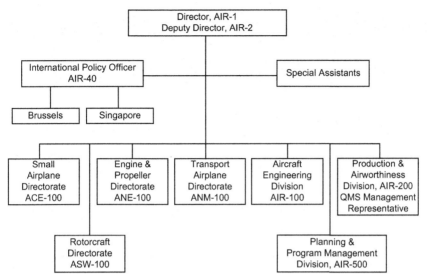

FIGURE 3.3 Structure of the Aircraft Certification Service

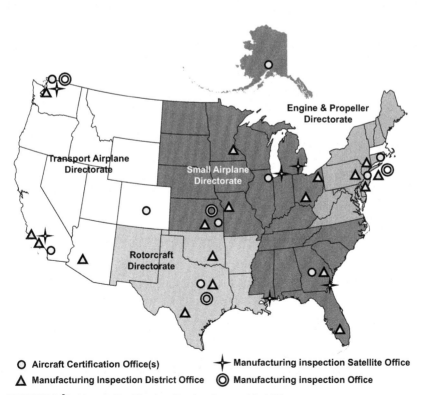

FIGURE 3.4 Aircraft Certification Service Geographical Directorates

Aircraft Certification Service – Products and services		
Design Approvals	**Production Approvals for Aircraft, Engines, and Propellers**	**Airworthiness Certification**
• Transport airplanes • Small airplanes • Engines and propellers (*including APUs*) • Rotorcraft • Airships • Manned free balloons	• Production certificate • Production under a type certificate only • Approved Production Inspection System (APIS)	• Standard airworthiness certificate • Special airworthiness certificate (amateur-built) • Approved Production Inspection system (APIS) • Special flight authorization • Export approvals • Import approvals
Design Modifications for Aircraft, Engines, and Propellers	**Design and Production Approvals for Parts/ Articles/Appliances**	**Representatives of the Administrator (Designees)**
• Amended type certificate • Supplemental type certificate • Field approval	• Parts Manufacturer Approval • Technical Standard Order authorization	• Designee resources • Designee process overview • Designee/FAA selection and appointment process • Designee training
Continued Operational Safety	**International Aviation**	
• Airworthiness Directives (AD) process • Alternate Method of Compliance (AMOC) • Design approval holder reporting requirements	• Bilateral agreements	

FIGURE 3.5 Main tasks of the Aircraft Certification Service

directorates share responsibility for the design and production approval, airworthiness certification, and continued airworthiness programs of all US civil aviation products (see Fig. 3.3).

The Aircraft Certification Service's responsibility is divided as follows.[17]

3.7.1.1. HEADQUARTER'S DIVISIONS
The **Aircraft Engineering Division** is responsible for overall policy and guidance for the engineering sector of the Aircraft Certification Regulatory Program (ACRP).[18]

---

[17] Details on the quoted FARs can be found in Chapter 4.

[18] The Federal Aviation Act of 1958 directs the FAA to promote safety of flight of civil aircraft in air commerce prescribing and revising minimum standards for design, materials, construction, and so on. The ACRP was developed to accomplish this goal.

Furthermore, it is responsible for Federal Aviation Regulations (FAR) 21, 39, and Special FARs[19] pertaining to type certification, and certification of restricted category and primary category aircraft.[20]

The **Production and Airworthiness Certification Division** is responsible for the regulations, policy, and guidance for manufacturing and airworthiness certification portions of the ACRP. It is also responsible for FAR 21, 43, 45, 183, and Special FARs pertaining to certification conformity, airworthiness certification, and production.

The **Planning and Program Management Division** is responsible for the coordination of the Service's strategic and tactical planning initiatives and processes. It is also responsible for Service's technical, general and managerial training requirements, administrative and program management guidance, coordination, and support for Service headquarter's organizations.

3.7.1.2. AIRCRAFT CERTIFICATION DIRECTORATES

The **Small Airplane Directorate** (Central Region) is responsible for FAR 23 and 31; technical guidance for restricted category small airplanes; airworthiness criteria for gliders and airships; technical guidance for primary category airplanes; FAR 23, glider and airship import TC projects; issuance of ADs for the above products; and participation in consensus standard development for light sport aircraft.[20]

The **Transport Airplane Directorate** (Northwest Mountain Region) is responsible for FAR 25 and technical guidance for restricted category transport airplanes[20]; FAR 25 import type-certification projects and issuance of ADs for the above products.

The **Rotorcraft Directorate** (Southwest Region) is responsible for FAR 27 and 29; technical guidance for restricted category rotorcraft, powered lift aircraft, and primary category rotorcraft; FAR 27 and 29 import TC projects; and issuance of ADs for the above products.

The **Engine and Propeller Directorate** (E&PD, New England Region) is responsible for FAR 33 and 35, and technical guidance on auxiliary power units (APUs); FAR 33 and 35 import TC projects; and issuance of ADs for the above products.

The Service also has other functions.

International Policy Office. This office, including the staff in Brussels and Singapore, is responsible for policy guidance on bilateral agreements, import and export of aeronautical products, and other international airworthiness issues, programs, and procedures.

---

[19] Special FARs establish additional airworthiness standards for aircraft to cope (normally) with particular operation. For instance, Special FAR 23 is for aircraft to be certificated in the Normal category for a reciprocating or turbopropeller multi-engine-powered small airplane that is to be certificated to carry more than 10 occupants and that is intended for use in operations under FAR 135.

[20] See Chapter 8.

Aircraft Certification Offices (ACOs). Each directorate incorporates three or more ACOs within their geographical areas issuing the actual certification of aircraft and products. They work directly with the applicant and provide the main interface between the public and the FAA.

Aircraft Evaluation Group. A Flight Standards group is colocated with each directorate and it is responsible for determining operational acceptability and continuing airworthiness requirements of newly certified or modified aircraft, engines, propellers, and parts.

Before describing in more detail the four Directorates mentioned above, we introduce some useful definitions.

Aircraft Certification Office (ACO). The aircraft certification directorate's engineering operational element. This office administers and secures compliance with agency regulations, programs, standards, and procedures governing the type design of aircraft, aircraft engines, or propellers. It offers certification expertise on investigating and reporting aircraft accidents, incidents, and service difficulties. The term "ACO" refers to the Engine Certification Office (ECO), the Rotorcraft Certification Office (RCO), the Special Certification Office (SCO), the Airplane Certification Office (ACO), and all other ACOs.

Manufacturing Inspection Office (MIO). The MIO oversees Manufacturing Inspection District Offices (MIDO) and Manufacturing Inspection Satellite Offices (MISO) in its geographic area and provides organizational leadership and technical guidance to these offices. The MIO manages all geographically located production facilities and designees. They administer the airworthiness certification policies, office staffing, and internal budget allocation.

Manufacturing Inspection District Office (MIDO). This is a subordinate office to the MIO in its geographical area. This office oversees production certification, airworthiness certification, approval holders (manufacturing facilities), and designees in its geographical area. MIDOs support ACOs during type-certification programs; they investigate and submit enforcement reports on noncompliance with applicable FARs. MIDOs investigate and ensure corrective measures for service difficulties, which are implemented as identified in the quality system.

Manufacturing Inspection Satellite Office. This subordinate geographically remotes office reports to an MIDO and is responsible for the same activities as of the MIDO.

3.7.2. The Small Airplane Directorate

The Small Airplane Directorate (Central Region) consists of the **Directorate headquarters** located in Kansas City; four **ACOs** located in *Anchorage, Atlanta, Chicago,* and *Wichita*; and seven **MIDOs**[21] located in *Atlanta, Cleveland, Kansas City, Minneapolis, Orlando, Vandalia,* and *Wichita*.

---

[21] MIDOs assist with production approval and certification (manufacturing); airworthiness certification; manufacturing facilities approval holder issues; manufacturing designee oversight; support to ACOs during design approvals.

The primary functions of the Directorate headquarters in Kansas City are to

(1) Provide administrative support and resource management for the Directorate field offices.

(2) Develop type-certification policies and regulations for *small airplanes, airships,* and *balloons,* and ensure standardized application of the policies and regulations.

(3) Administer type-certification of small airplanes, airships, and balloons in field offices outside the Directorate.

(4) Monitor *continued airworthiness information* and process airworthiness actions for small airplanes, airships, and balloons.

The Small Airplane Directorate is responsible for several aspects of aviation, such as:

(1) Continued airworthiness and general aviation safety

(2) Type certification

(3) Technical Standard Orders (TSO)

(4) Parts manufacturer approval (PMA)

(5) Field approval[22]

NOTE: FAR 1 defines a "small aircraft" as an aircraft of 12,500 lbs or less maximum certificated takeoff weight. Therefore, any airplane, including transport category airplanes, could be considered "small" by the Part 1 definition if the airplane is less than 12,500 lbs. However, as commonly used, and in the most basic meaning, small airplanes have generally been considered fixed-wing aircraft that are not transport category airplanes (i.e., fixed-wing airplanes type-certificated to standards other than FAR 25). Therefore, generally speaking, small airplanes are fixed-wing airplanes that are not transport category. Depending on the category, small airplanes can reach up to 19,000 lbs maximum takeoff weight.

A small airplane is not the same as a General Aviation (GA) aircraft, because GA aircraft are operated under FAR 91, which could be any category of airplane, including transport category and rotorcraft. Additionally, airplanes operated under FAR 121 and 125, which may include small airplanes, are not considered General Aviation aircraft when operated under these rules.

3.7.3. The Transport Airplane Directorate

The Transport Airplane Directorate (Northwest Mountain Region), functionally, has oversight responsibility for transport category airplane design approvals and modifications worldwide, as well as oversight responsibility for over 900 production approval holders. The Transport Airplane Directorate works closely with other FAA offices throughout the country and with foreign regulatory authorities to accomplish this mission.

Among the FAA offices working with the Directorate, it is worth mentioning

---

[22] Field approval is a maintenance performance approval for a major repair or major alteration that is performed by a Flight Standards Service, Aviation Safety Inspector.

1. The **Aircraft Certification Services** (**ACOs**; in Seattle, Los Angeles, and Denver).
2. The **MIDOs** (in Los Angeles, Phoenix, Seattle, and Van Nuys).
3. The **MIO** (in Seattle).

The Directorate relies on **Designated Representatives**[23] of the Administrator to act on behalf of the FAA. This designee force includes Engineering Designees, Manufacturing Designees, and Organization Designees.

The Directorate's three most important responsibilities are

1. Continued operational safety.
2. Regulations and policy for all transport category airplanes.
3. Design, production, and airworthiness certification.

3.7.3.1. CONTINUED OPERATIONAL SAFETY

1. Monitoring the transport category airplane fleet to ensure that airplanes continue to meet regulations and are safe throughout their operational life cycle.
2. Looking for conditions that affect the safety of airplanes. This is done by surveillance, inspection, review, investigation and analysis of service difficulties, incidents, and accidents.
3. If an unsafe condition is identified, this will trigger the following actions:
   a. Working with the manufacturers to mandate corrective action through ADs or
   b. Revision of regulations/policy or
   c. Issuing of new regulations/policy.
4. Performing surveillance and oversight of production approval holders.

3.7.3.2. REGULATIONS AND POLICY FOR ALL TRANSPORT AIRPLANES

1. Developing and establishing FAA type design and airworthiness standards for all transport category airplanes.
2. The type design standards are codified in Title 14, Code of Federal Regulations (14 CFR), Part 25. This is commonly referred to as Part 25 of the FAR.
3. These FAR 25 standards are applied to aircraft worldwide, working with other civil aviation authorities to "harmonize" these standards whenever possible.

3.7.3.3. DESIGN, PRODUCTION, AND AIRWORTHINESS CERTIFICATION

(1) The Directorate is responsible for the release of design, production, and airworthiness approvals of all aircraft and aircraft parts in Washington,

---

[23] A designee is an Administrator of the FAA authorized by law to examine, test, and/or make inspections necessary to issue airman or aircraft certificates.

Oregon, Idaho, Montana, Colorado, Wyoming, California, Arizona, Utah, Nevada, Hawaii, and the Pacific Rim countries.
(2) Determining and ensuring that each aircraft design meets the applicable regulations (**design approval**).
(3) Issuing a type certificate, when an applicant shows that its aircraft design meets the standards.
(4) Ensuring that each manufacturing facility is capable of producing aircraft to the approved design (**production certification**).
(5) Ensuring that each aircraft produced in the manufacturing facility is built to the approved design.
(6) Ensuring that each aircraft produced is in a condition for safe operation (**airworthiness approval**).
NOTE: Transport airplanes are either:
Jets with 10 or more seats or a Maximum Takeoff Weight (MTOW) greater than 12,500 lb or
Propeller-driven airplanes with greater than 19 seats or an MTOW greater than 19,000 lb.

3.7.4. The Rotorcraft Directorate

The Rotorcraft Directorate is responsible for
(1) FAA regulations and policy related to engineering certification of rotorcraft and powered-lift aircraft;
(2) FAA certification of rotorcraft worldwide, and both fixed- and rotary-wing aircraft within the FAA Southwest Region; and
(3) Approval of the design and production for all fixed- and rotary-wing aircraft manufactured or modified within the FAA Southwest Region.
In addition to certifying all aircraft, the Directorate has the responsibility for writing rules and policy for rotorcraft and working with all the **ACOs**— also outside the above-mentioned territory—to achieve standardized application of the rules for rotorcraft. Furthermore, it works with its counterparts in other countries to issue domestic approvals for foreign-manufactured rotorcraft.
The Rotorcraft Directorate has one ACO (in Fort Worth); three MIDOs (in Fort Worth, Oklahoma City, and San Antonio); and one MIO (in Fort Worth).
NOTE: The rotorcrafts are
— Normal Category Rotorcraft: 7000 pounds or less, and nine or less passenger seats.
— Transport Category Rotorcraft: Bigger/heavier rotorcraft (above 7000 lbs). Although it could be technically possible to certify a rotorcraft under 7000 lbs in the transport category, this is not historically done.

3.7.5. The Engine and Propeller Directorate

The E&PD (New England Region) is located in Burlington, MA. It is responsible for original type certification or changes to approved designs of aircraft engines and propellers in addition to Technical Standard Order (TSO) approvals of Auxiliary Power Units (APUs) and assuring that aviation parts are manufactured to approved standards.

The E&PD is responsible for developing rules, policy, and guidance for these products, and assures standardization across all FAA **ACOs** that perform certification work on these products. The E&PD Standards Staff is the working element of the E&PD that directly carries out these functions.

The **ECO** (in Burlington) and each of the **ACOs** (in Boston and New York) that perform E&PD-related certification work are accountable for planning, directing, and controlling engine and propeller type certification programs in addition to TSO approvals of APUs. Both the ECO's and ACOs' primary responsibilities are to find compliance to the applicable Airworthiness Standards (i.e., FAR 33 and 35 and TSO-C77B) and assure continued airworthiness of these products once in service.

There are five **MIDOs** (in Farmingdale, New Cumberland, Boston, Saddle Brook, and Windsor Lock) and one **MIO** (in Boston).

3.8. "ONE WORLD, ONE GOAL: AVIATION SAFETY"

In this chapter, in dealing with the JAA, we have emphasized the necessity of having in place a legally recognized European authority. In fact, despite a huge amount of work accomplished for unification of regulations and procedures in Europe, the JAA did not have the authority to impose these rules.

The EASA now has this power and can perform as a single authority. For instance, once an aircraft is type certificated by the EASA, this type certificate is valid for all the Member States, without being just a "recommendation" for the issue of a national type certificate. Today, we have a single European Agency instead of 31 national authorities, and a single certificate for aeronautical products instead of 31.

Another shortcoming of the JAA was the complexity of bilateral agreements[24] with authorities such as the FAA or Transport Canada. For example, an Airbus certificated by the JAA could be accepted in the United States only when it was in possession of a type certificate issued by a European Member State.

The JAA has carried out long and complex work with the FAA and Transport Canada for the release of new bilateral agreements, also relating to single European Member States.

The new legal reality requires European Member States to comply with European Law; they neither deviate from common European rules, nor impose additional requirements or conclude agreements with third countries. As a consequence, Member States are represented by the EASA. Furthermore, Member States are bound by and must reflect the Agency's decisions and positions when carrying out their representative roles in frameworks such as the ICAO and ECAC.

---

[24] See Chapter 5, "Type certification of imported products."

The Agency is committed to establishing proper relations with non-EU members of the ECAC and to pursue relationships with other international partners through special arrangements, associations, partnerships, and mutual recognition agreements. It must also recognize that, legally, bilateral safety agreements are a competence of the European Commission.

At present, the EASA has already agreed to some working arrangements with a certain number of non-EU states: Brazil, Canada, China, Israel, Japan, New Zealand, Russia, Singapore, United States, the Inter-States Aviation Committee of the Community of Independent States, United Arab Emirates, and several Civil Aviation Authorities of ECAC—non-EU Member States.

No bilateral agreement has been formalized. Therefore, from a strictly legal point of view, the existing bilateral agreements of the EU Member States are still in force.

In this context, the EASA is carrying on the tradition of an annual US—Europe International Aviation Safety Conference. The Europe—US Aviation Safety Conference has been taking place for 50 years to promote cooperation and mutual recognition of safety standards.

The US FAA and the EASA cooperate to improve aviation safety and to facilitate, when appropriate, reciprocal acceptance of certificates approvals by, whenever possible, harmonizing standards and implementing guidance. In this context, the US/Europe International Aviation Safety Conference provides a forum for open discussion with other civil aviation authorities and industry representatives on current initiatives and strategic directions. This conference also provides a forum for interested parties to participate in harmonization and safety enhancement activities and to present initiatives of their own to the global community.

The conferences interest aviation authorities and industry worldwide that are working on aircraft certification, maintenance, operations and aviation safety issues, programs, and projects.

The Europe—US International Aviation Safety Conference on 7—9 June 2005 was jointly organized by the EASA, the JAA, and the FAA of the United States. For many years, this event provided a forum for open discussion between the JAA and other civil aviation authorities and industry representatives on current initiatives and strategic directions. Today, this annual conference also provides a forum for interested parties to participate in harmonization and safety enhancement activities, and to present initiatives of their own to the global community.

More than 350 high-level aviation experts from all over the world came together in Cologne, Germany, to discuss future trends in aviation safety. Under the title "Aviation Safety Regulation—Setting the Sights for the Future," this conference hosted by the EASA focused on bilateral agreements and future regulation in aviation safety.

In opening the conference, Patrick Goudou, Executive Director of the EASA, said:

Our mission is to set and achieve the highest common standards of safety and environmental protection in civil aviation. I am confident we can achieve our goals through international cooperation and a strong partnership with the United States in particular.

The international cooperation for the global aviation safety is still of current interest and in the last years it was discussed during the EU/US International Safety Conference co-chaired by the EASA and the FAA. In June 2007, the conference was hold in Prague with the theme *"How can open data sharing contribute to global aviation safety?"*

Accident rates in the United States and Western Europe have dropped dramatically over the years, but the challenge is to drive them lower, toward the zero accident goal. According to Mr. Goudou's conference overview:

Global collaboration among all players and the availability of new technologies can give a new impulse to reach the zero accident goal that we all want to achieve. Open exchange of data, knowledge, and experience should also involve more actively aviation growth regions, like South-East and South Asia as well as Africa to really be efficient. This year's Conference agenda has been drafted with these ideas in mind.

The discussion of the main theme of the conference focused on the importance of the international sharing of data on a global reporting system. The right implementation of the ICAO Annex 13 principles with nonpunitive, confidential reporting system, is an effective tool that should be expanded around the world.

The 2008 EU/US International Safety Conference was hold in St. Petersburg, FL, in June and it was titled: "Global Safety Management: Revolution or Evolution?"

Among many items discussed in line with the main theme of the conference, authorities and industries discussed the issue of Safety Management System (SMS)[25] implementation.

In the closing remarks of the Conference, Nicholas Sabatini[26] stated:

You, the industry have asked that the authorities identify what actions or issues we are committed to moving forward on. We had the opportunity last night to discuss the plethora of views that have been aired on safety management this week, and with Patrick Goudou's support, I would like you to know that the FAA and EASA will move forward from this conference to work with ICAO towards changes in what ICAO has proposed to its signatory States. Understand that fundamentally, safety management is the right thing–we fully support the need for all parties to manage safety. But we need to do so in a manner where the burden on the industry is also considered. I hope that other authorities who have joined us this week, will also consider the views expressed by our customers and that collectively we can influence change in ICAO's basic requirements.

The 2009 EU/US International Safety Conference was held in Athens, Greece, in June and it was titled: "Global safety in challenging times" "How can we better achieve harmonized implementation?"

---

[25] See Section 9.6 of Chapter 9.

[26] Nicholas Sabatini was an associate administrator of the FAA for Aviation Safety.

Improving aviation safety in time of economic hardship requires the cooperation of all aviation players, regulators, and industry. The EASA and the FAA have therefore put this cooperation at the top of the agenda of the Conference.

We can have an idea of the themes treated reading an excerpt of the ***Closing remarks*** *from Patrick Goudou*

> *"Yesterday, during a meeting with the FAA, with John and his team, we have analyzed the main issues that have been raised during this conference in the various workshops and during the talks we had with you. We came to the conclusion, that, in fact, we have 4 main issues in front of us:*
>
> *SMS. The discussion in the plenary session and in the workshop panel has shown that we have already followed up on this issue but that it is certainly not finished; we have to continue working on it*
> *Data sharing. More work is needed to build more trust and confidence and to better ensure respect for the sensitivity of this area and confidentiality along with a just culture.*
> *New EASA rules. The impact of the new EASA regulations needs to be further explained in order to create the right environment and to foster mutual understanding of these regulations*
> *Training. We have also highlighted the importance of training, be it training pilots or training mechanics. If we look at the recent accidents and incidents, we come to the conclusion that training can help a lot to mitigate risks.*
>
> *I would like to reassure you that EASA together with the FAA will continue to cooperate hand in hand and that we have the firm intention to continue this in the future, especially at the technical level, for the sake of safety.*
>
> *This also means that harmonization is not finished; we will continue working on this together."*

Chapter four

Airworthiness Requirements

4.1. REQUIREMENTS, REGULATIONS, AND STANDARDS

Before dealing with EASA regulations, it is worth considering the JAA require-ments, which are the basis of these regulations, and their relationship with their FAA analogs. Even if all JAA requirements are to be superseded, it is necessary to start with them to establish continuity and gain an understanding of their origin.

Having already mentioned the **standards** as the technical documents issued to define design criteria, we now consider the "**requirements**" (in the JAA terminology) or "**regulations**," "**airworthiness standards**" (in the FAA termi-nology) or "**certification standards**" (in the EASA terminology): the compul-sory standards.

The Organisation Scientifique et Technique International du Vol à Voile (OSTIV),[1] for example, publishes a standard for the design of sailplanes and powered sailplanes entitled "OSTIV Airworthiness Standard." This document defines this organization's vision on this subject. However, if anyone applies for the certification of a sailplane in Europe, they must make reference to CS 22,[2] "Sailplanes and Powered Sailplanes," because this is the only set of sail-plane airworthiness standards with legal value, adopted by all EU Member States. This means that the OSTIV Standard[3] can only be a guide as well as a valuable reference point (also for further amendments of the CS 22).

4.2. JARs AND FARs

When the JAR requirements were first issued in the 1970s, several different stan-dards for aircraft certification were in force in different countries. If we consider the western world only, among the most renowned we can quote the Federal Aviation Regulations (FARs) issued by the FAA, adopted in the United States

---

[1] The OSTIV is an independent organization linked to the Fédération Aéronautique Inter-nationale. The organization's aim is to encourage and internationally coordinate the science and techniques of sailplane flight and design.
[2] See Section 4.5.4.
[3] Before the issue of JAR 22, this was adopted as a national requirement by some states.

as well as in many other countries. In the United Kingdom, for example, the Civil Aviation Authority which replaced the Air Registration Board (ARB) in 1972 made use of the British Civil Air Regulations (BCARs). In France, the *Direction Générale de l'Aviation Civile* (DGAC) had the *Régles AIR*. In Germany, the *Luftfahrt Bundesamt* had its own regulations for sailplanes. This situation posed many difficulties in aircraft exportation.

Finally, on 1 January 1992, the JARs became part of the regulations of the European Community, assuming legal status in the Community Countries (all existing equivalent regulations had to be superseded). At present, only JARs (now replaced by the EASA regulations, as we will see) and FARs (or derivative regulations) are in practical use.

4.3. LIST OF JARs AND FARs[4] DIRECTLY OR INDIRECTLY RELATED TO AIRWORTHINESS CERTIFICATION

4.3.1. JAR 1/FAR 1. Definitions and Abbreviations

These codes contain definitions and abbreviations of terms used in other JAR/FAR codes. JAR 1 is based partly on those definitions contained in ICAO Annexes and partly on FAR 1. FAR 1 also contains *rules of constructions*, that is, characterization of wording such as the use of *"shall," "may," "a person may not,"* and *"includes."*

4.3.2. JAR 11. JAA Regulatory and Related Procedures

This code contains the requirements applicable to the following:
(1) The retention by the central JAA of documents related to the development and production of JARs.
(2) The format and structure of JARs.
(3) The development of JARs and amendments to JARs until their publication by the JAA.
(4) The procedures for granting exemptions in the JARs.
(5) The procedures for consultation on special conditions.
(6) The development of Advisory Circulars—Joint (ACJ) until their publication by the JAA.

4.3.3. FAR 11. General Rulemaking Procedure

This part applies to the issuance, amendment, and repeal of any regulation for which the FAA follows public rulemaking procedures under the Administrative Procedure Act. In this context, the code prescribes requirements applicable to

---

[4] **The correct denomination of the FAA regulations should be 14 CFR Part XX (Ex. Part 11).** For the sake of practicality, and to clearly see the difference from JAA and EASA requirements, we use the denomination "FAR XX (Ex. FAR 11)."

(1) Procedures for issuing a rule, from the "advanced notice of proposed rule-making", to the "notice of proposed rulemaking", and to the "final rule".
(2) Petitions for exemptions (from individual or entity).
(3) Petitions for rulemaking (from individual or entity).
(4) Special conditions for issuing a rule.

4.3.4. JAR 21. Certification Procedures for Aircraft and Related Products and Parts

See relevant paragraph in this chapter.

4.3.5. FAR 21. Certification Procedures for Products and Parts

See relevant paragraph in this chapter.

4.3.6. JAR 22. Sailplanes and Powered Sailplanes[5]

See relevant paragraph in this chapter.

4.3.7. JAR-VLA. Very Light Aeroplanes[6]

See relevant paragraph in this chapter.

4.3.8. JAR 23. Normal, Utility, Aerobatic, and Commuter Category Aeroplanes

See relevant paragraph in this chapter.

4.3.9. FAR 23. Airworthiness Standards: Normal, Utility, Acrobatic, and Commuter Category Airplanes

See relevant paragraph in this chapter.

4.3.10. JAR 25. Large Aeroplanes

See relevant paragraph in this chapter.

4.3.11. FAR 25. Airworthiness Standards: Transport Category Airplanes

See relevant paragraph in this chapter.

---

[5] The FAA adopted JAR 22 as an acceptable standard for the certification of sailplanes and powered sailplanes in the United States.
[6] An equivalent FAA standard does not exist. The FAA adopted these requirements for the certification of very light aeroplanes in the United States. The acceptable criteria for the adoption of JAR-VLA are included in AC 21.17−2A. The FAA also issued adjunctive rules in AC 23−11 to authorize IFR and night flight of such aeroplanes.

4.3.11/a. FAR 26. Continued Airworthiness and Safety Improvement for Transport Category Airplanes

This part establishes requirements for support of the continued airworthiness of and safety improvements for transport category airplanes. These requirements may include performing assessments, developing design changes, developing revisions to Instructions for Continued Airworthiness (ICA), and making necessary documentation available to affected persons. Requirements of this part that establish standards for design changes and revisions to the ICA are considered airworthiness requirements.

4.3.12. JAR 26. Additional Airworthiness Requirements for Operations

This code prescribes specific additional airworthiness requirements with which operators must ensure that compliance has been established if operating in accordance with the Part of JAR-OPS relevant to the particular type of operations.

(1) Subpart B relates to Commercial Air Transportation (Aeroplanes).

(2) Subpart C (*reserved*) relates to General Aviation (Airplanes).

(3) Subpart D (*reserved*) relates to Commercial Air Transportation (Helicopters).

(4) Subpart E (*reserved*) relates to General Aviation (Helicopters).

4.3.13. JAR 27. Small Rotorcraft

See relevant paragraph in this chapter.

4.3.14. FAR 27. Airworthiness Standards: Normal Category Rotorcraft

See relevant paragraph in this chapter.

4.3.15. JAR 29. Large Rotorcraft

See relevant paragraph in this chapter.

4.3.16. FAR 29. Airworthiness Standards: Transport Category Rotorcraft

See relevant paragraph in this chapter.

4.3.17. FAR 31. Airworthiness Standards: Manned free balloons[7]

See relevant paragraph in this chapter.

---

[7] The JAA has not issued requirements for free balloons.

4.3.18. JAR-E. Engines

This code is based on the English BCAR Section C and contains the airworthiness requirements for engines. Subsections B and C deal specifically with piston engines; subsections D and E deal specifically with turbine engines.

4.3.19. FAR 33. Airworthiness Standards: Aircraft engines

This part prescribes airworthiness standards for the issue of type certificates for aircraft engines and changes to those certificates. Subparts C and D deal specifically with reciprocating aircraft engines, and Subparts E and F deal specifically with turbine aircraft engines.

4.3.20. JAR-APU. Auxiliary Power Units[8]

This code is based on FAA Technical Standard Order TSO-C77a and provides airworthiness requirements for the release of Joint Technical Standard Order (JTSO) authorizations for turbine-powered auxiliary power units for use in aircraft.

4.3.21. FAR 34. Fuel Venting and Exhaust Emission Requirements for Turbine Engine-Powered Airplanes[9]

The provisions of this subpart are applicable to all in-use aircraft gas turbine engines of the classes specified, certificated for operations within the United States.

As regards foreign airplanes, this FAR applies only to those foreign civil airplanes that, if registered in the United States, would be required by applicable FARs to have a US standard airworthiness certificate to conduct the operations intended for the airplane.

4.3.22. JAR-P. Propellers

The requirements of this code apply to propellers of conventional design.

4.3.23. FAR 35. Airworthiness Standards: Propellers

This part prescribes airworthiness standards for the issue of type certificates and changes to those certificates for propellers.

Each person who applies under FAR 21 for such a certificate or change must show compliance with the applicable requirements of FAR 35.

---

[8] The FAA rules for APU certification are contained in the TSO C 77 B.

[9] Exhaust emissions refer to substances emitted into the atmosphere from the exhaust nozzle of an aircraft engine. Fuel venting emissions refer to raw fuel, exclusive of hydrocarbons in the exhaust emissions, discharged from aircraft gas turbine engines during all normal ground and flight operations.

4.3.24. JAR 36. Aircraft Noise

JAR 34. Aircraft Engine Emission

JAR 36 consists of five subparts and reproduces the Standards agreed by the ICAO for Environmental Protection in Annex 16, Volume I: Aircraft Noise.

JAR 34 reproduces the Standards agreed by the ICAO for Environmental Protection in Annex 16, Volume II: Aircraft Emissions.

4.3.25. FAR 36. Noise Standards: Aircraft type and airworthiness certification

This part prescribes noise standards for the issue of the following certificates:

(1) Type certificates, and changes to those certificates, and standard airworthiness certificates for subsonic transport category large airplanes and for subsonic jet airplanes regardless of category.

(2) Type certificates, and changes to those certificates, standard airworthiness certificates, and restricted category airworthiness certificates for propeller-driven, small airplanes and for propeller-driven, commuter category airplanes, except those airplanes that are designed for "agricultural aircraft operations" (as defined in FAR 137.3, as effective on 1 January 1966) or for dispersing fire-fighting materials to which FAR 36.1583 does not apply.

(3) Type certificates, and changes to that certificate, and standard airworthiness certificates for Concorde airplanes.

(4) Type certificates, and changes to those certificates, for helicopters, except those helicopters that are designated exclusively for agricultural aircraft operations, for dispensing fire-fighting materials or for carrying external loads.

4.3.26. FAR 39. Airworthiness Directives

The regulations in this part provide a legal framework for the FAA's system of Airworthiness Directives.[10]

4.3.27. FAR 43. Maintenance, Preventive Maintenance, Rebuilding, and Alterations

See relevant paragraph in Chapter 9.

4.3.28. FAR 45. Identification and Registration Marking

This part prescribes the requirements for

(1) Identification of aircraft, and aircraft engines and propellers, which are manufactured under the terms of a type or production certificate.

(2) Identification of certain replacement and modified parts produced for installation on type-certificated products.

---

[10] The FAA's Airworthiness Directives are legally enforceable rules that apply to aircraft, aircraft engines, propellers, and appliances.

(3) Nationality and registration marking of US-registered aircraft.

4.3.29. JAR-TSO. Joint Technical Standard Orders

While the requirements for issue of JTSOs are found in JAR 21 Subparts O and N−O, the code provides the list of JTSOs as follows:
- **Index 1**: the JTSOs that are technically similar to FAA TSOs.
- **Index 2**: the JTSOs that are applicable only to JAR (different from FAA TSOs, or corresponding FAA TSOs not existing).

4.3.30. JAR-OPS 1. Commercial Air Transportation (Aeroplanes)

This code prescribes requirements applicable to operation of any civil aeroplane for the purpose of commercial air transportation by any operator whose principal place of business is in a JAA Member State, with exceptions indicated in the same code.

4.3.31. JAR-OPS 3. Commercial Air Transportation (Helicopters)

This code prescribes requirements applicable to any civil helicopter for the purpose of commercial air transportation by any operator whose principal place of business is in a JAA Member State, with exceptions indicated in the same code.

4.3.32. JAR-MMEL/MEL. Master Minimum Equipment List/Minimum Equipment List

See relevant paragraph in Chapter 5.

4.3.33. FAR 91. General Operating and Flight Rules

Except as provided in cases indicated, this part prescribes rules governing the operation of aircraft (other than moored balloons, kites, unmanned rockets, and unmanned free balloons, which are governed by FAR 101, and ultralight vehicles operating in accordance with FAR 103) within the United States, including the waters within three nautical miles of the US coast.[11]

4.3.34. FAR 101. Moored Balloons, Kites, Unmanned Rockets, and Free Balloons

This part prescribes rules governing the operation in the United States, of moored balloons, kites, unmanned rockets, and free balloons, whose

---

[11] Although this part is essentially operative, airworthiness is recalled for equipment, instrument, and certification requirements. The same applies to other operative parts such as FAR 121, 125, 129, 133, 135, JAR-OPS, and JAR-AWO.

characteristics and limitations (as applicable weight, gas capacity, quantity and quality of propellant, etc.) are defined.

4.3.35. FAR 103. Ultralight Vehicles

This part prescribes rules governing the operation of ultralight vehicles in the United States. For the purposes of this part, ultralights are defined in terms of maximum weight (powered and unpowered), maximum speed (powered), and maximum stalling speed; the operations are limited to a single occupant and their use to recreation or sport purposes only.

4.3.36. FAR 119. Certification: Air Carriers and Commercial Operators

This part applies to each person operating or intending to operate civil aircraft as an air carrier or commercial operator, or both, in air commerce or, when common carriage is not involved,[12] in operations of US-registered civil airplanes with a seat configuration of 20 or more passengers, or a maximum payload capacity of 6000 lb or more. This part prescribes in particular the certification requirements an operator must meet to obtain and hold a certificate authorizing operations under FAR 121, 125, or 135.

4.3.37. FAR 121. Operating Requirements: Domestic, Flag, and Supplemental Operations

This part prescribes rules governing (in particular):
(1) The domestic, flag, and operations of each person who holds an Air Carrier Certificate or Operating Certificate under FAR 119.
(2) Each person employed by a certificate holder conducting operations under this part, including maintenance, preventive maintenance, and alteration of aircraft.

4.3.38. FAR 125. Certification and Operations: Airplanes having a seating capacity of 20 or more passengers or a maximum payload capacity of 6000 pounds or more; and rules governing persons on board such aircraft

This part prescribes rules governing the operations of the above-mentioned US-registered civil airplanes when common carriage is not involved, unless they required to be operated under FAR 121, 129, 135, or 137, and unless other cases described in this part are applicable.

---

[12] See Chapter 8, Section 8.6.2.1 *"Definitions."*

4.3.39. FAR 129. Operations: Foreign air carriers and foreign operators of US-registered aircraft engaged in common carriage[12]

This part prescribes rules governing the operations within the Unites States of each foreign air carrier holding (defined) permits issued by the Civil Aeronautic Board of the US Department of Transportation.

4.3.40. FAR 133. Rotorcraft External Load Operations

This part prescribes airworthiness and operating certification rules for rotorcraft used in the above-mentioned operations in the United States by anyone, with the exceptions defined in the same document.

4.3.41. FAR 135. Operating Requirements: Commuter and on-demand operations and rules governing persons on board such aircraft

This part prescribes rules governing the commuter or on-demand operations of each person who holds, or it is required to hold an Air Carrier Certificate or Operating Certificate under FAR 119 and relevant items.

4.3.41. bis FAR 136. Commercial Air Tours[13] and National Parks Air Tour Management

This part applies to each person operating or intending to operate a commercial air tour in an airplane or helicopter and, when applicable, to all occupants of the airplane or helicopter engaged in a commercial air tour.

This subpart also clarifies the requirements for the development of an air tour management plan for each park in the national park system where commercial air tour operations are flown.

4.3.42. FAR 137. Agricultural Aircraft Operations

This part prescribes rules governing agricultural operations within the United States and the issue of commercial, and private agricultural aircraft operator certificates for those operations.

4.3.43. FAR 145. Repair Stations

This part describes how to obtain a repair station certificate. This part also contains the rules a certificated repair station must follow relating to its performance of maintenance, preventive maintenance, or alterations of an aircraft, airframe, aircraft engine, propeller, appliance, or component part to which

---

[13] *Commercial Air Tour* means a flight conducted for compensation or hire in an airplane or helicopter where a purpose of the flight is sightseeing.

FAR 43 applies. It also applies to any person who holds, or is required to hold, a repair station certificate issued under this part.

4.3.44. FAR 147. Aviation Maintenance Technician Schools

This part prescribes the requirements for issuing aviation maintenance technician school certificates and associated ratings and the general operating rules for the holders of those certificates and ratings.

4.3.45. JAR-AWO. All Weather Operations

This code prescribes requirements for
(1) Automatic landing systems.
(2) Airworthiness certification of aeroplanes for operations with decision heights of 60 m (200 ft) down to 30 m (100 ft)—Category 2 operations.
(3) Airworthiness certification of aeroplanes for operations with decision height below 30 m (100 ft) or no decision height—Category 3 operations.
(4) Directional guidance for takeoff in low visibility.

4.3.46. JAR/CS-VLR. Very Light Rotorcraft

See relevant paragraph in this chapter.

4.3.47. References for certification of parts of aircraft

(1) JTSO authorization (JAR 21 Subpart O).
(2) Technical Standard Order (TSO) (FAA AC 20−110).
(3) Joint Part Approval authorization (JAR 21 Subpart P).
(4) Part Manufacturer Approval (FAR 21.303).
(5) Military and industrial specifications.
(6) Specifications written in the aircraft certification process.
   NOTE: We will deal with this subject in more detail in Chapter 5, in the section "Parts and appliances approval."

4.3.48. General remarks

The standards dealing with the same products[14] have been put in sequence in the above list. The list shows the existence of **operational standards** in addition to the product type-certification standards. These operational standards contain airworthiness requirements that influence the aircraft configuration in relation to their particular operations.[15]

---

[14] According to JAR and FAR, products are aircraft, aircraft engines, and propellers.
[15] Obviously, the JAA and FAA operational standards are related to the aircraft registered in the country having those standards as legal operational rules.

A JAR/FAR 23 aeroplane, for example, can obtain a type certificate with the installation (as flight and navigation instruments) of an airspeed indicator, an altimeter, and a magnetic direction indicator only. However, to obtain a certificate of airworthiness (the document that authorizes the flight), other instruments and equipment that depend on the particular type of operation (e.g., tourism, aerial work) and on the flight conditions [Visual Flight Rules (VFR), Instrumental Flight Rules (IFR), night flight, etc.] must be installed as prescribed by the operational rules.

Furthermore, the environmental standards such as FAR 34 and JAR/FAR 36 must be considered. For the FAA and EASA, compliance with the environmental protection requirements is part of the type certification.[16]

The JAA requirements were adopted by the JAA Member States, the EASA requirements are now used by the EU Member States, and the FAA regulations are used in the United States[17] (Canada has almost equivalent rules).

Nevertheless, the manufacturing companies wanting to sell their products on both sides of the Atlantic must perform a double certification, with a substantial increase in costs, especially when the standards are different. For many years, the transport aeroplane industry has been penalized because the contents of JAR 25 and FAR 25 were not equivalent (even including the same paragraph numbering). JAR 25 originated under a strong English influence, with philosophies borrowed from BCAR Section D. It is also because of manufacturer's complaints that, for many years now, the JAA and FAA have carried out a harmonization process that is well advanced, but not yet totally accomplished. The situation is better for JAR/FAR 23 aeroplane standards and for the JAR/FAR 27 and 29 rotorcraft[18] standards, because these JAR requirements were produced with the cooperation of the FAA, with the common will of avoiding the situation that has penalized transport aeroplanes: these standards are now almost harmonized.

We will see in the subsequent chapters how it is possible to minimize the burden related to the acceptance of certifications made by different authorities.

4.3.49. Historical background of FAA aircraft airworthiness regulations

Figure 4.1, copied from the Order 8110.4C, gives a synthesis of the evolution of the FAA aircraft airworthiness regulations.[19]

---

[16] See Chapter 5.

[17] Many states adopt the FAA regulations as a basis for their national regulations.

[18] The term "rotorcraft" is not only limited to helicopters but also includes gyroplanes (even if they are less common).

[19] **Civil Air Regulations (CARs).** The CARs were part of the original certification basis for aircraft first certified in the 1940s, 1950s, and 1960s by the Civil Aeronautics Administration. As such, the CARs may still be needed as a reference for older aircraft, or as a standard for minor changes to older aircraft designs.

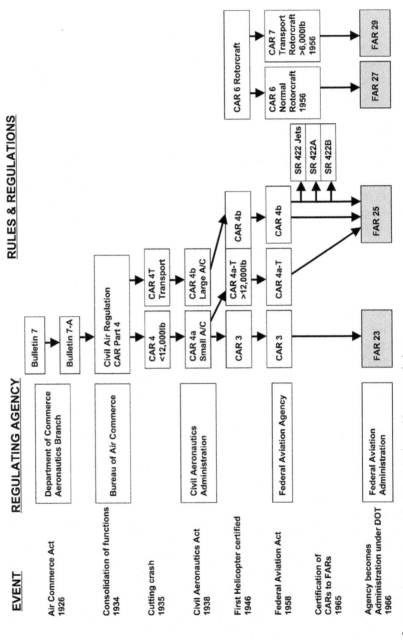

FIGURE 4.1 Historical background of aircraft airworthiness regulations

4.4. ADVISORY MATERIAL

Some rules can be interpreted in different ways. This is the reason why the authorities issue advisory material for the explanation of the rule or, in certain cases, suggest suitable procedures to perform a demonstration of compliance to the same rule.

The FAA publishes "Advisory Circulars" (ACs) as documents separate from the standards, while the JAA and EASA include similar documents at the end of the JAA/EASA standards.

If we look at the JARs, in Section 2, these standards contain the "ACJs" that are "Acceptable Means of Compliance (AMC), and Interpretations."[20] The ACJs provide a means, *but not the only means*, by which a requirement can be met.[21] A numbering system is adopted in which the ACJ uses the same number as the paragraph of the JAR to which it is related.

By the same approach, the EASA Certification Standards (CS) contain the AMC, with the same meaning as the ACJs.

For the Implementing Rules (IRs) of the EASA, such as Part 21, Part M, Part 145, and so on, documents containing the AMC and guidance material (GM) have been issued.

The AMC have the meaning already defined, whereas the GM helps to illustrate the meaning of a specification or requirements.

4.5. EASA REGULATIONS

Figure 4.2 depicts the EASA's regulation organizational structure.

4.5.1. The Basic Regulations

The Basic Regulation establishes common requirements for the regulation of safety and environmental sustainability in civil aviation. It gives the European Commission powers to adopt detailed rules for the Regulation's implementation.

The Basic Regulation was EC No. 1592/2002 already mentioned, now repealed by the Regulation (EC) No. 216/2008 of 20 February 2008, which outlines the tasks of the Agency starting from the necessity that "*a high and uniform level of protection of the European citizen should at all times be ensured in civil aviation, by the adoption of common safety rules and by measures ensuring that products, persons and organizations in the Community comply with such rules and those adopted to protect the environment.*"

---

[20] JAR 11 defines ACJ as "an accompanying text, containing explanations, interpretations or acceptable means of compliance, in order to clarify and to provide guidance for the application of requirements."

[21] This means that the designer (or better, the "applicant," as it is normally defined) can choose other means of compliance, but in this case has to convince the authority about the validity of the choice.

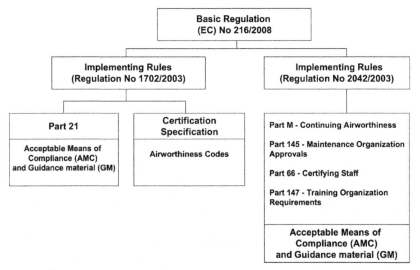

FIGURE 4.2 EASA regulation structure

While the Regulation EC No. 1592/2002 established competency only for the regulation of the airworthiness and environmental compatibility of products, the No. 216/2008 is extended to air operation, fly crew licensing, and third country aircraft.

During the next few years, this competency will also cover the safety regulation of airport and air traffic control service.

This Basic Regulation applies to:

a. the design, production, maintenance, and operation of aeronautical products, parts, and appliances, as well as personnel and organizations involved in the design, production, and maintenance of such products, parts, and appliances

b. personnel and organizations involved in the operation of aircraft.

The principal objective of this Regulation is to establish and maintain a high uniform level of civil aviation safety in Europe.

Additional objectives, listed in Article 2, set forth the environmental protection, simplification and effectiveness of common certification procedures, free movement of goods, persons, and services, and so on.

4.5.2. Implementing Rules

These IRs contain documents defined as **Parts**, which are divided into two sections: Section A, detailing the requirements to be satisfied by aeronautical subjects; and Section B, containing the procedures to be followed by the national authorities.

The IRs (EC) No. 1702/2003 for the airworthiness and environmental certification of aircraft and related products, parts, and appliances, as well as for the certification of design and production organizations, specify:

(1) The issuing of type certificates, restricted type certificates, and changes to those certificates.

(2) The issuing of certificates of airworthiness, restricted certificate of airworthiness, permit to fly, and authorized release certificates.

(3) The issuing of repair design approval.

(4) The showing of compliance with environmental protection requirements.

(5) The issuing of noise certificates.

(6) The identifying of products, parts, and appliances.

(7) The certifying of certain parts and appliances.

(8) The certifying of design and product organizations.

(9) The issuing of Airworthiness Directives.

Annex to this document is **Part 21, "Certification of aircraft and related products, parts and appliances, and design and production organizations."**

This document replaces JAR 21, which remains the core of the same document. The changes to the JAR document reflect the new legal status of the EASA toward the national authorities, and a full revision of the document in the light of the JAA certification experience.

The IRs (EC) No. 2042/2003 specify the continuing airworthiness of aircraft and aeronautical products, parts, and appliances, and the approval of organizations and personnel involved in these tasks.

The following are Annexes to this document:

(1) Annex I, Part M establishes the measures to be taken to ensure that airworthiness is maintained, including maintenance. It also specifies the conditions to be met by persons and organizations involved in such continuing airworthiness management.

(2) Annex II, Part 145 establishes the requirements to be met by an organization to qualify for the issue or continuation of an approval for the maintenance of aircraft and components.

(3) Annex III, Part 66 establishes the requirements for the issue of an aircraft maintenance license and conditions of its validity and use, for aeroplanes and helicopters.

(4) Annex IV, Part 147 establishes the requirements to be met by organizations seeking approval to conduct training and examination as specified in Part 66.

4.5.3. AMC and GM for Parts 21, M, 145, 66, and 147[22]

As already mentioned, AMC illustrate a means, but not the only means, by which a specification contained in an airworthiness code or a requirement in an IR can be met. GM helps to illustrate the meaning of a specification or requirement.

4.5.4. Airworthiness Codes

Almost all airworthiness codes are directly derived from the JARs. The JAR denomination has been changed in Certification Specification (CS).

---

[22] See also the "Advisory material" section in this chapter.

Currently, the airworthiness codes are as follows:

(1) CS-Definitions. Derived from JAR 1.

(2) CS-22: Sailplanes and Powered Sailplanes. Derived from JAR 22.

(3) CS-23: Normal, Utility, Acrobatic, and Commuter Aeroplanes. Derived from JAR 23.

(4) CS-25: Large Aeroplanes. Derived from JAR 25.

(5) CS-27: Small Rotorcraft. Derived from JAR 27.

(6) CS-29: Large Rotorcraft. Derived from JAR 29.

(7) CS-VLR: Very Light Rotorcraft. Derived from JAR-VLR.[23]

(8) CS-VLA: Very Light Aeroplanes. Derived from JAR-VLA.

(9) CS-E: Engines. Derived from JAR-E.

(10) CS-P: Propellers. Derived from JAR-P.

(11) CS-34: Aircraft Engine Emission and Fuel Venting. Derived from JAR 34.

(12) CS-36: Aircraft Noise. Derived from JAR 36.

(13) CS-APU: Auxiliary Power Units. Derived from JAR-APU.

(14) CS-ETSO: European Technical Standard Orders. Derived from JAR-TSO.

(15) CS-AWO: All Weather Operations. Derived from JAR-AWO.

(16) AMC-20: General AMC for Airworthiness of Products, Parts, and Appliances.

(17) CS-31 HB: Hot Air Balloons (issued 27 February 2009).

For the certification of parts of aircraft, the references are the following:

(1) European Technical Standard Order (ETSO) authorization (Part 21 Subpart O).

(2) Specifications written in the aircraft certification process.

(3) Standard parts in accordance with officially recognized standards.

4.6. GENERAL CONSIDERATIONS ON AIRWORTHINESS STANDARDS

Before considering the single standards (at least the more representative ones relating to the information this book is aimed to provide), it is worth considering the "philosophies" that are the basis of their compilation.

4.6.1. Publication

The standards are made by Working Groups that are responsible for their compilation and amendments. Before publication, the Authorities concerned (the JAA, FAA, or EASA) submit the standards to public evaluation, allowing interested people and organizations to express their opinions on the matter. The relevant rules and the procedures for these phases are contained in JAR 11, "Regulatory and Related Procedures," and in FAR 11, "General Rulemaking Procedures."

---

[23] See the "JAR-VLR" section in this chapter. At the end of 2002, it was still in a status of NPA.

The EASA does not have a similar standard, but since 2003 has a standard rulemaking procedure (EASA Management Board Decision 07/2003).

This procedure has now been amended and replaced by the EASA Management Board Decision 08/2007.

4.6.2. Special conditions

As mentioned earlier, the standards do not anticipate aeronautical progress. Therefore, in several cases, a "nonconventional aircraft" is the object of the certification, or one with some peculiarities for which the "applicable" airworthiness requirements of the relevant JAR/FAR/CS do not contain adequate or appropriate safety standards. As we have also considered that a "blocked" standard might prevent aeronautical progress, what should be done in such situations? JAR/FAR 21, Paragraph 16, and EASA Part 21, Paragraph 21A.16B, provide an answer mentioning **"special conditions."** It is a matter of adding such safety standards as the authority finds necessary to establish a level of safety equivalent to that established in the applicable JAR/FAR/CS. The special conditions are issued in accordance with JAR/FAR 11 and, for EASA, according to the "Products certification procedure"-Decision 02/2004.

We will return to the "level of safety" concept. However, to mention just one of the numerous possible examples, special conditions were issued for turbine engine installations on FAR 23 aircraft when FAR 23 did not yet contain safety standards for this kind of installation. It is not difficult to imagine the number of special conditions issued for the certification of "Concorde" in the 1960s.

In many cases, if design peculiarities that require special conditions become commonplace in the aeronautical field, for example "winglets," such special conditions are included (after discussions and evaluations according to JAR/FAR 11 and EASA rulemaking procedure) in the JAR/FAR/CS standards via amendments.

4.6.3. Severity of the airworthiness standards

The "level of safety" concept is a matter of serious concern regarding the compilation of the standards. The authorities could be tempted to play safe by issuing very restrictive standards. The immediate result would be to make it impossible for an aircraft to be certified for technical or simply for economical reasons.[24] Within airworthiness standards, it is therefore necessary to balance criteria of **acceptability** (from the safety point of view) and the **practicability** of the same criteria.

The application of a rule involves expense. Increase of safety is not always proportional to the severity of the rule, even before considering the expense: at and beyond a certain point, negligible safety increases incur great expenditure. At this point, the rule is no longer "practicable" (Fig. 4.3).

---

[24] It used to be said that the limit trend of the airworthiness standards was to make aircraft certification impossible!

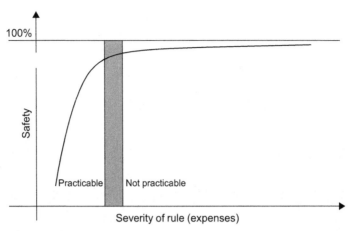

FIGURE 4.3 Airworthiness rules

As a "golden" rule in airworthiness rulemaking, a proposal should be
a. economically reasonable,
b. technologically practicable, and
c. appropriate to the particular type of aircraft.

Various airworthiness standards have been produced for different types of
aircraft (aeroplanes, rotorcraft, etc.) and also for different categories of the
same type of aircraft (for weight, passenger numbers, etc.). An attempt has
been made to arrange the aircraft in groups that are as "homogeneous" as
possible. Neglecting the obvious necessity to distinguish free balloons from
transport aeroplanes, a distinction was made, for example, among the categories of aeroplanes named normal, utility, aerobatic, and commuter in
JAR/FAR 23 and the Large Aeroplanes/Transport category airplanes in
JAR/FAR 25. We must not infer that the airworthiness standards are different
because the transport aeroplanes should be safer than other types of aircraft.
Safety must be maximized for all aircraft, taking into account the criteria of
"practicability" mentioned above. As a fundamental concept, simple aircraft
should have simple airworthiness standards to comply with.

It is certainly less easy to understand why a 19,000-lb commuter certified
according to JAR/FAR/CS-23, if it "adds on weight" by some pounds then
comes under the JAR/FAR/CS-25 regulations. However, it is clear that the
creation of classes implies that significant parameters are established consequently involving precise numbers.[25] The designer should be capable of
choosing the right airworthiness standard in light of the possible development
of the project.

In any case, it is worthwhile considering that airworthiness, such as medicine, is not an exact science!

---

[25] The same also applies for boxers' categories or for tax brackets.

4.6.4. Stalling speed for single-engine aeroplanes

In cases of "acceptability," "practicability," and examples of "philosophies," we will now see how single-engine aeroplanes are regulated from a particular point of view: the stalling speed. A single-engine aeroplane, in case of engine failure, by definition can only glide. If this condition is not manageable in safety terms, it should "never" happen. In fact, despite the great progress made in engineering techniques, the engine that "never"[26] fails does not exist. We can also add that the present engine failure rate should not be compatible—and therefore not acceptable—with safety if any engine failure were to cause an accident. It is then necessary that the gliding and (especially) the power-off landing of a single-engine aeroplane be managed by a pilot of average skill.[27] It is evident that the result of an out-of-field landing is mainly influenced by the approach speed. However, the minimum gliding approach speed in the landing configuration is a function of the power-off stalling speed in the same configuration; hence, a limitation of this speed is required. As a result, the stalling speed of single-engine aeroplanes in landing config-uration (V_{so}) is limited to 61 knots. The same limitation exists for twin-engine aeroplanes that cannot meet a certain minimum rate of climb with an inoper-ative engine.

For all other twin-engine aeroplanes (even with an engine failure probability that is double that of a single-engine aeroplane), the proba-bility of a twofold engine failure in the same flight is considered close to "never," and therefore acceptable, so that no stalling speed limit is prescribed.[28]

It is interesting to note that, on the basis of the above-mentioned principles, JAR-VLA[29] contains a speed limitation in landing configuration of 45 knots, because it allows the installation of JAR 22 powered sailplane engines that, at least in principle, are considered less reliable than the engines installed on JAR/FAR 23 aeroplanes.

A stalling speed limitation in landing configuration was also introduced in JAR 22, because the trend toward the increase of water ballast quantity for speed contests was producing such an increase in wing loading, and therefore in the stalling speed, to jeopardize the possibility of a safe landing in the case, for example, of an aborted takeoff or breaking of the tow rope; normally, in these cases, there is insufficient time to dump the water ballast.

---

[26] We will see later the meaning of "never" in flight safety assessment.

[27] As a basic concept, an aircraft must be manageable in all foreseen conditions, by a crew of average skill (for that class of aircraft), and not necessarily by over-skilled people.

[28] We can understand the lack of single-jet engine aeroplanes in the civil market, because they have performance normally incompatible with the above stalling speed limitation. We will mention this issue again with regard to "crashworthiness."

[29] As we will see, this airworthiness standard concerns aeroplanes up to 750 kg maximum weight.

4.6.5. Crashworthiness

We have mentioned a stalling speed limitation of 61 knots. However, is it really true that such a limitation could produce a safe power-off landing?

When limitations of this type are introduced, generally they are the result of experience and analysis of accidents that occurred in relevant situations. They are certainly not chosen at random. Nevertheless, the limitation cannot take into consideration all the conditions of the area where the aeroplane is likely to land (or crash if the ground is particularly uneven). Then, the possibility of a crash must be considered, for whatever reasons and not only for single-engine aeroplanes. The airworthiness standards have become more and more stringent from this point of view. This is what we call **crashworthiness**.

JAR/FAR/CS-23 contains appropriate safety standards for emergency landing conditions. It deals with structural rules for the occupants' protection, also requiring expensive static and dynamic tests for the seat/restraint system, the seats, and the fuselage structure supporting the same.

FAR 23 prescribed something more. To allow the certification of speedy single-engine aeroplanes (turbine engine-powered), the design of which is severely penalized by the 61-knot limitation, these regulations enable an increase of stalling speed to be "exchanged" with the additional severity of the crashworthiness regulations; we again encounter the acceptability/practicability balance. EASA issued the Notice of Proposed Amendment (NPA) No. 2008-08[30] of 30 April 2008 to amend CS 23 and harmonize with the FAR 23.

Crashworthiness concerns all types of aircraft.

The JAR 22 Study Group tried to avoid dynamic tests for aircraft such as sailplanes and powered sailplanes. These aircraft are normally produced in such small numbers that it becomes economically difficult for the manufacturers to sustain the cost of dynamic crash tests. Nevertheless, the problem does exist and it is a serious one because these machines, for which an out-of-field landing is not even an emergency, tend to crash land frequently. We therefore have to consider the classical configuration of these aircraft that, in principle, does not offer suitable protection for the occupants.

We could think of a "survival cage" able to bear some tens of **g**, but this is not the solution because, even if the cage does not break, the occupants could still sustain extended or maybe fatal injuries.

According to an FAA definition, a survivable crash is "*one where the survivable human tolerance acceleration limit has not been exceeded in any of the principal aircraft axes, where the structure and structural volume surrounding occupants remain sufficiently intact during and after impact to permit survival and where an item of mass does not become unrestrained and create a hazard to occupants.*"

(Of course, this definition is related to dynamic effects only, and not other effects such as fire, smoke, etc.).

---

[30] A-NPA: Advanced Notice of Proposed Amendment. This refers to a document used to seek early advice from interested parties on a possible future NPA.

The OSTIV Sailplane Development Panel (SDP) studied this problem for some time, appointing a Crashworthiness Panel and reaching solutions recalling (to a certain extent) those adopted in Formula 1. The criteria they adopted could be summarized as "stiff cage and soft nose," that is, a sufficiently strong structure to protect the occupants but with a yielding front part, able to absorb impact energy. The OSTIV also provided advice on the seat design; these should be devised as "energy absorbing."[31] Standards for headrests were introduced, very effective items in the rebound phase after impact. The seat profile and the safety harness configuration were studied; the accident analysis pointed out the possibility of spine damage due to sliding under the safety harness in the impact phase, a movement defined as "submarining." Furthermore, the accident analysis showed that the landing gear standards did not offer sufficient energy absorption, with consequences for the occupants' spine. Therefore, these standards were improved.

The criteria coming from the OSTIV SDP were very often introduced in JAR 22, after evaluation of the relevant Study Group.

The NPA-2007-12[30] *"Cockpit crashworthiness"* was published on August 2007 on the EASA webpage. The NPA is based on a proposal of the OSTIV SDP for an amendment of the CS-22 to improve the crashworthiness design of sailplanes' cockpits. The CS-22 was consequently amended on 1 October 2008.

JAR-VLA contains a paragraph dealing with "emergency landing conditions," which has not been updated since the original publication in 1990 and could be considered in need of modernization.

Based on NPA 2008-11, an Amendment of 5 March 2009 provides requirements for rapid escape in normal and crash attitude.

Dynamic crash tests should also be avoided for these aeroplanes, but an update of the crashworthiness criteria is reasonable—for instance, taking into consideration the studies performed for sailplanes.

The airworthiness standards for transport aeroplanes (JAR/FAR/CS-25) and for rotorcraft (JAR/FAR/CS-27 and -29) contain paragraphs on crash landing inclusive of dynamic crash tests.

4.6.6. Fire protection

An aircraft has engines, electrical installations, and other components, making it subject to fire hazard. First, the "fire zones" of the aircraft, that is, those in which a fire can develop, must be located—an engine compartment, for example. There are essentially three methods of protecting the occupants from fire: (a) abandoning the aircraft,[32] (b) passive protection to contain the

---

[31] It may seem trivial, but how many people know that foam rubber cushions can be dangerous? They could indeed return most of the absorbed impact energy.

[32] Independent of the fire emergency, the abandonment of the aircraft is considered in limited categories of civil aircraft such as sailplanes/powered sailplanes and aerobatic aeroplanes. This is necessary for the hazard of flight collisions, especially during thermal flights of sailplanes, and for the hazard of the structure overloading and the critical situation that can occur during aerobatic operations. The applicable airworthiness standards provide suitable rules for this type of emergency.

fire for the time necessary for landing, and (c) active protection by means of extinguishers. Of course, the combination of these last two means is possible. For military aircraft, normally carrying explosive material, abandoning the aircraft is favored (unless the fire is so limited that it can be put out by means of extinguishers), the active or passive protection being limited to the time necessary for the acknowledgment of the situation by the crew and their bailing out.

This cannot imply that fire protection for military aircraft is "optional." If, for example, we consider MIL-HDBK-516B, which establishes the airworthiness certification criteria to be used in the determination of airworthiness of all manned and unmanned (military) aircraft, the constant reference to FAA documents such as FARs and ACs can be noted. However, although the FAR requirements are mandatory in the case of civil aircraft, they are considered as useful airworthiness criteria in the case of military aircraft, not necessarily applicable. Various types of operational missions can bring to a certification basis tailored on the particular type of aircraft, for which, moreover, can also be applied to various documents such as Military (MIL) Specifications, Joint Service Specification Guides, and so on.

In the case of civil aircraft, passive protection is prescribed to allow a safe emergency landing whenever possible. This is achieved by suitable isolation of the fire zones so that essential structures and installations can be protected for the time necessary for landing. The use of extinguishers is not excluded, but they are not considered as primary protection.

Active protection, by means of portable or fixed extinguishers, is prescribed in some categories of aircraft (e.g., transport and commuter aeroplanes), for accidental fires in the cockpit, the cabin, and the baggage or cargo compartments.

The airworthiness standards also provide rules for materials used for the cabin interiors, from the points of view of flammability and noxious smoke emissions.

Because the requirements must normally be substantiated by tests, the certification standards provide acceptable procedures for such tests. To give an idea of the content of these documents, an example can be found in Appendix F to FAR 23, "Test Procedure," of which an extract is reported here.[33]

4.6.6.1. ACCEPTABLE TEST PROCEDURE FOR SELF-EXTINGUISHING MATERIALS FOR SHOWING COMPLIANCE WITH PARAGRAPHS 23.853, 23.855, AND 23.1359

(1) **Conditioning.** Specimens must be conditioned to 70°F, 65°F, and at 50–65 percent relative humidity until moisture equilibrium is reached, or for 24 hours.
(2) **Specimen configuration.** Except as provided for materials used in electrical wire and cable insulation and in small parts, materials must be tested either as a section cut from a fabricated part as installed in the airplane or as a specimen simulating

---

[33] JAR and EASA standards contain equivalent procedures.

a cut section, such as a specimen cut from a flat sheet of the material or a model of the fabricated part. The specimen may be cut from any location in a fabricated part; however, fabricated units, such as sandwich panels, may not be separated for a test. The specimen thickness must be no thicker than the minimum thickness to be qualified for use in the airplane, except that (1) thick foam parts, such as seat cushions, must be tested at 1/2-inch thickness; (2) when showing compliance with Paragraph 23.853(d)(3)(v) for materials used in small parts that must be tested, the materials must be tested at no more than 1/8-inch thickness; (3) when showing compliance with Paragraph 23.1359(c) for materials used in electrical wire and cable insulation, the wire and cable specimens must be the same size as used in the airplane. In the case of fabrics, both the warp and fill directions of the weave must be tested to determine the most critical flammability conditions. When performing the tests prescribed in Paragraphs (d) and (e) of this appendix, the specimen must be mounted in a metal frame so that (1) in the vertical tests of Paragraph (d) of this appendix, the two long edges and the upper edge are held securely; (2) in the horizontal test of Paragraph (e) of this appendix, the two long edges and the edge away from the flame are held securely; (3) the exposed area of the specimen is at least 2 inches wide and 12 inches long, unless the actual size used in the airplane is smaller; and (4) the edge to which the burner flame is applied must not consist of the finished or protected edge of the specimen, but must be a representative of the actual cross-section of the material or the part installed in the airplane. When performing the test prescribed in Paragraph (f) of this appendix, the specimen must be mounted in a metal frame so that all four edges are held securely and the exposed area of the specimen is at least 8 × 8 inches.

(3) **Vertical test.** A minimum of three specimens must be tested and the results are averaged. For fabrics, the direction of weave corresponding to the most critical flammability conditions must be parallel to the longest dimension. Each specimen must be supported vertically. The specimen must be exposed to a Bunsen or Tirrill burner with a nominal 3/8-inch internal diameter (ID) tube adjusted to give a flame of 1.5 inches height. The minimum flame temperature measured by a calibrated thermocouple pyrometer in the center of the flame must be 1550°F.

(4) **Horizontal test.** A minimum of three specimens must be tested and the results averaged. Each specimen must be supported horizontally. The exposed surface when installed in the airplane must be faced down for the test. The specimen must be exposed to a Bunsen or Tirrill burner with a nominal 3/8-inch ID tube adjusted to give a flame of 1.5 inches height. The minimum flame temperature is measured.

(5) **Forty-five-degree test.** A minimum of three specimens must be tested and the results averaged. The specimens must be supported at an angle of 45° to a horizontal surface. The exposed surface when installed in the aircraft must be faced down for the test. The specimens must be exposed to a Bunsen or Tirrill burner with a nominal 3/8-inch ID tube.

(6) **Sixty-degree test.** A minimum of three specimens of each wire specification (make and size) must be tested. The specimen of wire or cable (including insulation) must be placed at an angle of 60°.

(7) **Burn length.** Burn length is the distance from the original edge to the furthest evidence of damage to the test specimen due to flame impingement, including areas of partial or complete consumption, charring or embrittlement, but not

including areas sooted, stained, warped or discolored, or areas where material has shrunk or melted away from the heat source.

4.6.7. Safety assessment

Let us consider the control system of a light aeroplane: cables, pulleys, perhaps some rods. These items are very often in view and easy to inspect. For such systems, if designed according to good design practice and applicable airworthiness standards and maintained following the maintenance manual instructions (providing the replacement of worn parts), no particular studies will be needed to assure the system's safety during the entire operating life of the aeroplane. We can therefore talk of a system that "never" fails. It is quite different if, considering a more sophisticated aircraft, the control system depends on the electrical and hydraulic systems, or even the mechanical transmissions are eliminated, as for fly-by-wire systems, with computers playing an important part.

The above example on control systems can obviously be extended to all aircraft systems and equipment.

In this case, the safety assessment would require more refined rules and instruments. The essentially informative nature of this book cannot provide a thorough discussion on this very specific topic. Nevertheless, it is worth outlining some basic concepts.

The rules for safety assessment are contained in different aircraft airworthiness standards at Paragraph XX.1309,[34] and advisory material in the respective ACJs/ACs/AMC&GM. As specified by the title of Paragraph 1309, they are related to "Equipment, Systems, and Installations."

As a consequence, these rules do not apply to performance, flight qualities, and structural load and strength of Subparts B, C, and D.[35] However, they do apply to any system on which compliance with the requirements of Subparts B, C, D, and E is based. As a typical example (contained in FAA AC 23-1309-1D), Paragraph 23.1309 does not apply to the stall characteristics of Paragraph 23.201, but nevertheless it applies to a stick pusher (stall barrier) installed to satisfy the latter paragraph.

That being said, if we were to ask a layman (better still, a passenger) what kind of reliability a vital aircraft system should have, the answer would immediately be 100 percent. It has nevertheless been demonstrated that such reliability is an impossibility. As an example, setting in parallel "n" items (redundancy), 100 percent reliability can be obtained for n tending to infinity!

A system with a high degree of redundancy would be heavy, expensive, and complex: so subject to drawbacks that it would make such redundancy questionable. It is then more convenient to design such systems with a minimum

---

[34] JAR 22 does not contain this paragraph; JAR-VLA provides general indications only to minimize hazards in case of failure. This is consistent with the (generally) simple systems of the relevant aircraft.

[35] See the "Structure of aircraft airworthiness standards" section in this chapter.

degree of redundancy (the reliability of the single components can be increased), in order that its reliability, even if not amounting to 100 percent, is such as to ensure **an acceptable safety level**.

The definition of an acceptable safety level implies the definition of an **acceptable accident rate**; this cannot be defined as abstract wishful thinking, but on the basis of what is **practicable**.

What is practicable for the future can be forecast by the analysis of past accident rates. Therefore, after taking into consideration, the accident rate in commercial (occidental) aviation in the 10-year period from 1970 to 1980, a rate of catastrophic accidents[36] a little less than 1×10^{-6} flight hours was detected. From this accident analysis, it was also found that about 10 percent of the catastrophic accidents could be attributed to system failures. Hence, the portion of catastrophic accidents attributed to systems was of the order of 1×10^{-7} flight hours.

Starting from the arbitrary hypothesis that a commercial large aircraft could present some 100 hazards (potential failure conditions) leading to a catastrophic effect, it follows that, for each system, the acceptable probability of a catastrophic failure is less than 10^{-9} flight hours.

This is the basic concept for "the maximum probability of a catastrophic effect for a single system"[37] of a transport aeroplane.

The general intention is that effects of a catastrophic nature should virtually never occur in the fleet life of a particular type of aircraft. This would mean, for example, that in the case of a fleet of 100 aircraft of a particular type, each flying 3000 hours per annum, one or more of the various catastrophic effects might be expected to occur once in 30 years, which is close to the concept of "virtually never,"[38] a situation near to that never we have already considered.

We have to bear in mind that there are some systems operating constantly and others operating in a certain flight phase only (the latter could make up as much as 80 percent of the total: e.g., a landing gear system). Hence, a probability failure per flight hours of such systems can be established by dividing the probability by the average flight duration estimated for the particular type of aircraft.

4.6.7.1. FAILURE CONDITIONS

Failure conditions are defined as effects on the aircraft and its occupants, both direct and consequential, caused or contributed to by one or more failures, considering relevant adverse operational or environmental conditions. Failure conditions may be classified according to their severity as follows (AMJ 25.1309):

---

[36] A "multifatality" accident, normally leading to the loss of the aircraft.

[37] Accident analysis for other types of aircraft leads to different values. For example, for JAR 23 single engines, it becomes 10^{-6}.

[38] Total hours per year was 3×10^5. In 30 years, 9×10^6, near to 10^7, which could imply a catastrophic accident (considering all aircraft systems).

(1) Minor. Failure conditions that would not significantly reduce aeroplane safety, and which involve crew actions that are well within their capability.

(2) Major. Failure conditions that would reduce the capability of the aeroplane or the ability of the crew to cope with adverse operating conditions to the extent that there would be, for example, a significant reduction in safety margins or functional capabilities, a significant increase in crew workload or in conditions impairing crew efficiency, or discomfort to occupants, possibly including injuries.

(3) Hazardous. Failure conditions that would reduce the capability of the aeroplane or the ability of the crew to cope with adverse operating conditions to the extent that there would be

(a) A large reduction in safety margins or functional capabilities

(b) Physical distress or higher workload such that the flight crew cannot be relied on to perform their tasks accurately or completely, or

(c) Serious or fatal injury to a relatively small number of the occupants.

(4) Catastrophic. Failure conditions that would prevent continued safe flight and landing.

An **inverted relationship** between the **severity** of the failure conditions and the **probability** of occurrence is established.[39] Hence,

1	Minor failures	Become	Probable
2	Major failures	Become	Remote
3	Hazardous failures	Become	Extremely remote
4	Catastrophic failures	Become	Extremely improbable

Each of the above probabilities has a maximum value assigned, which depends on the type of aircraft considered—for example, for large aircraft, extremely improbable is 10^{-9}, as we have already seen; extremely remote is 10^{-7}; remote is 10^{-5}, and so on.

Figures 4.4 and 4.5, extracted from Book 2 of CS-25, show the above criteria.

We can gain a better indication of the safety levels relating to the above figures through another example. A single aircraft might fly a total of 5×10^4 hours and a large fleet of 200 aircraft (same type) might then accumulate a fleet total of 10^7 hours. Thus,

(1) A catastrophic failure condition (at worst 10^{-9}) would be unlikely to arise in the whole fleet's life.

(2) A hazardous failure condition (at worst 10^{-7}) might arise once in the whole fleet's life.

(3) A major failure condition (at worst 10^{-5}) might arise once in an aircraft's life and would arise several times in the whole fleet's life.

(4) A minor failure could arise several times in the aircraft's life.

---

[39] Where the effects are less hazardous, they are "permitted" to occur more frequently.

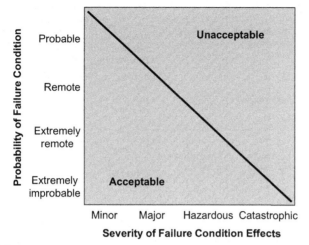

FIGURE 4.4 Classification of failure conditions

The safety assessment of equipment, systems, and installation is a very important (and fascinating) part of aircraft design. It is of paramount importance to start the assessment from the very beginning of the design. A late assessment could bring unpleasant surprises, leading to expensive design changes.

As mentioned before, the techniques of safety assessment are a specialist matter.

4.6.8. Fatigue strength

To remain within the limits of general information and guidance, leaving the rules, the advisory material, and the numerous treatises dealing with the subject as direct reference for further studies, we can see in summary how the airworthiness standards confront the structural fatigue that has caused so many air crashes, especially in the past.[40]

The airworthiness standards essentially consider two types of structure:

(1) Single load path structures, where the applied loads are eventually distributed through a single member, the failure of which would result in the loss of the structural capability to carry the applied loads.[41]

(2) Multiple load path structures, identified with redundant structures in which (with the failure of an individual element) the applied loads would be safely distributed to other load-carrying members.[42]

In the first case, the structure must result in **safe-life**, that is, be able to sustain a certain number of events such as flights, landings, or flight hours,

---

[40] Example, the "Comet" crashes in the 1950s, due to fatigue caused by fuselage pressurization.

[41] Example, a wing-fuselage attachment made by a single structural element. Such an arrangement is common in light aircraft.

[42] Example, a multiplex wing-fuselage attachment, made by several structural elements. Such an arrangement is classical in large aeroplanes.

Effect on aeroplane	No effect on operational capabilities or safety	Slight reduction in functional capabilities or safety margins	Significant reduction in functional capabilities or safety margins	Large reduction in functional capabilities or safety margins	Normally with hull loss
Effect on occupants excluding flight crew	Inconvenience	Physical discomfort	Physical distress, possibly including injuries	Serious or fatal injury to a small number of passengers or cabin crew	Multiple fatalities
Effect on flight crew	No effect on flight crew	Slight increase in workload	Physical discomfort or a significant increase in workload	Physical distress or excessive workload impairs ability to perform tasks	Fatalities or incapacitation
Allowable qualitative probability	No probability requirement	<...Probable...>	<...Remote...>	<.................> Extremely Remote	Extremely improbable
Allowable quantitative probability: Average probability per flight hour on the order of:	No probability requirement	$<10^{-3}$ Note 1	$<10^{-5}$	$<10^{-7}$	$<10^{-9}$
Classification of failure conditions	No safety effect	<...Minor...>	<...Major...>	<...Hazardous...>	Catastrophic

Note 1: A numerical probability range is provided here as a reference only. The applicant is not required to perform a quantitative analysis, nor substantiate by such an analysis, that this numerical criteria has been met for minor failure conditions. Current Transport category aeroplane products are regarded as meeting this standard simply by using current commonly-accepted industry practice.

FIGURE 4.5 Relationship between probability and severity of failure conditions

during which there is a low probability that the strength will degrade below its design ultimate value due to fatigue cracking.

In the second case, the structure must be of **damage-tolerance** design, that is, be able to retain its required residual strength for a period of unrepaired use after the failure or partial failure of a principal structural element due to fatigue, corrosion, accidental damage, and bird strikes.[43] Such a structure is defined as **fail-safe**.

For large aeroplanes and large rotorcraft, the relevant airworthiness standards require fail-safe structures,[44] unless this entails such complications that

---

[43] The bird strike is considered for large aeroplanes only, but it could be the object of special conditions for other types of aircraft.

[44] Rotorcraft structures include rotors, rotor drive systems between the engines, and rotor hubs, controls, fuselage, fixed and movable control surfaces, engine and transmission mountings, landing gear, and their related primary attachment.

an effective damage-tolerant structure cannot be reached within the limitations of geometry, inspection, or good design practice. Under these circumstances, a design that complies with the safe-life fatigue evaluation requirements is used. A typical example of a structure that might not be conducive to damage-tolerance design is the landing gear and its attachments.

The divided opinions within the National Transportation Safety Board (NTSB)[45] at the end of 2002, as described by *Flight International*, give an idea of the importance of the choice between fail-safe and safe-life. The question was whether a component with exceptional reliability needs to be fail-safe.

A report on a January 2000 MD-83 crash led to the conclusion that the stabilizer was jammed fully leading edge up, pitching the aircraft nose-down so strongly that elevator forces could not counteract it. The failure of the screw-jack assembly controlling the horizontal stabilizer pitch angle, because of inadequate lubrication, was also ascertained. Because the screw-jack mechanism is the sole component that sets and limits stabilizer pitch, its failure becomes catastrophic. Therefore, by some NTSB staff's opinion, the mechanism should have been redesigned according to a "more rational" fail-safe criterion. Other NTSB staff argued that the component was reliable (the retrofit would have involved more than 800 civil aircraft), because

(1) The same basic mechanism has been used in all MD DC-9, MD-80, and -90 aircraft since 1965.

(2) This kind of accident was the only one in more than 100 million flying hours.

(3) The accident was caused by inadequate maintenance (the carrier was fined by the FAA).

In the end, common sense prevailed and the issue was resolved without modifications to the mechanism. The FAA issued an Airworthiness Directive[46] (AD 2000-15-15) providing inspection, check, and test "*to prevent loss of pitch trim capability due to excessive wear of the jackscrew assembly of the horizontal stabilizer, which could result in reduced controllability of the airplane ....*"

The AD prescribes in particular the replacement of the jackscrew assembly with a new or serviceable assembly in case of metallic parts deterioration, corrosion, pitting, or distress.

The section "Airworthiness Limitations" of the "Instructions for Continued Airworthiness" must contain the inspections, replacement of parts, and other procedures necessary to prevent accidents caused by structural failures.

For JAR/FAR/CS-23 aeroplanes, it is possible to choose between the two philosophies safe-flight/fail-safe. Exceptions are made for composite airframe structures that must be designed according to fail-safe criteria, excluding the already-mentioned cases of impracticability.

---

[45] NTSB: the US organization dealing with aircraft accidents and issuing recommendations based on the investigation performed.

[46] Airworthiness Directives: documents issued by the authority making particular mandatory actions (changes, inspections, etc.).

The previous remarks about continued airworthiness are also applicable to these aircraft.

For JAR/FAR/CS-27 rotorcraft that are prone, similar to all rotorcraft, to particular fatigue problems, mixed criteria are generally followed, based mainly on time change items (parts to be replaced according to prearranged and approved schedules). Everything must obviously be clear in the "Instructions for Continued Airworthiness."

For JAR/CS-VLA aeroplanes and JAR/CS-22 sailplanes, the airworthiness standards contain very little information on fatigue,[47] similar to the old airworthiness standards for light aeroplanes. As a matter of fact, the low average utilization of these aircraft (100–200 flight hours/year) was not worrisome. However, after several years, and the intensive use of the machines in certain kinds of operation (e.g., school, aerial work, air taxi), fatigue problems became one of the causes of accidents, so much so, that step by step the fatigue airworthiness standards have been amended in JAR/FAR 23. Similar problems also occurred for sailplanes (perhaps less for VLAs as they are "younger"), so that no sailplane or VLA is now certified without fatigue assessment made by the manufacturers and authorities. For instance, LBA (Luftfahrt-Bundesamt), in Germany, a long time ago issued standards for fatigue assessment of sailplane composite airframe structures.

In the case of loads and loading spectra, the assumptions made for fatigue assessment are as follows:

(1) For rotorcraft, it is explicitly required that for each portion of the flight structure of which failure could be catastrophic, loads or stresses must be verified (or determined) by in-flight measurements; the same must be done for the loading spectra being considered. Then, the identification of these "critical" structural items is of paramount importance.

(2) For large aeroplanes, the principal loads that should be considered in establishing a loading spectrum are flight loads (gust and maneuver), ground load, and pressurization loads. The loading spectra are based on measured statistical data derived from government and industry load history studies and, where no sufficient data are available, on a conservative estimate of the anticipated use of the aeroplane. In assessing the possibility of serious fatigue failures, the design is examined to determine probable points of failure in service. In this examination, consideration is given, as necessary, to the results of stress analysis, static and fatigue tests, strain gauge surveys, tests of similar structural configurations, and service experience.

(3) For JAR/FAR/CS-23 aeroplanes, criteria similar to the above are adopted.

(4) For sailplanes and VLAs, apart from the general design recommendation to avoid stress concentration areas as far as possible, fatigue tests are performed, but only if they are essential, for economic reasons. If possible, reference is made to data resulting from fatigue tests performed on similar

---

[47] JAR-VLA offers some simplified criteria, but these must be considered carefully.

structures and service experience. Another way to avoid fatigue tests is the design of critical structures with stress levels under the fatigue limit of the material involved. Obviously, this must be properly demonstrated by static tests and strain gauge surveys.

If fatigue tests are necessary, the technical literature provides typical load spectra and programs for the repeated application of loads.

Apart from the consideration made for sailplanes and VLAs, the fatigue life assessment is performed through analysis, and fatigue tests on structures or single parts, according to criteria that are detailed in airworthiness standards and ACJs/ACs/AMC&GM. All analysis and test schedules are normally agreed with the authorities.

Fatigue test programs for large aeroplanes can last some years; hence, it is not generally possible to complete them before the aeroplanes' type certification. It is therefore required that at least 1 year of safe operations must be demonstrated when the type certificate is issued. Subsequently, to maintain the validity of the type certificate, the fatigue life substantiation must always exceed the number of cycles/flight hours reached by the "oldest" aeroplane (lead aeroplane).

4.7. JAR/FAR 21

JAR/FAR 21 contain the "Certification Procedures for Aircraft and Related Products and Parts" for JAA and FAA certification, respectively. JAR 21 deals with

(1) Procedural requirements for the issue of type certificates and changes to those certificates, the issue of standard certificates of airworthiness, and the issue of Export Airworthiness Approvals.

(2) Procedural requirements for the approval of certain parts and appliances.[48]

(3) Procedural requirements for the approval of organizations related to the subject of the previous points.

(4) Rules governing the holders of any certificate or approval specified in the previous points.

In a similar way, FAR 21 deals with

(1) Procedural requirements for the issue of type certificates and changes to those certificates, the issue of production certificates, the issue of airworthiness certificates, and the issue of Export Airworthiness Approvals.

(2) Rules governing the holders of any certificate specified in Paragraph (a)(1) of Paragraph 21.1 (Applicability).

(3) Procedural requirements for the approval of certain materials, parts, processes, and appliances.

---

[48] Appliance means any instrument, mechanism, equipment, part, apparatus, appurtenance or accessory, including communications equipment, that is used or intended to be used in operating or controlling an aircraft in flight, is installed in or attached to aircraft, and is not a part of an airframe, engine, or propeller. JAR 21 normally uses "parts and appliances" together, to include also the "parts" of airframes, engines, and propellers.

JAR/FAR 21 are therefore the rules upstream of the airworthiness standards, dictating, so to speak, the "rules of the game." The relationship between authorities and enterprises for certification of design and production of aeronautical materials is established. We will return to these issues in subsequent chapters (**see Note 49**).

4.8. EASA PART 21 (FIRST ISSUE)

As previously mentioned, this document replaced JAR 21, which remains the core of the same document. The changes to the JAR document reflected the new legal status of the EASA toward the national authorities and a full revision of the document in light of the JAA certification experience. To understand the evolution of this fundamental document, a comparison is made between the **first issue of EASA Part 21** (September 2003) and **JAR 21 Amendment 5** (June 2003).[49]

Without attempting a full comparison between the two documents, it is worth making the following observations.

4.8.1. Type certificates[50]

Subpart H of Part 21 (Article 21A.184) includes the "restricted type certificates"[51] missing in JAR 21.

---

[49] Following the establishment of the EASA in September 2003 and the adoption of EASA IRs, Certification Specifications (CS), and AMC and GM (AMC) the Joint Aviation Authorities Committee made the decision that in future the JAA would publish amendments to the airworthiness JARs by incorporation of reference to EASA IRs, AMC, and CS. Such publications would have a JAA cover letter with reference to the relevant EASA document, as well as any differences to it agreed by the JAA.

JAR 21 Amendment 6 of November 2004, was converted with reference to the above mentioned IRs and then replaced by **JAR 21 Amendment 7** dated February 2007. Then, the Article 1 dealing with the applicability of JAR 21 Amendment 7 as follows:

(1) This Regulation lays down, in accordance with Article 5(4) and 6(3) of the basic Regulation, common technical requirements and administrative procedures for the airworthiness and environmental certification of products, parts, and appliances specifying:
  (a) the issue of type-certificates, restricted type-certificates, supplemental type-certificates, and changes to those certificates;
  (b) the issue of certificates of airworthiness, restricted certificates of airworthiness, permits to fly authorized release certificates;
  (c) the issue of repair design approvals;
  (d) the showing of compliance with environmental protection requirements;
  (e) the issue of noise certificates;
  (f) the identification of products, parts, and appliances;
  (g) the certification of certain parts and appliances;
  (h) the certification of design and production organizations;
  (i) the issue of airworthiness directives.

[50] See Chapter 5, "The type certificate."

[51] See Chapter 8, "Restricted certificates of airworthiness," which defines and comments on the restricted type certificates.

4.8.2. Airworthiness certificates[52]

Subpart H of Part 21 (Article 21A.173) classifies the airworthiness certificates as follows:

(1) A certificate of airworthiness shall be issued to aircraft that conform to a type certificate that has been issued in accordance with this Part.

(2) Restricted certificates of airworthiness shall be issued to aircraft that

    (a) conform to a restricted type certificate that has been issued in accordance with this Part, or

    (b) have been shown to the Agency to comply with specific certification specifications ensuring adequate safety.

(3) Permits to fly shall be issued to aircraft that do not meet, or have not been shown to meet, applicable certification specifications but are capable of safe flight under defined conditions.

The certificates in 1 are equivalent to the Standard certificates of airworthiness of JAR 21.[53]

The certificates in 2 are consequent to the restricted type certificates and do not exist in JAR 21 (Amendment 5).

The certificates in 3 have the characteristics of the Special certificates of airworthiness[54] currently issued by national authorities and are not included in JAR 21 (Amendment 5).

As explained in Chapter 8, the Special certificates of airworthiness (present, e.g., in FAR 21) constitute a very complex matter for which there were no harmonization amongst EU Member States.

In 2006, the EASA NPA No. 09-2006 addressed this matter, proposing amendments to the Part 21 to distinguish permits to fly for various purposes such as the experimental certificates of airworthiness from special flight permits of FAR 21.

Part 21, as amended on 30 March 2007, contains a new Subpart P "**Permit to Fly**" (see Chapter 8).

4.8.3. Environmental protection

Part 21, in Subpart B, includes the designation of applicable environmental protection requirements and certification specifications, missing in JAR 21 (Amendment 5).

4.9. STRUCTURE OF AIRCRAFT AIRWORTHINESS STANDARDS

If we look at the airworthiness standards for aircraft certification (JAR/CS-22, JAR/CS-VLA, JAR/CS-VLR, JAR/FAR/CS-23, -25, -27, and -29), we note

---

[52] See Chapter 8.
[53] See Chapter 8, "Standard certificates of airworthiness."
[54] See Chapter 8, "Special airworthiness certificates."

a common structure that entails a certain unity and uniformity. Apart from the forewords, the lists of pages, and so on, we find "subparts" and "appendices." As mentioned previously, the JARs/CS also contain advisory material. Each subpart contains paragraphs under a title (e.g., "Ground Loads," "Control Systems," etc.), and it is of interest to see that, in all the above standards, the same topics are generally dealt with in paragraphs bearing the same number (e.g., "Weight limits" Paragraph XX.25; "Materials and workmanship" Paragraph XX.603; etc.). This makes it easier to pass from one standard to another, and to define comparisons when that is needed.

Some details of this structure are as follows:

(1) **Subpart A: General.** This Subpart provides information about the types and categories of aircraft to which the standard is applicable.

(2) **Subpart B: Flight.** This Subpart deals with the flight tests to be carried out to show compliance with the requirements for performance, controllability and maneuverability, stability, and so on. It is worth stating that this Subpart does not exclusively cover certification flight tests; other Subparts contain some requirements that must be complied with through flight tests.

(3) **Subpart C: Structure.** This Subpart contains the requirements for flight and ground load assessment, and for structural design of airframes, control systems, landing gears, and other components. Crashworthiness and fatigue requirement parameters are also provided.

(4) **Subpart D: Design and Construction.** This Subpart deals with the design technique, materials, safety factors, control system and landing gear design, structural tests to be carried out, cockpit and passenger cabin design, fire protection and flutter requirements, and so on.

(5) **Subpart E: Power Plant.** This Subpart contains the requirements for power plant installations and related systems (such as fuel, oil, exhaust systems, etc.). Power plant controls, accessories, and fire protection are also considered.

(6) **Subpart G: Operating Limitations and Information.** This Subpart provides requirements for all the information that must be available to the pilot and other personnel for correct aircraft operations: from marking and placards, to the flight manual content.

(7) **Appendices.** These documents are of various natures; they can provide simplified design load criteria, test procedures for assessment of material flammability, instructions for continued airworthiness, and other information.

NOTE:

(a) **Aircraft Category.** The term "category," as used with respect to the certification of aircraft, means a grouping of aircraft based on their intended use or operating limitations, for example, normal, utility, acrobatic, or primary.

(b) **Aircraft Classification.** The term "classification," as used with respect to the certification of aircraft, means a broad grouping of aircraft having similar characteristics of propulsion, flight, or landing, that is, airplane, rotorcraft, glider, or balloon.

4.10. AIRCRAFT AIRWORTHINESS STANDARD APPLICABILITY

As mentioned above, Subpart A of aircraft airworthiness standards defines types and categories of specific aircraft. We consider this in more detail.

4.10.1. JAR/CS-22. Sailplanes and Powered Sailplanes

(1) Sailplanes with a maximum weight not exceeding 750 kg.

(2) Single-engine (spark or compression ignition)-powered sailplanes with a design value W/b^2 (weight to span^2) not greater than 3 (W in kg, b in m), and maximum weight not exceeding 850 kg.

The maximum number of occupants for both sailplanes and powered sailplanes must not exceed two.

The term "powered sailplane" includes those powered sailplanes that may be incapable of complying with the minimum rate of climb required by Paragraph 22.65 and a maximum takeoff distance required by Paragraph 22.51, and which must consequently be prohibited from taking off solely by means of their own power (so they are launched similar to sailplanes). These powered sailplanes are referred to as "self-sustaining powered sailplanes," and additional requirements of Appendix 1 are applicable to them.

JAR/CS-22 contains Subparts H and J with standards for engines and propellers to be installed on powered sailplanes (engines and propellers certificated according to JAR-E and JAR-P as relevant can, of course, be installed).

NOTE: In JAR/CS-22, the term "sailplane" is related both to sailplanes and powered sailplanes. In JAR 22, the requirements applying only to powered sailplanes are annotated with the letter P in the margin.

Sailplane categories are as follows:

(1) Utility. Sailplanes intended for normal soaring flight and some aerobatic maneuvers (listed in the requirements) if demonstrated during type certification.

(2) Aerobatic. Sailplanes intended for aerobatic maneuvers in addition to those permitted in the utility category. The permitted aerobatic maneuvers must be established during type certification.

NOTE: Powered sailplanes are considered as "sailplanes with an auxiliary engine." That is why the installation of power plants complying with less severe requirements is allowed. Furthermore, unlike aeroplanes, powered sailplanes are allowed to fly with the engine off (and retracted if that is possible), similar to sailplanes. Flight tests must be carried out both with power on and power off—and power plant retracted when this is possible.

To explain the strange-looking parameter W/b^2, we can say that, when the first standards for powered sailplanes began to appear, some manufacturers envisaged the possibility of producing aeroplanes "disguised" as powered sailplanes. The intent was to comply with requirements that were less severe—at that time. It is clear that for an 850-kg powered sailplane, the formula W/b^2 not greater than 3 leads to a minimum wing span of 16.8 m, which gives the aircraft an appearance more like a sailplane than an aeroplane.

4.10.2. JAR/CS-VLA. Very Light Aeroplanes

This includes aeroplanes with a single engine (spark or compression ignition) having no more than two seats, with a maximum certificated takeoff weight of not more than 750 kg and a stalling speed in the landing configuration of not more than 45 knots (CAS). The approval must be given only for day-VFR.

NOTE: JAR-VLA was issued as a simplification of FAR 23 (JAR 23 did not yet exist). Unfortunately, while other requirements have been continuously updated by the relevant study groups, JAR-VLA has remained practically unchanged since publication; this is why some inaccuracies are still present as well as some old-fashioned concepts (e.g., in crashworthiness matters). An increase in maximum takeoff weight would also be advisable, because 750 kg is somewhat insufficient, especially in the case of composite airframes (avoiding expensive carbon fibers).

In the United States, where the requirements have been adopted (see Note 6), the VLA certification for night flight and IFR is possible, in compliance with the additional requirements of AC 23−11. In Europe, this possibility has not yet been approved.

4.10.3. JAR/FAR/CS-23. Normal, Utility, Aerobatic, and Commuter Category Aeroplanes

(1) Aeroplanes in the normal, utility, and aerobatic categories that have a seating configuration, excluding the pilot seat(s), of nine or fewer and a maximum certificated takeoff weight of 5670 kg (12,500 lb) or less.

(2) Propeller-driven, twin-engine aeroplanes in the Commuter category that have a seating configuration, excluding the pilot seat(s), of 19 or fewer and a maximum certificated takeoff weight of 8618 kg (19,000 lb) or less. Aeroplane categories are as follows:

(1) **Normal.** The normal category is limited to nonaerobatic operations. Nonaerobatic operations include stalls (except whip stalls) and some simple maneuvers (listed in the requirements) in which the angle of bank is not more than 60°.

(2) **Utility.** The utility category is limited to the operations of the Normal category, spins (if approved for the particular type of aeroplane), and some aerobatic maneuvers (listed in the requirements) in which the angle of bank is between 60° and 90°.

(3) **Acrobatic.** The acrobatic category has no restrictions other than those shown to be necessary as a result of required flight tests.

(4) **Commuter.** The commuter category is limited to any maneuver incident to normal flying, stalls (except whip stalls), and steep turns in which the angle of bank is 60° or less.

NOTE: Historically, smaller FAR 23 airplanes were typically simple and slow while bigger airplanes were more complex and faster. Consequently, the existing approach to standards based on **weight and engine type** was effective.

Although the existing approach has produced safe airplanes for decades, technological advances have changed the original assumptions of the FAR 23. The new small turbine engines, composite airframes, and lightweight digital electronics offer FAR 23 airplanes the operational capability and performance of traditionally larger FAR 25 airplanes.

FAR 23 standards have evolved beyond their original intent to address the increasing performance and complexity.

Unfortunately, the slow, simple FAR 23 airplanes have suffered as the standards have shifted toward more complex airplanes.

Although not impossible, certifying a simple, two-place airplane is cumbersome and expensive. At the same time, FAR 23 does not completely address very complex high-performance turbine products.

Today, the certification authorities use special conditions to address the certification requirements of complex high-performance turbine airplanes.

This kind of remark has conducted the FAA to make a serious thought on the matter with the institution of a dedicated and well-qualified team to make recommendations for future changes of FAR 23.

The FAA has recently published a "**Part 23—Small Airplane Certification Process Study**" (Recommendations for General Aviation for the next 20 years).

The first team's recommendation is to reorganize the FAR 23 based on **airplane performance and complexity**, versus the existing weight and propulsion divisions.

The study proposes the following:

- **Part 23 Category A**. Low complexity, low performance
- **Part 23 Category B**. Medium complexity, medium performance
- **Part 23 Category C**. High complexity, high performance.

The FAA document gives plenty of details about these categories.

The study is not limited to certification standards. Study team members reviewed other areas affecting general aviation, such as pilot training, operations, and maintenance. The study offers a variety of short-term and long-term recommendations.

4.10.4. JAR/CS-25. Large Aeroplanes FAR 25. Transport Category Airplanes

These comprise the following:

- Large turbine-powered aeroplanes (JAA/EASA)
- Transport category airplanes (FAA).

NOTE: There are no limitations with regard to weight, number of engines, and number of occupants.

Maximum weights corresponding to the airplane's operating conditions (such as ramp, ground or water taxi, takeoff, en route, and landing), environmental conditions (such as altitude and temperature), and loading conditions (such as zero fuel weight, center of gravity position, and weight distribution) are established so that they are not more than

(1) The highest weight selected by the applicant for the particular conditions.

(2) The highest weight at which compliance with each applicable structural loading and flight requirement is shown.

(3) The highest weight at which compliance is shown with the certification requirements of FAR 36.

JAR/CS-25 take into account turbine-powered aeroplanes only. Actually, large aeroplanes powered by reciprocating engines have not been designed for many years. FAR 25, issued before JAR 25 as a derivation of older regulations, does not have this limitation. In effect, transport aeroplanes powered by reciprocating engines are still flying in some parts of the world, with valid type certificates.

4.10.5. JAR/CS-27. Small Rotorcraft FAR 27. Normal Category Rotorcraft

Rotorcraft with a maximum weight of 3175 kg (7000 lb) or less and nine or less passenger seats.

Multiengine rotorcraft may be type certificated as **Category A** provided the requirements referenced in Appendix C are met.

NOTE: For Category A definition, see the Notes on JAR/FAR/CS-29.

4.10.6. JAR/CS-29. Large Rotorcraft FAR 29. Transport Category Rotorcraft

Rotorcraft categories are as follows:

(1) Rotorcraft must be certificated in accordance with either the **Category A** or **Category B** requirements of JAR/FAR/CS-29. A multiengine rotorcraft may be certificated as both Category A and Category B, with appropriate and different operating limitations for each category.

(2) Rotorcraft with a maximum weight greater than 9072 kg (20,000 lb) and 10 or more passenger seats must be type certificated as Category A rotorcraft.

(3) Rotorcraft with a maximum weight greater than 9072 kg (20,000 lb) and nine or less passenger seats may be type-certificated as Category B rotorcraft, provided the Category A requirements of Subparts C, D, E, and F of JAR/FAR/CS-29 are met.

(4) Rotorcraft with a maximum weight of 9072 kg (20,000 lb) or less but with 10 or more passenger seats may be certificated as Category B rotorcraft provided the Category A requirements of JAR/FAR/CS-29.67(a)(2), 29.87, 29.1517 and of Subparts C, D, E, and F of JAR/FAR/CS-29 are met.

(5) Rotorcraft with a maximum weight of 9072 kg (20,000 lb) or less and nine or less passenger seats may be type certificated as Category B rotorcraft.

NOTE: **Category A** means a multiengine rotorcraft designed with engine and system isolation features specified in JAR/CS-27/JAR/CS-29 and capable of operations using takeoff and landing data scheduled under a critical engine failure concept, which assures adequate designated surface area and adequate performance capabilities for continued safe-flight or safe-rejected takeoff in the event of an engine failure.

Category B means a single-engine or multiengine rotorcraft that does not meet Category A standards. Category B rotorcraft have no guaranteed capability to continue safe flight in the event of an engine failure, and unscheduled landing is assumed.

4.10.7. FAR 31. Manned Free Balloons

(1) Captive gas balloons deriving lift from a captive lighter than air gas.

(2) Hot-air balloons deriving lift from heated air.

NOTE: There are no limitations on weight and number of occupants. The certification maximum weight is the highest weight at which compliance with each applicable requirement of this part is shown.

4.10.8. JAR/CS-VLR. Very Light Rotorcraft

Very light rotorcraft with a single engine (spark or compression ignition) having no more than two seats, with a maximum certificated takeoff weight of not more than 600 kg. The approval has to be for day-VFR only.

NOTE: The Italian RAI-ENAC issued a standard for this type of rotorcraft in the 1990s (it was approved on 22 April 1996), to allow the certification of ultralight helicopters that were limited, by the relevant low in Italy, to a maximum weight of 450 kg. Such a maximum weight was considered inadequate and, furthermore, these ultralights had no certification standards.

RAI-VLR (this was the title of the standard) was issued as a simplification of JAR 27, following a path resembling that adopted for the issue of JAR-VLA (simplification of FAR 23). RAI-VLR is now a national standard that could lead to a type certification and the issue of Standard certificates of airworthiness.[55] Nevertheless, because the RAI was bound to the Cyprus agreement, an Elementary Aircraft category was created in Italy, for which a special certification was required.

Then, the standard was "offered" to the JAA, which set up a study group for the evaluation and eventual revision of this document, to finally issue a JAR-VLR.

The JAR-VLR was issued in September 2003 and then adopted by the EASA as CS-VLR.

4.11. AIRWORTHINESS STANDARDS FOR UNMANNED AIRCRAFT

These aircraft are internationally known as **Unmanned Aerial Vehicles (UAVs)** or **Uninhabited Aerial Vehicles (UAVs)**, and also as **Remotely Piloted Vehicles** and **Remotely Operated Aircraft**.

---

[55] See Chapter 8.

Lately, it has been agreed to adopt the term **Unmanned Aircraft Systems (UAS)**[56]; consequently, this is now the internationally accepted official term we will refer to.

UAS have been mainly used by the world's armed forces for wartime operations for more than 60 years, for battlefield observations in the past, and more recently as a wartime means of attack. We can therefore argue that UAS have already reached a technical maturity and this is going to evolve as for any other kind of aircraft. However, up to the present, UAS missions have been normally limited on battlefields, to restricted flying areas, outside the zones open to civil aircraft operations.

Today, when the great potential of this type of machine has been recognized, the global industry has requested the opportunity of also using them commercially in civil airspace. This possibility is also of interest to the defense industry, because they can achieve better operational flexibility—for example, in the case of transfer flights.

We have mentioned the potential of UAS for civil applications. We now consider what kind of applications these might be.

As a first example, thousands of rotary wing UAS are used for agricultural purposes in Japan (crop spraying—pesticides and fertilizers). These machines, all built in Japan, carry a useful load of 25–150 kg.

Some classifications have been drafted and the list below includes just a few examples taken from the multitude of possible UAS uses:

- Forestry services—fire control and other kinds of surveillance.
- National weather services—atmospheric sampling, meteorology, and so on.
- Agriculture and wildlife—agricultural monitoring, river and estuary surveys, illegal waste disposal surveys, crop dusting, mapping, fishing law enforcement, and so on.
- Electricity authorities—monitoring nuclear facilities, power line verification, and so on.
- Postal services—urgent package delivery in remote areas.
- Coastguards—surveillance for counter narcotics, illegal alien intrusion detection, illegal fishing control, search and rescue missions, and so on.
- Civil aviation—noise measurement for aircraft certification purposes.
- Telecommunications—as telecom relays (replacing satellites), local TV coverage, and so on.

It is clear from these examples that, in many cases, the scope of UAS is to carry out the "dirty jobs," that is, dangerous tasks, or tasks too long or too tedious for a crew.

Can UAS be legally defined as "aircraft"? An answer has been given by the ICAO EURNAT Office: UAS are aircraft as defined by Annex 2 of the ICAO. Furthermore, the Chicago Convention in Article 8 declares that

---

[56] As explained in Section 4.11.1, the Air Vehicle Station (AVS)—the ground station—and the vehicle, should be considered as a unique system. The term UAS better conveys this idea.

No aircraft capable of being flown without a pilot shall be flown without a pilot over the territory of a contracting State without the special authorization by that State and in accordance with the terms of that authorization. Each contracting State undertakes to ensure that the flights of such aircraft without pilot in regions open to civil aircraft shall be so controlled as to obviate danger to civil aircraft.

Therefore, the real problem is now to develop concepts for the safe integration of UAS in general air traffic. It is then necessary to develop rules harmonized with the existing rules for air traffic control.

The issues concerning the above rules can be easily classified, as for "manned" aircraft, into three basic segments:

(1) Personnel licenses,

(2) Air traffic management (ATM), and

(3) Airworthiness.

Hence, we return to the main safety factors discussed in Chapter 1: man, the environment, and the machine.

Studies and conferences on the above subjects have been taking place for many years. In Europe, there are some institutes and associations dealing with these issues. One is the European UVS International (formerly EURO UVS), equivalent to the Association for Unmanned Vehicle System International in the United States. A great contribution to this discussion has also been made by EUROCONTROL, particularly concerning ATM matters. Other initiatives have been taken all over the world.

4.11.1. Airworthiness standards

We should not be misled by the title: at the time of writing (June 2010), no official airworthiness standards exist for UAS.

In the 1990s, at the request of the national industry, the Italian RAI-ENAC issued a draft of a UAV airworthiness standard. This document was presented at the annual EURO UVS conference in June 1999, triggering great debate on the subject. It was indeed probably the first attempt to define some sort of airworthiness standards for civil UAS. Instead of attempting to invent everything from scratch, the JAR-VLA standard was chosen as a basic standard to be adapted to fixed-wing UAVs up to 750 kg.

It could be argued that, to transform a standard for manned aircraft into a UAS standard, it would be sufficient to delete all requirements inherent to the occupants, such as the cockpit and the passenger cabin requirements. But it is not so simple because the airworthiness "philosophies" we have considered in the previous chapters would not be fully utilized in doing so.

It is therefore necessary to set up new philosophies specific for UAS before trying to convert them into new standards.

The definition of "airworthiness" given at the beginning of Chapter 2 is perfectly suitable to UAS ("requirements" and "allowable limits" should also exist for these machines), provided that we clarify the meaning of "safe condition." In other words, what constitutes "safety" for UAS? This is a topic that requires debate and validation. Because a UAS is an aircraft, any UAS's requirements should, as far as practicable, be consistent with ICAO Annex 8 that states in its Foreword that *"the objective of international airworthiness standards is to define for application by the competent nations authorities, the minimum level of airworthiness constituting the international basis for the recognition by States, under Article 33 of the Convention, of certificates of airworthiness for the purpose of the flight of aircraft of other States into or over their territories, thereby achieving, among other things, protection of other aircraft, third parties and property."*

If we consider the various airworthiness standards, we clearly understand that they are written having in mind the **occupant's protection**. The protection of people and properties on the ground is an added purpose of the safety obtained through compliance with the standards. For some categories of aircraft (aerobatic aeroplanes, sailplanes, and powered sailplanes), the abandonment of the aircraft in emergency cases is even considered. (The presence of a pilot on board could, in certain cases, avoid or limit the damage on the ground, but this is not guaranteed, as demonstrated by various accident reports.)

On the other hand, it is always very difficult to establish exactly what to do to avoid damage on the ground when considering air accidents: the most reasonable way to approach this problem is by trying to prevent the accident from occurring.

From the above considerations, we can logically assume that we need to approach UAS standards, which by definition have no occupants, from a totally different perspective.

If we start from a general, but fundamental, safety principle of the protection of human beings, we can state that the UAS standards should aim to avoid any damage to people (and properties) in the UAS range of action. This can only mean one thing: **to avoid in-flight collisions and uncontrolled ground impact**.

This can be achieved by applying the system safety assessment concepts that we have already mentioned in this chapter, and the standards for flight, structural strength, and so on, which can be acquired from the current standards for manned aircraft. This also leads to an additional parameter, **mission effectiveness**, which applies equally to UAS and manned aircraft. This can be achieved by taking from the current standards, based on a century of experience, everything that might be applicable to UAS.

In the case of UAS safety assessment, it is clear that the **severity of failure conditions** for UAS will be very different compared with manned aircraft.

For example, for manned aircraft, a **catastrophic failure condition** is one that **would prevent continued safe flight and landing**. For a UAV, this situation would be not at all catastrophic if the aircraft has a "Flight Termination System" (FTS) capable (using parachutes, for instance) of bringing the machine back to the ground. An FTS failure could instead become catastrophic and there are other numerous examples supporting this argument.

This is just an example because it is not certain that such described FTS will be considered acceptable in future requirements.

The current parachute recovery for little manned aircraft or ultralights is usually considered acceptable on a "no hazard"[57] basis only; therefore, it is not able to cope with the FAR/JAR/CS XX.1309, requiring the aircraft systems to be designed to assure *"continued safe flight and landing."*

It is then possible that, while an FTS with parachute could be acceptable (also for economic reasons) only for very light UAS (LUAS), more sophisticated FTS providing automatic flight guidance for a normal landing will be required for other, more complex, UAS.

We can infer that a new setting of standards taken from Paragraph XX.1309 has to be arranged for the determination of **severity failure conditions** and **probability of occurrence**. But in the case of CS-VLA, for example, we have seen that Paragraph 1309 has a small number of applicable requirements.[58] Therefore, we have to provide something different for a UAS standard based, for example, on CS-VLA.

Furthermore, the installation of anticollision systems, or similar devices certainly not installed on VLAs, could become compulsory.

Avoiding the risk of in flight collisions (currently known as **sense and avoid**) is one of the most challenging problems to solve to integrate UAS into civil (nonsegregated) airspace.

Another peculiarity of UAS standards should be the incorporation of requirements for the "Air Vehicle Station" (AVS)—the ground guidance station—that have to be considered as an integral part of the flying material and should be consistent with it.

In conclusion, we can argue from the analysis made so far that, to produce UAS airworthiness standards, many difficulties must be overcome; these difficulties are not only related to UAS technology, which already exists and is evolving, but also associated with the creation of the related airworthiness philosophies and their correct transfer into the standards to be issued for the different UAS categories and kind of operations (also to be defined).

4.11.1.1. BASIC CRITERIA FOR AIRWORTHINESS STANDARDS

In the last paragraph, we wrote that a UAS standard could be developed from appropriate standards already existing for manned aircraft.

As we shall see in Chapter 8 (Section 8.6), the type certificate of an aircraft can be obtained independently from the kind of operations to be carried out.

---

[57] "No hazard": see Chapter 5, Section 5.2.
[58] JAR-VLA deals with a simple aeroplane and the safety of two occupants. A UAS of the same weight is a sophisticated machine if we look at the systems, and we have also to consider the lives of the 300 or 400 people aboard a large aircraft that could be brought down.

Actually, the basic airworthiness standards, such as FAR/JAR/CS 23, 25, 27, ... and so on, are not directly linked to the purposes for which the aircraft will be flown. This means that the aircraft has to satisfy additional airworthiness requirements for each particular kind of operation.

This is what normally happens for civil aircraft.

Another philosophy, very often adopted for military aircraft, is to produce airworthiness certification basis for each particular type of aircraft having defined its characteristics, performance, type of missions, and so on with different criteria.

The **MIL-HDBK-516B** establishes the same airworthiness certification criteria for the airworthiness of all manned and unmanned aircraft, considering that they have the same specific *safety-of-flight* (*SOF*) system requirements.

Therefore, specific criteria are included for these types of systems to ensure the establishment of minimum levels of design for safe operation and maintenance.

In the case of UAS, the document only established that, being this a case of unmanned systems, SOF risks associated with loss of aircrew may not apply. Of course, as with manned air vehicles, SOF risk associated with personnel, damage to equipment, property, and/or environment must be considered.

However, all this can be applied as, while civil air vehicles must satisfy precise and established standards (albeit with special conditions whenever necessary), in the case of military air vehicles the certification basis could be tailored, as mentioned, to satisfy particular types of operational missions, and it could make reference to FARs, and also to many other military specifications.

This "**safety target approach**," as it is normally defined, can work for military aircraft for which the State is both customer and responsible for the safety of a relatively limited number of types of aircraft.

Such an approach is clearly not viable for civil aircraft for which it should be necessary to produce different airworthiness standards even for the same type of aircraft for different kinds of operations, without a rational and internationally recognized airworthiness basis. Furthermore, these standards could end up being not consistent with the ICAO Annex 8.

In conclusion, it is logical to argue that the civil UAS airworthiness standards will be produced with a "**conventional approach**" philosophy similar to the one adopted for manned aircraft.

Of course, if these standards are developed from the existing ones, suitable criteria must be established first for UAS classification, to set a comparison with the different classes and categories of manned aircraft.

The conventional versus safety target approach criteria for certification is discussed in the EASA A-NPA 16-2005[30] based on a report of the joint JAA/EUROCONTROL UAV Task-Force that is attached as an appendix to the document.

This A-NPA, which is a preliminary step to develop a policy for UAS certification can be really considered as a step in the right direction.

Among the different topics—such as UAS classification, design organization approval, environment, safety analysis, certificate of airworthiness, and so on—the A-NPA takes an important position toward the problem of "*sense and avoid*," which is considered an operational matter to be defined by the authorities responsible for air navigation services. Therefore, these positions should lead to the production of additional requirements as operational rules for the certification of the relevant equipments.

Thus, if sense and avoid is not taken into consideration—being generally regarded as irremissible for entering non-segregate airspace—the airworthiness certification will contain limitations for operation in segregated airspace only—physically, or through suitable agreements with the competent air traffic authorities.

Actually, the problem of sense and avoid related to anticollision purposes is the most difficult problem to be solved. The fundamental issue making this matter difficult is the (correct) statement that the risk of in-flight collisions for UAS should not be higher than the risk for corresponding manned aircraft. Therefore, the regulatory standards must not be less demanding than those currently applied to comparable manned aircraft.

Furthermore, an ATC must not be put in condition to apply different rules to UAS or manned aircraft.

The technological fulfillment of these constraints is a matter of a period of time judged between 10 and 15 years. Also for this reason, the orientation of the A-NPA for the establishment of a "basic" set of airworthiness regulations—as for manned aircraft—is the right choice.

Many airworthiness authorities have already issued special certificates of airworthiness for UAS for special purposes, based on the case-by-case certification criteria. However, the development of UAS utilization requires, as for manned aircraft, a set of basic airworthiness regulations, possibly harmonized at international level.

Coming back to the EASA A-NPA, the Comment Response Document (CRD-16-2005) was published in December 2007 and received many comments from authorities, organizations, and individuals, with good agreement on the main options described in the A-NPA.

As a consequence, on August 2009 EASA issued a "*policy statement: airworthiness certification of Unmanned Aircraft Systems (UAS).*"

The policy represents a first step in the development of comprehensive civil UAS regulation, and it may be regarded as an interim solution to aid acceptance and standardization of UAS certification procedures and will be replaced in due course by AMC and GM to EASA Part 21 when more experience has been gained.

The overall objective of this policy is to facilitate acceptance of UAS civil airworthiness applications, while upholding the Agency's principle objective of establishing and maintaining a high uniform level of civil aviation safety in Europe.

With no persons onboard the aircraft, the **airworthiness objective** is primarily targeted at the protection of people and property on the ground. A civil UAS must not increase the risk to people or property on the ground compared with manned aircraft of equivalent category.

The protection of other airspace users dependent on ATC/ATM separation procedures and defined "detect and avoid" criteria are considered outside of airworthiness. However, there will be an airworthiness function to verify that equipment designed to meet such criteria, together with the unmanned aircraft's performance, are satisfactory.

Where applicable, a UAS must comply with the essential requirements for **environmental protection objective** as stipulated in Article 6 of the Basic Regulation.

Here is copied the **Table of Contents**:

1. Introduction
2. UAS definition
3. Policy scope
4. Policy objectives
5. Procedure for UAS type certification
6. Specific guidance in complying with Part 21 Subpart B
7. Guidance on special conditions
    7.1 Emergency recovery capability
    7.2 Command and control link
    7.3 Level of autonomy
    7.4 Human machine interface
    7.5 Control station
    7.6 Due to type of operation
    7.7 System safety assessment
8. Other issues
    8.1 Certificate of airworthiness
    8.2 Noise certificate
    8.3 Permit to fly
    8.4 Continuing airworthiness
    8.5 "Detect and Avoid"
    8.6 Security
    Appendix 1: Methodology for selecting the applicable airworthiness code(s)
    Appendix 2: Methodology for tailoring the selected airworthiness code(s).

4.11.2. The state-of-the-art

At the beginning of this review of airworthiness standards for UAS, it was stated that "at the time of writing (June 2010) no official airworthiness standards for UAS exist."

When *Aeronavigabilità* was written at the end of 2002, it was emphasized that

> *In any case, we must be able to discuss such requirements only when these problems will be faced with determination by authorities like the FAA and JAA (and subsequently the EASA).*

So, why bring up this issue now?

The answer is that, even without having achieved common and approved rules, much work has been carried out in the last 8 years.

During the last few years, not only the above-mentioned authorities but a great number of government agencies, military authorities, international organizations, and stakeholders groups have operated in this new UAS reality.

The *"Global Perspective 2008/2009 and 2009/2010,"* published by the UVS International, presents a general photograph of the current worldwide UAS situation.

It is not possible, in the limits of this book, to offer an exhaustive summary of this situation, but only some information in order to have an idea of the main current activities.

In Europe:

- **EASA** has established contacts with ICAO, FAA, EUROCONTROL, EUROCAE,[59] NATO, European Defence Agency, and so on to cooperate in UAS activities. Patrick Goudou, the EASA Director General, has declared the Agency's interest in UAS and the intention to cooperate and contribute to the development of a comprehensive UAS regulatory framework. We have already underlined the importance of the EASA A-NPA 16-2005 and the consequent *"Policy statement: airworthiness certification of UAS."*

- **EUROCONTROL**, since the beginning, has led work on ATM aspects of UAS in the European airspace. A UAV Task Force was established as a result of a joint initiative between the JAA and EUROCONTROL in 2002. In May 2004, they issued a final report called *"A concept for European regulations for civil unmanned aerial vehicles"* to develop a concept for the regulation of civil UAVs, with reference to safety, security, airworthiness (including continued airworthiness), operational approval, maintenance, and licensing. EUROCONTROL is not working in isolation, but it cooperates with national and international civil and military bodies for the UAS integration into nonsegregated airspace. The mission of EUROCONTROL is the harmonization and integration of air navigation services in Europe by the creation of a satisfactory and uniform ATM system for both civil and military users. For this reason, in 2008, the Agency created the "EUROCONTROL UAS ATM integration Activity" for the coordination of an UAS ATM integration work program.

- **EUROCAE**[59] deals with airborne and ground systems in cooperation with organizations such as ICAO, EASA, EUROCONTROL, European National Authorities, FAA, Radio Technical Commission for Aeronautics (RTCA),

---

[59] EUROCAE "European Organization for Civil Aviation Equipment" is a nonprofit making organization formed in 1963 to provide a European forum for resolving technical problems with electronic equipment for air transport. EUROCAE deals exclusively with Aviation standardization (Airborne and Ground Systems and Equipments) and related documents as required for use in the regulation of aviation equipment and systems.

major aeronautical industries, and others. EUROCAE WG-73 was launched in April 2006 following earlier work by EUROCONTROL, NATO, and JAA.

Similar to the European UAS expert group, WG-73 will work with EASA in the development of airworthiness criteria and Special Conditions to supplement EASA A-NPA-16 *Policy for UAV Certification*. WG-73 cooperates with other bodies with the main objective to deliver standards and guidance for UAS operating in nonsegregated airspace and is recognized as the European UAS expert group to assist EASA for additional airworthiness criteria and/or special conditions.

- **ASTRAEA**[60] stands for Autonomous Systems Technology Related Airborne Evaluation and Assessment. The program seeks to research, develop, and validate the necessary technologies, systems, facilities, and procedures to promote and enable safe, routine, and unrestricted use of UAS. The ASTRAEA program is divided into a series of projects, each managed by a private sector partner. The projects fit within three different themes,

  (1) *Regulatory Framework.* ASTRAEA was specifically designed to look at the differences between manned and unmanned environments; typically, the differences are set in two identified streams: technical and operations. In essence, the Regulatory Projects are looking at what might be an acceptable interpretation of the existing regulations, when applied to unmanned flight and to take care of the differences.

  (2) *Technology.* The technology being developed by ASTRAEA will cope with issues such as Ground Operations and Human Interaction; Communications and Air Traffic Control; UAS Handling; Routing; Collision Avoidance; and so on.

  (3) *Demonstration.* This is a key part of the program. Experiments and demonstrations are being performed both for individual technologies and for integrated systems. After a first phase in which technology systems, procedures and facilities to allow UAS to operate safely in the airspace have been examined, ASTRAEA is now ready to move in a second phase aimed to see the possibility of commercial UAS operating in nonsegregated airspace toward 2012.

- **Civil Aviation Authority (CAA).** In April 2008, the CAA has issued the third edition of the CAP 722 "Unmanned Aircraft System Operations in UK Airspace—Guidance" (amended 14 April 2009).

  It is intended to assist those who are involved in the development of UAS to identify the route to certification, to ensure that the required standards and practices are met by all UAS operators.

  Overall, the purpose of the document is to highlight the safety requirements that have to be met, in terms of airworthiness and operational standards,

---

[60] Half of the funding for ASTRAEA is being provided by public sector organizations and the rest from a consortium of UK companies. Six of the UK leading universities are partners in the ASTRAEA program, engaged in UAS-related projects.

before a UAS is allowed to operate in the United Kingdom. While UAS flights beyond the limits of visual control are currently restricted to segregated airspace, the ultimate aim is to develop a regulatory framework that will enable the full integration of UAS activities with manned aircraft operations throughout UK airspace.

The CAP 722 obviously applies to UAS operating in the United Kingdom. Nevertheless, the document includes criteria that can be discussed and taken into account for the formulation of new international UAS requirements.

Outside Europe:

- **ICAO.** After a first ICAO exploratory meeting on UAVs in Montreal in May 2006, with the objective of determining the potential role of ICAO in UAV regulatory development work, a meeting with EUROCAE and an informal ICAO meeting in January 2007, it was decided to establish an ICAO study group.

  The role of this UAS Study Group (UASSG) is to assist the Secretariat in developing a framework for regulatory development, guiding the Standards and Recommended Practices development process within ICAO, and to support a safe, secure, and efficient integration of UAS into nonsegregated airspace and aerodromes.

  Sixteen contracting states and eight international organizations have nominated experts to the Study Group.

  We have to remember that ICAO has an international role, harmonizing procedures, and terminology for all the civil aviation around the world. The main general task is to issue rules for the UAS such as to be treated like the other aircraft.

  For a list of priorities, one of the first is a terminology that could be universally valid, as a revision of the current terminology for manned aircraft. A priority list of terms has been identified by the UASSG and is going to expand.

  Another very big task is the amendment of the 18 Annexes to the Convention as to the applicability to UAS. Practical solutions should be found to avoid undue limitations to the commercial use of UAS.

  The UASSG will issue an "ICAO Unmanned Aircraft System Circular" that should be available early in 2010 containing extensive background information for the states willing to develop a regulatory framework.

- **FAA.** To address the increasing civil market and the desire by civilian operators to fly UAS just like any other aircraft, the FAA is developing new policies, procedures, and approval processes. At FAA Headquarters in Washington, DC, a team of experts from various parts of the agency is working on guidance that will increase the level of access to airspace for UAS in a step-by-step fashion without being overly restrictive in the early stages.

  Developing and implementing this new UAS guidance is a long-term effort and is still a "work in progress."

  More immediately, the FAA is reviewing certification requests from several manufacturers. The FAA has already issued a consistent number of airworthiness certificates in the "experimental" category—for research and development, crew manufacturers training, or UAS market survey).[61]

The FAA established an Aviation Rulemaking Committee (ARC) on April 2008 to develop recommendations to allow the operation of small UAS (sUAS) within the National Airspace System (NAS). One year later, the sUAS ARC provided a comprehensive set of recommendations to the FAA for various categories of sUAS for the maximum weight of 25 kg. If these recommendations are accepted, the sUAS in compliance with the weight restrictions will be able to operate in the US according to the FAA specifications.

Because the sUAS could have many useful applications, the outstanding work of the sUAS ARC can be considered a good step forward.

- **RTCA.** It is a private, not-for-profit corporation that develops consensus-based recommendations regarding communications, navigation, surveillance, and ATM system issues. RTCA functions as a Federal Advisory Committee. Its recommendations are used by the FAA as the basis for policy, program, and regulatory decisions and by the private sector as the basis for development, investment, and other business decisions.

Many federal agencies and commercial operators are currently operating or seeking authority to operate UAS in the NAS. The Committee SC-203 was established in 2004 to develop recommended UAS Minimum Aviation System Performance Standards (MASPS) necessary for the safe integration of UAS in the NAS. The SC-203 terms of reference include: (1) MASPS for UAS systems level; (2) MASPS for UAS control and communication; and (3) MASPS for UAS sense and avoid.

---

[61] The FAA issued the Order 8130.34 *"Airworthiness Certification of Unmanned Aircraft Systems"* on March 2008 to establish procedures for issuing a special airworthiness certificate in the experimental category for the purposes of research and development, market survey, or crew training to UAS (see Chapter 8, Section 8.5.2.5).

Here, we report some basic principles:

The airworthiness certificate authorizes an operator to use defined airspace and includes special provisions unique to each operation. For instance, the certificate includes a requirement to operate only under VFR and during daylight hours.

It is required coordination with an appropriate air traffic control facility and the UAS and to have a transponder able to operate in well-defined modes.

To make sure, the aircraft will not interfere with other aircraft, a ground observer or an accompanying "chase" aircraft must maintain visual contact with the UAS.

Flight termination must be initiated at any point that safe operation of the aircraft cannot be maintained or if hazard to persons or property is imminent.

In the event of lost link, the UA must provide a means of automatic recovery that ensures airborne operations are predictable and that the UA remain within the flight test area. The chase aircraft or observer, all other UAS control stations, and the appropriate ATC facility will be immediately notified of the lost link condition and the expected aircraft response.

To release initial GM and qualitative considerations, the document *"Guidance Material and Considerations for Unmanned Aircraft Systems"* was published in March 2007.

Since February 2009, SC-203 and EUROCAE WG-73 collaborate for the development of a pilot project for initial UAS safety assessment.

- **Civil Aviation Safety Authority (CASA).** The Australian Authority issued Part 101 of *Civil Aviation Safety Regulations* to consolidate the rules governing all unmanned aeronautical activities into one body of legislation. This part sets out the requirements for the operation of unmanned aircraft (including model aircraft). These rules require the operation of a large UAV with a launch mass greater than 150 kg to be issued with either an Experimental Certificate or a Certificate of Airworthiness in the restricted category.

The AC 21-43(0) of June 2006 *"Experimental certificate for large unmanned vehicle (UAV)"* provides guidance to applicants seeking an Experimental Certificate for a large UAV, which is an aeroplane with a launch weight of 150 kg or above. After this summary list of initiatives for the UAS integration into the airspace, we can conclude that even if the full integration in nonsegregated airspace is too far to be achieved, a great number of UAS engaged in civil operation are already a reality, even with all the limitations that the authorities are trying to solve on a case-by-case basis. Even if the integration in nonsegregate space will require a number of years to become a reality, the initiative of some authority trying to ensure the limited operational use of UAS, will allow the international community to achieve that knowledge that only practical experience can produce.

This will produce a basic set of airworthiness regulations internationally harmonized that the UAS community has long been waiting for.

4.11.2.1. THE LIGHT UAS

In Europe, unmanned aircraft are divided into two major groups, each of which is regulated by different authorities:

UA with a maximum takeoff mass of more than 150 kg. These systems are regulated by the EASA and

UA with a **maximum takeoff mass of less than 150 kg**, commonly designated as **Light**.

Unmanned Aircraft Systems

These systems are regulated by the national civil aviation authorities.

The European Commission, Directorate General Energy and Transport (now renamed Directorate General Mobility and Transport), organized a hearing on LUAS in Brussels, Belgium, on 8 October 2009. This was the first hearing on UAS that was ever organized by the European Commission.

A report of this hearing has been released for publication on 1 April 2010, and it is interesting to quote some excerpts of the first part of this document.

The main objectives of the hearing were

To understand the current European LUAS industrial base and the current LUAS applications in Europe.

To identify potential obstacles and best practices in Europe.

To exchange directly with the European LUAS community views and assess the future potential role of EC for the insertion of LUAS.

Current applications based on the effective usage of LUAS

The hearing showed that LUAS are already used in Europe for a large spectrum of governmental and nongovernmental applications.

The use of LUAS is significant for civil security operations, in particular for supporting the fight against building fires, post fire investigations, motorway road traffic collision monitoring, chemical cloud release monitoring, searching frozen lakes for missing persons (thermal). UAS greatly improve the preintervention situational awareness of the authorities, which can be of prime importance in the case of dangerous environments such as collapsed buildings (earth quakes), chemical clouds, floods, and so on.

LUAS are also widely used for the monitoring of wildlife and nature observation, and reveal excellent capabilities in support of the meteorological domain (better capabilities/manoeuvrability than balloons). The following applications were also highlighted at the hearing: atmospheric and climate research, land monitoring (vegetation, fauna, hydrology, salt water infiltration), and ocean monitoring (sea state, algae, sea ice, and icebergs).

Potential advantages and benefits of LUAS for citizens

LUAS provide authorities with new possibilities that did not exist before with manned aircraft. They limit physical risks for civil servants in dull, dirty, and dangerous environments, due to the absence of crew on board and the non-necessity to be physically involved on site.

LUAS are easy to transport, relatively simple to deploy, easy to launch and recover, and show real advantages in terms of durability, modularity, silence, substantial autonomy, and high degree of controllability.

In the absence of pilots onboard, the air vehicle brings new potentialities in terms of protection of the environment, noise abatement, reduced fuel consumption, and CO_2 emission.

LUAS present a high level of mobility and reactivity, supplying authorities with a rapid response capability in support of outdoor and indoor operations. Simpler than any manned aircraft systems developed for similar activities in terms of deployment and use, LUAS have low-cost operations and are less demanding in terms of resource allocated.

LUAS allow long-time surveillance, modularity through fusion of data coming from multiple onboard sensors (electrooptic, infrared, radar, etc.), and operations under extreme conditions.

The user base for LUAS is very large, enabling the use of these systems for all types of missions and by a large customer base. They also offer possibilities for operations run by public and private entities, thereby creating new business opportunities for the sector.

Most current nonmilitary LUAS applications take place within visual line-of-site and at altitudes inferior to 150 m and are therefore outside airspaces used by manned aircraft. Consequently, a significant number of applications could rapidly be fulfilled with the existing LUAS technology.

Current obstacles to LUAS development

In Europe, no harmonized rules and standards exist for the insertion of unmanned aircraft.

As already seen, the certification and operational requirements for LUAS with a minimum takeoff mass of **less than 150 kg** are the responsibilities of the European National Aviation Authorities (NAA).

Because of the complexity of the task, very few states have developed *ad hoc* legislation and certification processes, and currently no harmonization has taken place between national regulations.

The ICAO has engaged activities related to the insertion of unmanned aircraft, but the development of ICAO rules is not foreseen before a long time. Additionally, ICAO does not seem to consider itself competent in the field of LUAS.

Common certification processes and standards:
A must for the industry

In Europe, no harmonized technical airworthiness code has been developed for LUAS, and no type approval/certification process is in place. The UAS sector below 150 kg is composed of aerial vehicles of very different types, capabilities, size, and weight.

Therefore, adaptation shall be required to accommodate them on the basis of their intrinsic relevant risk levels.

The LUAS community needs a single certification process applicable in all EU states that should provide national authorities with a single set of safety rules applied uniformly in all states and a set of rules allowing minimum segregation for operations.

In many states, the grant of an aerial work license to a UAS operator is almost impossible, as no appropriate framework for certification of the UAS exists. This affects the development of professional activities based on LUAS utilization for governmental and commercial use.

Conclusions and recommendations

This first hearing has been a real success and a fruitful exercise. The LUAS community has provided the Commission with a great number of elements of appreciation of the current situation relative to LUAS, allowing a better understanding of their requirements and permitting to define the line of action required for the introduction of LUAS in European airspace.

The hearing demonstrated that LUAS are already used by a significant number of governmental authorities, in particular for police, customs, border control, fire fighting, natural disasters, and search and rescue missions.

Once a legal framework exists, a totally new aerial work service supply industry should sprout rapidly.

Nonmilitary LUAS operations are currently mainly conducted at altitudes inferior to 150 m above ground level and within the visual line-of-sight. In that condition, the operational environment does not conflict with flights of manned aircraft. This call for the development of specific rules for LUAS, simpler than those existing for manned aircraft, or that will be required for unmanned aircraft with a mass of more than 150 kg.

A single set of rules for Europe would favor the creation of an open and fair European market.

It is necessary to harmonize the requirements and limitations for LUAS certification and operations within Europe, and also to harmonize the requirements with a number of non-European Union regulators such as the FAA, Transport Canada, and CASA Australia.

Chapter five

Type Certification

5.1. TYPE CERTIFICATION OF AIRCRAFT, ENGINES, AND PROPELLERS

5.1.1. The type certificate

The **type certificate** is a document by which the authority states that an applicant has demonstrated the compliance of a **type design** to all applicable requirements. This certificate is not in itself an authorization for the operation of an aircraft, which must be given by an **airworthiness certificate**.[1]

5.1.2. The type design

The type design of a product,[2] which must be adequately identified according to EASA Part 21 (Paragraph 21A.31) and FAR 21 (Paragraph 31), consists of the following[3]:

(1) The drawings and specifications, and a listing of those drawings and specifications. They are necessary to define the configuration and the design feature of the product shown to comply with the applicable type certification basis and environmental protection requirements.

(2) Information on materials and processes and on methods of manufacture and assembly of the product needed to ensure the conformity of the product.

(3) An approved Airworthiness Limitations section of the instructions for continued airworthiness[4] as defined by the applicable airworthiness code.

(4) Any other data necessary to allow, by comparison, the determination of the airworthiness, the noise characteristics, fuel venting, and exhaust emission (where applicable) of later products of the same type.

In other words, the type design "freezes" not only the product configuration but also the production methods. Every deviation from the type design becomes a "change" which must be approved, as we will see. This is to make sure that the series products are not inferior to the prototype identified by the type design, in terms of flight safety.

---

[1] See Chapter 8.

[2] Products are aircraft, engines, and propellers.

[3] The text is that of EASA Part 21. FAR 21 has slightly different wording, but with the same meaning. JAR 21 is similar, without reference to the environmental protection requirements.

[4] Continued airworthiness. This can be defined as the airworthiness of products during their operational life. Hence, the relevant information gives a description of the product and its characteristics, servicing information and maintenance instructions, and so on.

Airworthiness: An Introduction to Aircraft Certification.

5.1.3. Environmental protection

EASA Part 21 and FAR 21, for type certification, include the designation of applicable environmental protection requirements and certification specifications, missing in JAR 21 (till Amendment 5).

According to Annex 16 of the Convention of Chicago, the environmental protection includes noise requirements and emission requirements (prevention of intentional fuel venting and emissions of turbojet and turbofan engines).

Subpart I of EASA Part 21 include the instructions for the issue of **noise certificates**. Such documents do not exist in the FAA certification.[5]

An example of the influence that the environmental protection requirements can have on the design of an aircraft is the case of **supersonic business aeroplanes (SSBJ)**. Supersonic transport (SST) ended with the withdrawal of "Concorde." The big aerospace companies at present are not likely to produce new SST, struggling as they are to find new markets for more efficient and economic transport aeroplanes. The competition between Boeing and Airbus, with their new models B 787 and A 350 still on certification phase, is an example.

Nevertheless, the supersonic aeroplane is still attractive in the market of business jets. "Time is money ..." *Flight International* of October 2004 stated about some initiatives and ideas related to SSBJ projects.

One of the thorniest issues for the operation of (civil) supersonic aeroplanes is how to persuade regulators and legislators to change the rules banning supersonic overland flight. It is clear that an SSBJ being forced to fly subsonically over land is not so appealing. On the other hand, the only way to change the rules is **a reduction in the sonic boom** to an acceptable level for people on the ground.

The idea of quiet supersonic transport (QSST) is not new. One of the pioneers of this concept was Allen Paulson, the founder of Gulfstream. He pursued the dream of an SSBJ until his death in 2000. At his father's bequest, his son Michael Paulson engaged the notorious Skunk Works[6] to design an SSBJ using an innovative airframe shape to reduce the sonic boom.

Other studies and research have been carried out in the United States, all with the aim of reducing the sonic boom. NASA, of course, is involved in this.

If the research for low-boom technology is to be validated, some prototypes will have to be built, adding costs of many millions of dollars.

In any case, it is worth reading the cautiously optimistic conclusions of *Flight International*:

> *... it is now more likely than at any time in recent history that a supersonic business jet will become a reality within the next 10 years. And if an SSBJ enters service it will only be a matter of time before a larger aircraft— possibly a 50-seat transatlantic jet to replace Concorde, perhaps a 300-seat transpacific airliner—takes to the skies.*

---

[5] The noise certification is part of the type certificate.
[6] See Chapter 6, "Construction of prototypes and test articles."

In 2010, there still are a few SSBJ in the development phase and, before being able to reach a true QSST, a possible compromise could be the production of an aircraft to be operated just below the speed of sound over land and at supersonic speed over water. Such a solution does not require a change of regulation to allow supersonic flights over land.

5.1.3.1. DESIGNATION OF APPLICABLE ENVIRONMENTAL PROTECTION REQUIREMENTS

For EASA Part 21, the applicable noise and emissions requirements are included in the ICAO Annex 16 with a different applicability for the various categories of aeroplanes and helicopters.

FAR 21 normally refers to *applicable aircraft noise, fuel venting, and exhaust emissions requirements.*

From Chapter 3 of this book, we can quote:

- — FAR 36. Noise Standards.
- — FAR 34. Fuel venting and exhaust emission requirements for turbine-engine-powered airplanes.

NOTE: According to FAR 34:

Exhaust emissions means substances emitted into the atmosphere from the exhaust discharge nozzle of an aircraft or aircraft engine.

Fuel venting emissions means raw fuel, exclusive of hydrocarbons in the exhaust emissions, discharged from aircraft gas turbine engines during all normal ground and flight operations.

5.1.3.2. A LOOK INTO THE FUTURE

A current estimate of the aircraft's contribution to the total global emission from combustion of fossil fuels CO_2 is $2-3\%$.

With reference to the ICAO Annex 16 in Chapter 3, we have already discussed the effect of CO_2 and other aircraft emissions on the environment.

Aircraft emission could appear not important compared with the total global emission, but we must consider its rapid increase due to a dramatic forecast increase of the air travel in the next years (aircraft flights are expected to double by 2020 and triple by 2030).

The necessity of a drastic reduction in the emissions leads to the need of reducing the amount of fuel burned.

Fuel burning depends—apart from the engines; efficiency—on the thrust necessary to fly. At cruise speed, the thrust (T) should equal the overall drag (D), and the lift (L) of the overall weight (W).

$T = W \times D/L$; then, T is proportional to the weight and in inverse relation to L/D, the glide ratio.

Empty weight reduction can be obtained employing new materials, better structural and cabin furniture design, and so on.

The increase in the glide ratio depends on the aerodynamic design of the aircraft and can be obtained, for example, by increasing the wing span (without undue weight penalization), reducing the lift-induced drag by winglets, the parasitic drag by a good design of the fuselage and other nonlifting parts, lowering the skin friction, and so on. Good aerodynamic design is also important in reducing the wave drag, which can be a problem at high subsonic speed.

All this is not new of course, and during the last decades the airliners have really progressed in this sense. But the rapid increase of air travel, the growing concern about climate change, and the dramatic increase in fuel price have convinced all the concerned stakeholders to research into alternative solutions to cope with this new challenge.

Among many programs on the subject, we can quote **Clean Sky initiative.**

Proposed by the European Commission, Clean Sky aims to create a radically innovative Air Transport System centered on the reduction of the environmental impact of air transport through the reduction of noise and gaseous emissions, and improvement of the fuel economy of aircraft for the benefit of society at large. Clean Sky will embody a new approach to research financing at the European level, bringing together public and private funds, involving industry and nonprofit research institutions.

Clean Sky aims to develop advanced technologies for the next generation of aircraft to establish an innovative and competitive Air Transport System. Through the development of full-scale demonstrators, Clean Sky will perform an overall assessment of individual technologies at the fleet level, thus ensuring earliest possible deployment of its research results. The activity will cover all main flying segments of the Air Transport System and the associated underlying technologies identified in the Strategic Research Agenda for Aeronautics developed by the Aeronautics Technology Platform ACARE.[7]

5.1.4. Design organization

So far, we have dealt with airworthiness authorities and their commitment. Now, we will consider the designer's perspective,[8] that is, the "person" defined as the **applicant** becoming the **type certificate holder (TCH)** once the type certificate is issued. It goes without saying that designing and demonstrating compliance with the applicable requirements needs a technical organization adequate for this kind of project; this could range from very few to several hundred technicians.

JAR 21, Paragraph 21.13, states that the applicant must hold (or have applied for) an appropriate[9] Design Organization Approval (DOA). The requirements for a **JAA DOA** are contained in Subpart JA of JAR 21.

---

[7] Advisory Council for Aeronautical Research in Europe (ACARE) proposed some challenging targets for aerospace manufacture, including a 50% reduction in CO_2 emissions, an 80% reduction in NOx and a 50% reduction in noise nuisance for aircraft entering into service from 2020, relative to their year 2000 counterparts.

[8] We do not say the "manufacturer's perspective" because the manufacturer and the designer could be different "entities" (in a legal sense).

[9] That is, adequate to the design which is the object of the certification.

In a similar way, EASA Part 21 states in Paragraph 21A.14 that "any organization applying for a type certificate or restricted type certificate shall demonstrate its capability by holding a **DOA** issued by the Agency in accordance with Subpart J of Part 21."

By way of derogation, as an alternative procedure to demonstrate its capability, an applicant may seek Agency agreement for the use of procedures setting out specific design practices, resources, and sequence of activities necessary to comply with Part 21 when the product is one of the following:

(1) A very light aeroplane or rotorcraft, a sailplane or a powered sailplane, a balloon, a hot-air ship or

(2) A small aeroplane meeting all the following elements:

- **(a)** Single piston engine, naturally aspirated, of not more than 250 HP maximum takeoff power
- **(b)** Conventional configuration
- **(c)** Conventional material and structure
- **(d)** Flights under VFR, outside icing conditions
- **(e)** Maximum of four seats including the pilot and maximum takeoff mass limited to 3000 lb (1361 kg)
- **(f)** Unpressurized cabin
- **(g)** Nonpower-assisted controls
- **(h)** Basic acrobatic flights limited to +6/−3 g or
- **(i)** A piston engine or
- **(j)** An engine or a propeller type certificated under the applicable airworthiness code for powered sailplanes or
- **(k)** A fixed or variable pitch propeller.

Alternative procedures are acceptable means to demonstrate design capability for type certification in the above-mentioned cases, approval of a major design change to the type design under Supplemental type certificate (STC), and a major repair design.

This concept is the implementation, in the context of specific projects, of procedures required in Subpart J DOA, to ensure that the applicant will perform relevant activities as expected by the Agency, but without the requirements on the organization itself that can be found in Subpart J. The establishment of these alternative procedures may be seen as a starting phase for a Subpart J DOA, allowing at a later stage, at the discretion of the applicant, to move toward a full Subpart J DOA by the addition of the missing elements.

As an alternative to DOA, a manual of procedures must set out specific design practices, resources, and sequence of activities relevant for the specific projects, taking into account Part 21 requirements.

The EASA has an internal working procedure called "*Alternative Procedures to Design Organization Approval (ADOAP)*," describing how the Agency will internally handle the investigation of an applicant's alternative procedures in the absence of DOA.

EASA decisions related to design organizations having demonstrated their capability for design through alternative procedure to DOA are published in its official publication.

The FAA has a different approach. FAR 21 does not mention a formal approval of a design organization. In this chapter, we further consider the FAA type certification procedures in more detail.

5.1.5. Design Organization Approval (DOA)—JAA and EASA

We have already quoted that the requirements for acquiring this approval are contained in Subpart JA of JAR 21[10] and in Subpart J of EASA Part 21. It may be useful to illustrate the main characteristics of the DOA.[11]

The main duties and responsibilities of a design organization are as follows:
(1) To design.
(2) To demonstrate compliance with the applicable requirements.
(3) To independently check the statements of compliance.
(4) To provide items for continued airworthiness.
(5) To check the job performed by partners/subcontractors.
(6) To independently monitor the above functions.
(7) To provide the authority with the compliance documentation.
(8) To allow the authority to make any inspection and any flight and ground tests necessary to check the validity of the statements of compliance.

A crucial point, besides the normal design organization, is the institution of a **Design Assurance System (DAS)** for control and supervision of the design and design changes to the product covered by the application. This includes all the activities for the achievement of the type certificate, the approval of changes, and the maintenance of continued airworthiness.

In particular, the DAS should include an organizational structure to (Fig. 5.1)
(1) Control the design.
(2) Show compliance with the applicable certification standard and environmental requirements.
(3) Show compliance with protection requirements.
(4) Independently check this compliance.
(5) Liaise with the Agency.
(6) Continuously evaluate the design organization.
(7) Control subcontractors.

All these functions are essentially accomplished through the action of

---

[10] JAR 21 also contains a Subpart JB, which is a DOA for design organizations designing parts and appliances. The authority accepts such applications if it is agreed that the approval is appropriate for the purpose of assisting applicants for or holders of type certificates or STCs in showing compliance with the applicable requirements. The JB DOA is issued with reference to the above-mentioned applicants or holders. The JB DOA does not have privileges.

[11] Detailed explanations are contained in AMC&GM for Part 21.

Relationship Between Design, Design Assurance and Type Investigation

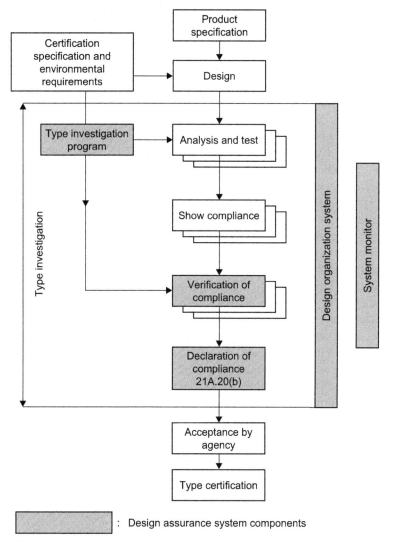

FIGURE 5.1 Relationship between design, design assurance, and type investigation

- A staff of **Certification Verification Engineers (CVEs)** responsible for checking and signing all the documents of compliance with the applicable requirements. The CVEs may work in conjunction with the individuals who prepare compliance documents, but may not be directly involved in their creation (this is to ensure independent checking).
- **System Monitoring**, which has the task of ensuring that all the responsibilities of the DAS are properly discharged, proposing corrective and preventive measures for continuous effectiveness. Normally, this is done through targeted audits. The System Monitoring could be a functional

emanation of the applicant's Quality Assurance System. The person responsible for the System Monitoring always reports to the **Head of Design Organization**.

A peculiarity of the organization is the **Office of Airworthiness** that, among its main tasks, ensures liaison between the design organization and the authority with respect to all aspects of type certification. This office carries out a true coordination action within the design organization; moreover, it issues and updates the **DOA Handbook**, which is the basic document of the organization, containing its description, the object of the certification, staff functions, all the procedures concerning design activities, tests, and others.

JAR 21 (Paragraph 21.20) and EASA Part 21 (Paragraph 21A.20) require that the applicant must declare, at the end of the type investigation that he or she has shown compliance with all applicable requirements. The **declaration of compliance** must be signed by the Head of Design Organization.

From a legal point of view, the declaration of compliance is issued by the authority through the type certificate, after the inspections, flights, and ground tests necessary to check the validity of the declaration of compliance.

An important feature of the DOA is shown by the **privileges** stated in Paragraph 21.A263. This states the possibility that the compliance documents for the applicable requirements may be accepted[12] by the Agency without further verification; furthermore, the design organization may obtain, following the prescribed investigations and within its terms of approval:

(1) The approval of flight conditions required for a permit to fly or

(2) A type certificate or approval of a major change to the type design or

(3) An STC or

(4) An ETSO authorization or

(5) A major design approval.

The holder of the DOA shall be entitled, within its terms of approval and under the relevant procedures of the DAS to

(1) Classify changes to the type design and repairs as "major" or "minor" (we will discuss this further in the present chapter).

(2) Approve minor changes to type design and minor repairs.[13]

(3) Issue information or instructions containing the following statement: "The technical content of this document is approved under the authority of DOA nr..."[14]

(4) Approve documentary changes to the aircraft flight manual, and issue such changes containing the following statement: "Revision no. xx to AFM ref. yyy, is approved under the authority of DOA no......"

---

[12] The design organization shall allow the Agency to review any report and make any inspection and perform or witness any flight and ground test necessary to check the validity of the compliance statements submitted by the applicant.

[13] That is, without direct intervention by the authority.

[14] They must contain a statement making reference to the DOA privilege.

(5) Approve the design of major repairs to products for which it holds the type certificate or the supplemental type certificate.[15]

(6) Approve the conditions under which a permit to fly can be issued [21A.263 (c)(6)] in accordance with 21A.710 (a) (2).[17]

    (i) except for initial flights of:

      — a new type of aircraft or

      — an aircraft modified by a change that is or would be classified as a significant major change or significant STC or

      — an aircraft whose flight and/or piloting characteristics may have been significantly modified;

    (ii) except for permits to fly to be issued for the purpose of 21A.701(a) (15).[16]

(7) Issue a permit to fly in accordance with 21A.711(b)[17]

---

[15] See "The Supplemental type certificate (STC)" section in this chapter.

[16] That is for noncommercial flying activity on individual noncomplex aircraft or types for which a certificate of airworthiness or restricted certificate of airworthiness is not appropriate.

[17] An appropriately approved design organization may issue a permit to fly under its privileges when the flight conditions have been approved in accordance with:

21A.710 Approval of flight conditions (*See also Figs 8.1 and 8.2 in Chapter 8*).

  (a) When approval of the flight conditions is related to the safety of the design, the flight conditions shall be approved by

    (1) the Agency or

    (2) an appropriately approved design organization, under the privilege of 21A.263 (c)(6).

  (b) When approval of the flight conditions is not related to the safety of the design, the flight conditions shall be approved by the Competent Authority, or the appropriately approved organization that will also issue the permit to fly.

  (c) Before approving the flight conditions, the Agency, the Competent Authority, or the approved organization must be satisfied that the aircraft is capable of safe flight under the specified conditions and restrictions. The Agency or the Competent Authority may make or require the applicant to make any necessary inspections or tests for that purpose.

For 21A.708, flight conditions include:

  (a) the configuration(s) for which the permit to fly is requested;

  (b) any condition or restriction necessary for safe operation of the aircraft, including

    (1) the conditions or restrictions put on itineraries or airspace, or both, required for the flight(s);

    (2) the conditions and restrictions put on the flight crew to fly the aircraft;

    (3) the restrictions regarding carriage of persons other than flight crew;

    (4) the operating limitations, specific procedures, or technical conditions to be met;

    (5) the specific flight test program (if applicable);

    (6) the specific continuing airworthiness arrangements including maintenance instructions and regime under which they will be performed;

  (c) the substantiation that the aircraft is capable of safe flight under the conditions or restrictions of subparagraph (b);

  (d) the method used for the control of the aircraft configuration, to remain within the established conditions.

for an aircraft it has designed or modified, and when the design organization itself is controlling under its DOA, the configuration of the aircraft, and is attesting conformity with the design conditions approved for the flight [21A.263(c)(7)].

The holder of the DOA has the following obligations:

(a) Maintain the handbook in conformity with the DAS;

(b) Ensure that this handbook is used as a basic working document within the organization;

(c) Determine that the design of products, or changes or repairs thereof, as applicable, comply with applicable requirements and have no unsafe feature;

(d) Except for minor changes or repairs approved under the privilege of 21A.263, provide to the Agency statements and associated documentation confirming compliance with Paragraph (c);

(e) Provide to the Agency information or instructions related to required actions under 21A.3(b) (reporting to the Agency);

(f) Where applicable, under the privilege of 21A.263(c)(6), determine the conditions under which a permit to fly can be issued;

(g) Where applicable, under the privilege of 21A.263(c)(7), establish compliance with 21A.711(b)[17] and (d)[18] before issuing a permit to fly (EASA Form 20), to an aircraft.

DOA can be considered as a significant improvement in the relationship between the applicant and the authority.[19] Many authorities have, for a long time, performed surveillance on designs and aeronautical material that can be defined as a "control of the control." All the aircraft were inspected and also checked in flight.[20] This kind of surveillance was expensive from the human resources point of view and could only be justified to compensate the lack of organization in an enterprise. The control of the control is a philosophically incoherent praxis because, to be effective, it should involve other levels of control (the control of the control of the control, that is, *quis custodiet ipsos custodes*? who guards the guardians?) until safety is assured.[21] The impossibility, but also the poor efficiency, of such a system is evident.

Hence, it is necessary that the applicant assumes the whole responsibility of safety, without the caveat that "if there is something wrong, the authority will correct it."

---

[18] 21A.711(d): "The permit to fly shall specify the purpose(s) and any conditions and restrictions approved under 21A.710."[17]

[19] We will use the term "authority" in a general sense; of course, the Agency is intended as an authority.

[20] In the United States, this problem was overcome a long time ago through different forms of organization. Thousands of aircraft per year were built before the crisis of general aviation, so that the FAA could not cope using "traditional" surveillance.

[21] This is something similar to the safety assessment of control.

But where is the real interest of the authority? The authority, through certification processes such as DOA [and Production Organization Approval (POA) in the case of production], promotes the enterprise to a condition of self-control leading to the creation of a product that is safe independent of the authority's surveillance. Hence, there is a transfer of responsibilities for the authority from the control of the **product** to the control of the **organization**; this is being ensured by means of audits of products[22] and audits of systems.[23]

Furthermore, the DOA privileges allow a more efficient authority's involvement, because the authority can choose what to see and what to approve, with focused interventions. This is also an advantage for the authority's technicians as they do not lose contact with aeronautical materials and tests, an indispensable prerequisite for training and updating.

From a certain point of view, the DOA privileges also become the authority's privileges.

Unfortunately, the alternative procedures replacing the DOA do not allow the above privileges. It should then be reasonable to prompt small organizations to instigate a DOA too, even if they normally deal with products for which the DOA is optional. This is rather difficult considering the way in which Subparts JA of JAR 21 and J of EASA Part 21 are now written, clearly with medium/large organizations in mind. The JAA have, for a long time, discussed the possibility of issuing advisory material that, without distortion of the basic philosophy, could make the DOA certification of small organizations easier. This would be an improvement in terms of both safety and efficiency of the authority.

5.1.6. Changes in type design

We have previously seen that all deviations from a type design are "changes" that have to be approved by the authority (in a direct or indirect way). Because these deviations can range, for example, from a simple correction of a drawing to the opening of a large door in the fuselage of an aircraft for conversion in a cargo aircraft, JAR/FAR 21/EASA Part 21 considers two kinds of changes:

(1) **Minor changes**, that is, those that have *no appreciable effect* on the mass, balance, structural strength, reliability, operational characteristics (noise, fuel venting, exhaust emission),[24] or other characteristics affecting the airworthiness of the product.

(2) **Major changes**, that is, all other changes.

FAR 21 has the same classification with some difference in wording and definition of **acoustical change** for different types of aircraft.

---

[22] Audit of product: checks performed on single tests or single-test articles to ensure the correct realization of the actions required to demonstrate compliance with the applicable requirements.

[23] Audit of system: checks performed on the applicant's organization, personnel, and procedures to ensure compliance with the applicable requirements.

[24] Noise, fuel venting, and exhaust emissions are in EASA Part 21 only.

The classification of changes is important because it makes a difference in the authority's involvement in the approval phase (we will also see its importance for establishment of the "certification basis"). We have already considered that an organization having a DOA can make a minor change approval without direct verification from the authority. But even without a DOA, the authority's attitude is less severe toward such changes. Nevertheless, the classification of changes is a delicate problem because, when the changes are not clearly minor or major as in the above example, that *appreciable effect* in the minor change definition can lead to a range of uncertainties. This is the reason why design organizations must have approved procedures for this classification, and why only design organizations with DOAs are allowed to make such classifications without further authority verification.

In any event, minor changes in a type design are approved:

(1) EASA—by the Agency or by appropriately approved design organization under a procedure agreed with the Agency.

(2) FAA—by a method acceptable to the Administrator.

GM 21A.91 of EASA Part 21 provides guidance on the classification of major changes (as opposed to minor changes as defined in Paragraph 21A.91). Furthermore, to make the classification easier, it provides a few major change examples per discipline: structure, cabin safety, flight, systems, propellers, engines, rotors and drive systems, environment, and power plant installations.

Figure 5.2 presents an outline of the change classification process.

We could question how much a certificated type design could be changed without the application for another type certificate. As an example: can a single-engine aeroplane be converted to a twin-engine aeroplane as a change to the same TC? The answer **used** to be provided by Paragraph 21.19 of JAR 21 (Amendment 5, an approach derived from the **now amended** FAR 21) and was negative. This JAR 21 paragraph listed other cases for which the application for a new TC was required, as follows.

For **aircraft**, an application for a new TC was required if the proposed change was:

(1) In the number of engines or rotors.[25]

(2) To engines or rotors using different principles of operation.[26]

For an **engine**, an application for a new TC was required if the proposed change was in the principle of operation.

For a **propeller**, an application for a new TC was required if the proposed change was in the number of blades or principle of pitch change operation.

The paragraph also prescribed the following general principle: "*Any person who proposes to change a product must apply for a new Type Certificate if the*

---

[25] Normally, with reference to an increase in number. Nevertheless, in some cases, a reduction in number has been accepted in the same TC (e.g., a three-engined aircraft converted to a twin-engined aircraft).

[26] Examples are a reciprocating engine replaced by a jet engine and a mechanically driven rotor replaced by a jet rotor.

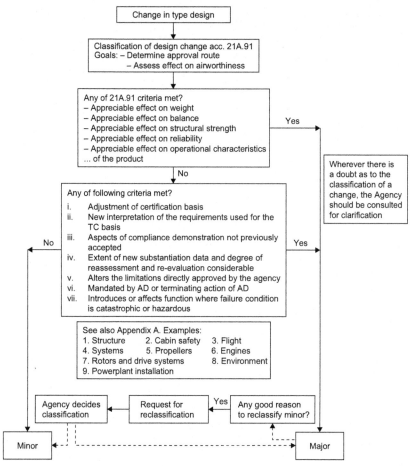

FIGURE 5.2 Classification of changes in type design

authority finds that the change in design, power, thrust, or weight is so extensive that a substantially complete investigation of compliance with the applicable requirements is required." This general principle can also be found in Paragraph 21A.19 of EASA Part 21 and Paragraph 21.19 of FAR 21, with slightly different wording.

The list of the particular cases requiring a new TC does not exist anymore and this allows the authorities a better opportunity to evaluate on a case-by-case basis.

5.1.7. Designation of the type certification basis:[27] the derivative prototype

During the operational life of an aircraft, many changes, minor or major, are normally introduced (after the authority's approval) for various reasons.

---

[27] The applicable airworthiness code as established in Paragraphs 17 and 101 of JAR/FAR 21 and EASA Part 21A as appropriate, special conditions, equivalent level of safety findings, and exemptions applicable to the product to be certificated.

It also happens that the TCH, after type certification, needs to differentiate the type design (normally for commercial reasons) in so-called "derivative" aircraft.[28]

The changes could be a different maximum takeoff weight, replacement of the engine type, a different fuselage length to contain a major (or minor) number of passengers—there are numerous examples.

The introduction of changes or the design of a derivative prototype are sometimes realized several years after the type certification of the product (which is called the "basic product"), and the applicable requirements may have been changed substantially in this period. The first thing to do, in any case, is to establish whether it is possible to type certificate the changed product as a **change to a TC** or whether there is a need for an application for a **new TC**.

As mentioned above, the FAR/EASA, Paragraphs 21.19/21A.19, establish when an application for a new TC is required. Nevertheless, that generic wording, leaving the final decision to the authority, has very often caused contention with the applicant. In fact, applicants usually prefer to start from a basic product because, if an application for a new TC is made, they have to start over again, and with the most recent basis for certification.

As a general rule (exceptions are given), the certification of a type design change should comply with the requirements applicable **at the date of the application** for the change.

There is also the possibility of adopting an earlier amendment (compared with the one existing at the date of application for change approval)[29] as follows.

Paragraph 21.101 of FAR 21 and Paragraph 21A.101 of EASA Part 21 introduce the concept of changes considered by the Administrator/Agency as **nonsignificant**.

Changes that meet one of the following criteria are automatically considered **significant**:

(1) The general configuration or the principles of construction are not retained.

(2) The assumptions used for certification of the product to be changed are no longer valid.

Having introduced these criteria, the above-mentioned Paragraph 101 states that an applicant may show that a changed product *complies with an earlier amendment* for any of the following:

(1) A change that the Administrator/Agency finds *not to be significant*.

(2) Each area, system, component, equipment, or appliance that the Administrator/Agency finds is *not affected by the change*.

(3) Each area, system, component, equipment, or appliance that is affected by the change for which the Administrator/Agency finds that compliance with a regulation applicable at the date of the application *would not contribute*

---

[28] One example among many others: Airbus aircraft of series A340-200, 300, 500, and 600.
[29] The earlier amendment may not precede the corresponding regulation incorporated for reference in the type certificate.

materially to the level of safety of the changed product or would be impractical.

Furthermore, an application for a change to an aircraft (other than a rotorcraft) of 2722 kg (6000 lb) or less maximum weight, or to a nonturbine rotorcraft of 1361 kg (3000 lb) or less maximum weight may show that the *changed product complies with the regulations incorporated by reference in the type certificate.* However, if the Agency/Administrator finds that the change is significant in an area, the Agency/Administrator may designate compliance with an amendment to the regulation incorporated by reference in the type certificate that applies to the change and *any* regulation that the Agency/Administrator finds is directly related, unless the Agency/Administrator also finds that compliance with that amendment or regulation would not contribute materially to the level of safety of the changed product or would be impractical.

This last provision of the above-mentioned paragraphs is less stringent for the approval of type design changes concerning general aviation aircraft.

The possibility of adopting earlier requirements for a new type certification is currently known as a "grandfather right."

With regard to the certification basis for a changed product, it is obvious that, with the same criteria used for the basic product type certification, if the Administrator/Agency finds that the regulations in effect on the date of the application for the change do not provide adequate standards with respect to the proposed change because of a novel or unusual design feature, the applicant must also comply with **special conditions**, and amendments to those special conditions, to provide a level of safety equal to that established by the regulations in effect on the date of the application for the change.

At this point, it is also important to introduce the concept of **substantial change**,[30] requiring the application for a new TC.

5.1.8. Advisory material

As we have illustrated so far, the definition of a type certification basis is a complex matter, involving a multitude of different cases and requiring experience and common sense. We may, for instance, consider that a series of step-by-step changes to a type design can lead to a cumulative effect such as to create a substantial change. Then, it may be necessary to go through the family "history" of related products case-by-case.

FAR/JAR 21 and EASA Part 21 define the basic criteria, and it would be impossible to make these criteria operational on a uniform basis without advisory material. After years of discussion, this advisory material is provided by **EASA GM 21A.101** and **FAA AC 21.101-1**, which give guidance for establishing the type certification basis for a product and identifying the conditions

---

[30] **Substantial change**: a design change of an extent sufficient to require a substantially complete investigation of compliance with the applicable requirements, and consequently a new TC in accordance with JAR/FAR 21.19/EASA Part 21A.19.

under which an applicant for a design change is required to apply for a new type certificate.

The GM/AC explains the criteria of 21A.19 and 21A.101, and their application.

They provide guidance as to the assessment of "**significant**" versus "**not significant**" changes to the type-certificated product. These documents also provide guidance for the determination of "**substantial**" versus **significant** changes.

The GM/AC is applicable to all major changes to the type design of aircraft, engines, and propellers.

Minor changes are considered to have no appreciable effect on airworthiness and are therefore by definition not significant.

These documents are also applicable to all significant changes to aircraft (other than rotorcraft) of 6000 lb or less maximum weight or to nonturbine rotorcraft of 3000 lb or less maximum weight (already mentioned above).

The GM/AC is full of examples to make practical application of a very complicated matter easier, and difficult items such as the influence of "service experience" are discussed to demonstrate that the introduction of the last amendment could be unnecessary.

Another valuable feature of these documents is the fact that all products (large and small aeroplane, rotorcraft, engines, etc.) are considered.

Figure 5.3, extracted from the GM (the FAA AC includes a very similar figure), shows the establishment of the type certification basis for changed products.

5.1.9. The supplemental type certificate

We have so far implied that changes are designed by the TCH. Nevertheless, another possibility does exist, and it is provided by Subpart E of JAR/FAR 21/EASA Part 21: any person who alters a product by introducing a **major** change, not sufficient to require a new application for a type certificate (see previous paragraph), shall apply to the authority for an STC.

To provide just a couple of the countless possible examples: a design organization (other than the TCH) can design an agricultural system for crop spraying to be installed on a type-certificated aircraft; in a similar way, a passenger transport aeroplane can be transformed into a cargo aeroplane.

Any organization applying for a JAR/EASA STC shall demonstrate its capability by holding a DOA or, by the way of derogation, alternative procedures setting out the specific design practices, resources, and sequence of activities necessary to comply with the applicable requirements.

EASA GM 21A.112B provides guidance to establish cases in which alternative procedures can be accepted.

For applications concerning an FAA STC, as we have mentioned dealing with the product type certification, there is no formal approval of the design organization.

Another peculiarity of the requirements governing the JAA/EASA STC, which cannot be found in the analogous FAA regulations, requires justifications

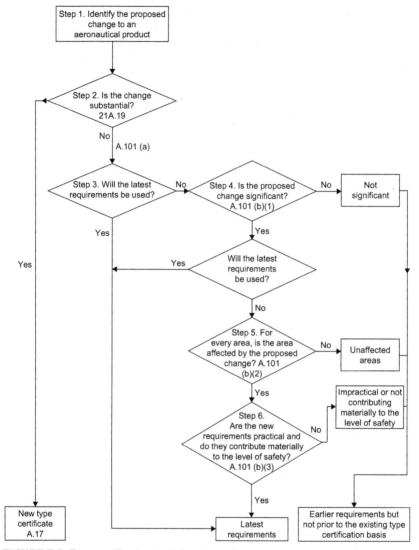

FIGURE 5.3 Type-certification basis for changed products

that the applicant is provided with all the information on the type design necessary to design the change, based either on the applicant's own resources or
through an arrangement with the TCH.

In the last case, the TCH shall not have technical objection to the above-
mentioned information, and shall collaborate with the STC holder to ensure
discharge of all obligations for continued airworthiness of the changed
product.

Because the STC is the approval of a major change in type design, what we
have discussed about the "change in type design" in this chapter is entirely
applicable.

The certification process of an STC is similar to the type-certification process of a product (aircraft, engine, and propeller).

Nevertheless, to cope with the peculiarity of this process, the EASA issued an Internal Working Procedure "Supplemental Type Certification Procedure," while the FAA issued AC 21-40A, "Application Guide for Obtaining a Supplemental Type Certificate."

Parts and appliances produced under an EASA STC shall permanently and legibly be marked according to Paragraph 21A.804 inclusive of the letters EPA (European Part Approval).

In relation to the practical introduction of an STC, according to FAR 21.120, an STC holder who allows a person to use the supplemental type certificate to alter an aircraft, aircraft engine, or propeller, must provide that person with written permission acceptable to the FAA.

The FAA considers the following privileges for a holder of an STC:

(1) In the case of aircraft, obtain airworthiness certificates.

(2) In the case of other products, obtain approval for installation on certificated aircraft.

(3) Obtain a production certificate for the change in type design that was approved by that STC.

5.2. PARTS AND APPLIANCES APPROVAL

All parts and appliances installed on certificated products (aircraft, engines, and propellers) must be approved. According to Subpart K of JAR/FAR/EASA Part 21, compliance with applicable requirements may only be made:

(1) Where applicable, under the **JTSO/TSO/ETSO authorization** procedures of Subpart O of JAR/FAR 21/EASA Part 21 or

(2) In conjunction with the type-certification procedures for the product (or its change) in which it is to be installed or

(3) In the case of **Standard Parts**,[31] in accordance with officially recognized standards or

(4) Where applicable (and JAA certifications) under the **Joint Part Approval authorizations (JPA auth.)** procedures of Subpart P of JAR 21 or

(5) Where applicable (and FAA certification) under the **Parts Manufacturer Approval (PMA)** issued under FAR 21.303 or

(6) According to FAR 21.303(a)(2), "Parts produced by an owner or operator for maintaining or altering his own product"

(7) In any other manner approved by the Administrator (the FAA).

---

[31] **Standard Part.** A part manufactured in complete compliance with an established government- or industry-accepted specification that contains design, manufacturing, and uniform identification requirements. The specification must include all information necessary to produce and conform the part, and must be published so that any person may manufacture the part. Examples of specifications include, but are not limited to, NAS, Air Force-Navy Aeronautical Standard (AN), Society of Automotive Engineers (SAE), SAE Aerospace Standard, and MS.

According to EASA Part 21A.307, Release of parts and appliances for installation:

> *No part or appliance (except a standard part), shall be eligible for installation in a type-certificated product unless it is: (a) Accompanied by an authorized release certificate (EASA Form 1)[32] certifying airworthiness; and (b) Marked in accordance with Subpart Q.*

JAR 21.307 is similar, quoting the Authorized Release Certificate (JAA Form 1).

The FAA prescribes an Authorized Release Certificate, FAA Form 8130-3[32] for airworthiness approval, export approval, conformity determination of prototype products/parts, and so on.

We will now try to describe this concept more thoroughly (the bold numbers below correspond to the list above).

(1) Subpart O of the JAR/FAR 21/EASA Part 21 prescribes procedural requirements for the issue of JTSO authorizations/FAA TSO/ETSO authorization, to mark parts (defined "articles" in Subpart O) with the appropriate JTSO/TSO/ETSO marking. The authorization is an approval of the design and for the production of an article that has been found to meet a specific JTSO/TSO/ETSO.

A Technical Data document must be produced according to the applicable JTSO/TSO/ETSO.

Furthermore, a **Declaration of Design and Performance (DDP)** has to be issued, containing the information for the definition of the article type design, the rated performance of the article, a statement of compliance certifying that the applicant has met the appropriate JTSO/TSO/ETSO, reference to relevant test reports, and reference to the appropriate maintenance, overhaul, and repair manual.

An article manufactured under a JTSO/TSO/ETSO authorization is, in principle, acceptable for installation in an aircraft. Nevertheless, some other technical requirements may be applied to the article by the participating authorities in accordance with the type certification regulations applicable to the aircraft in which the article is fitted (e.g., JAR/FAR/CS-25), and the compatibility with the characteristics of the product must be ascertained.[33]

---

[32] The certificate is to be used for import purposes, as well as for domestic and intra-Community purposes, and serves as an official certificate for the delivery of items from the manufacturer to users. Appendix I of EASA Part 21 includes a facsimile of the document and the completion instructions. The NPA 2007-13 is on the way for the introduction of new instructions for a better harmonization with other authorized release certificates such as the FAA Form 8130-3, the Transport Canada Form 24-0078, and others. FAA order 8130-21E describes the procedures for completion and use of the Authorized Release Certificate, FAA Form 8130-3, Airworthiness Approval Tag.

[33] A simple example: if an altimeter is limited to 30,000 ft, it cannot be installed in an aircraft with a maximum operating altitude of 50,000 ft.

Applicants for an ETSO authorization shall demonstrate their capabilities as follows:

(a) **For production**, by holding a POA[34] or through compliance with Subpart F procedures.[35]

(b) **For design**, for an auxiliary power unit (APU) by holding a DOA; for all other articles, by using procedures setting out the specific design practices, resources, and sequences of activities necessary to comply with the applicable requirements.[36]

For an FAA TSO authorization:

(a) **For production**, a quality control organization is required in compliance with FAR 21.143.

(b) **For design**, a formal DOA is not required as mentioned in the "Design organization" section in this chapter.

The code **JAR-TSO** contains the agreed common comprehensive and detailed aviation requirements for obtaining a JTSO authorization by showing compliance with the requirements of JAR 21. In Subpart B, there are two indexes:

- Index 1 lists all those JTSOs that are technically similar to FAA TSOs.
- Index 2 lists all those JTSOs that are applicable only to JAR. Examples are
  (a) Jointly agreed deviations from an FAA TSO or
  (b) When an FAA TSO does not exist for a particular application.

The code **CS-ETSO** contains the technical conditions an article should comply with to obtain an ETSO Authorization pursuant to Part 21, Subpart O. The technical conditions are contained in the respective ETSOs and are part of this code. Subpart B of CS-ETSO contains two indexes with the same criteria as JAR-TSO.

The **AC 20-110L** "Index of Aviation Technical Standards Order" describes the public procedures the FAA uses to develop and issue TSOs. Furthermore, the AC presents an index of the FAA TSOs that contain minimum performance standards for specific materials, parts, processes, and appliances used on civil aircraft.

(2) The JTSO/TSO/ETSO articles are only a part of what is installed on an aircraft and sometimes, as described previously, are not compatible with the same aircraft. It is therefore necessary to obtain the approval of parts and appliances specially designed for the product to be certificated or for a change approval. The production (and also the design) of parts and appliances for a product to be certificated or for a change approval can also be performed by external organizations, but the applicant is solely responsible for the airworthiness of such items.

The qualification procedure is generally similar to the JTSO/TSO/ETSO Authorization. In this case, a Technical Specification and a DDP

---

[34] See Chapter 7.

[35] See Chapter 7, "Production without Production Organization Approval."

[36] JAR 21 is more generic on this point.

(Declaration of Design Performance) must also be issued. A classification of the equipments, based on their criticality (the consequences of their failure), is performed, and according to this classification the authority's intervention is normally established.

The qualification procedure has to also consider compliance with Paragraph XX.1309 of the relevant JAR/FAR/CS, if applicable to that part of the product. It may be useful to remember that, while JTSO/TSO/ETSO parts and appliances can be installed in any product (except in the cases we have previously considered), the parts and appliances approved in conjunction with the type certification process of the relevant product can only be installed in products of the same type.

A misunderstanding could arise at this point: is an applicant compelled to install a JTSO/TSO/ETSO article when available?

This has often been cause of contention between applicants and authorities, especially for certification of small aircraft of General Aviation. It must be clear that this obligation does not exist, provided the above-mentioned rules are followed. The contention can become harsher when the applicant wants to install items originating from car production or other noncertificated production (e.g., ultralights). Why not? Everyone can testify about the reliability of instruments and various accessories in his or her own car (often proving to be serviceable and reliable for many years).

Nevertheless, direct transfer of a part from the counter of a car dealer to an aircraft is not possible. A qualification procedure similar (as far as possible) to that mentioned above must be put into action, ranging from knowledge of the item to its compatibility with the aircraft's operational conditions and installations (e.g., environmental and electromagnetic compatibilities); an acceptance procedure also needs to be established (vendors generally are unable to issue certificates of conformity). In conclusion, the qualified equipment should have its own part number so as not to be automatically interchangeable with the commercial equivalent.

All this costs time and money, in contrast to the installation of a JTSO/TSO/ETSO article, whose sole disadvantage is that it is normally more expensive: its cost may be a few times higher than the equivalent car or ultralight equipment. Hence, a choice has to be made. An investment in non-JTSO/TSO/ETSO equipment will be beneficial as regards the savings obtained in series production, allowing for a lower selling price and thus a more favorable position in the market.

Nevertheless, if the series production is uncertain, and there is an urgent need for a type certificate, the installation of JTSO/TSO/ETSO articles may be more convenient.

Another cause of contention in "light" aviation is when, say, Applicant A pretends to install in his own aeroplane noncertificated equipment already adopted by Applicant B. Part of the above-mentioned consideration is that equipment certificated with a product is valid for that product only; Applicant A does not have Applicant B's knowledge of the equipment, and does not know the changes that may have been made to that equipment and the acceptance procedures. In this case, Applicant A has to carry out equipment qualification for his product, similar to applicant B.

Of course, common sense should prevail in certification activities too. There are many noncritical parts (most parts) for which the authority, from a general point of view, could accept simplified qualification procedures, based on previous experience, technical evaluations, and so on.

We have previously mentioned that all parts and appliances installed on certificated products (aircraft, engines, and propellers) must be approved. Nevertheless, we have to consider the case of installation of equipment not specifically required for the aircraft's airworthiness, a case that occurs frequently. This equipment can be identified as follows:

(a) Entertainment equipment

(b) Domestic equipment

(c) Aerial work installations

(d) Experimental installations

(e) Instruments for additional information.[37]

In these cases, the **No Hazard** criterion is adopted, with the aim of ensuring that the above-mentioned equipment is not dangerous in itself, and that its presence on board will not jeopardize the performance and function of the aircraft's systems and, in general, the aircraft's airworthiness. It must be clear that the above criteria provide no guarantee for the correct function and nominal performance of this equipment that, from an airworthiness point of view, we can define as "tolerable."

If the equipment is a source of radio emission, the applicant has the responsibility to obtain the pertinent authorizations.

(3) This case applies to parts in accordance with standardization norms [e.g., Military Standards (MS), Society of Automotive Engineers Inc., Electronic Industries Association Standards Institute, American National Standards Institute, AIA-NAS], or with norms issued by the manufacturers of parts or products, and accepted by the authority as different from standardization norms. The above-mentioned norms are technical specifications that become parts of the product type design or change of the same.

(4) Subpart P of JAR 21 prescribes procedural requirements for the issue of a JPA auth. for replacement and modification parts (only minor changes are allowed) for installation on a type-certificated product. These parts are manufactured by people other than the TCH, holding or having applied for a suitable POA. As a consequence of the JPA auth., the parts are identified with a JPA marking.

Subpart P (of JAR 21) is "not applicable" for EASA Part 21.

According to EASA Part 21A.804(a)(3), all parts and appliances produced in accordance with approved design data not belonging to the TCH of the

---

[37] Information not required for aircraft operation.

related product (e.g., an STC), except for ETSO articles, need to be marked with the letters EPA (European Part Approval).

This approach is different from the JPA marking because this referred to parts and appliances produced in accordance with design data belonging to the TCH.[38]

(5) The FAA describes procedural regulations for the PMA, which is similar to JPA. These replacement parts are important especially for operators' fleet maintenance. In fact, these parts are generally less expensive than the original ones.

(a) *For the design* of these parts, the applicant must produce test reports and computations necessary to show that the design meets the airworthiness requirements of the Federal Aviation Regulations applicable to the product on which the part is to be installed, unless the applicant shows that the design of the part is identical to the design of a part that is covered under a type certificate. If the design of the part was obtained by a licensing agreement, evidence of that agreement must be presented.

(b) *For the production* of these parts, each holder of a PMA shall establish and maintain a fabrication inspection system that ensures that each completed part conforms to its design data and is safe for installation on applicable type-certificated products.

(6) The possibility that the US owners or operators have to produce their own parts is mainly related to old and "orphan" (no one even knows who owns the type certificate) aircraft for which it is difficult to find replacement parts. The FAA Memorandum of August 1993 explains how an owner- or operator-produced part can become an FAA-approved part:

- A part does not have to be solely produced by the owner to be considered an Owner-Produced Part.
- The aircraft owner must participate in the manufacture of the part in at least one of the five ways for it to be considered an Owner-Produced Part:

  (1) The owner provides the manufacturer of the part with the design or performance data.

  (2) The owner provides the manufacturer of the part with the materials.

  (3) The owner provides the manufacturer with fabrication processes or assembly methods.

  (4) The owner provides the manufacturer of the part with quality control procedures.

  (5) The owner personally supervises the manufacture of the new part.

---

[38] An EASA Part 145 approved organization can only fabricate parts for its own use in accordance with approved design data [Paragraph 145A.42(c)]. If those data come from the TC holder, Paragraph 21A.804(a)(3) would not be applicable and those parts will not need EPA marking. If the data come from an STC holder, minor change approval holder, or repair approval order, the parts will have to be marked as prescribed in the applicable data, which should include an EPA marking.

The key point is that the aircraft owner must participate in the part's manufacture and, if the Owner-Produced Part has all the characteristics of an approved part, it is only installed on the owner's aircraft and is not for sale, it would be considered as an FAA-approved part.

The characteristics of an approved part are as follows:

(1) *The part must be properly designed.* A properly designed part means that the part's design is FAA approved.

(2) *The part must be produced to conform to the design.* A properly produced part means the part conforms to the FAA-approved design.

(3) *The part's production should be properly documented.* A properly documented part provides evidence that the part was produced under an FAA approval and memorializes the production of the part.

(4) *The part must be properly maintained.* A properly maintained part means that the part is maintained in accordance with the rules prescribed under FAR Part 43.

(7) "In any other manner approved by the Administrator" is a general FAA statement for the approval of materials, parts, processes, or appliances outside the methodologies illustrated above.

5.3. THE MASTER MINIMUM EQUIPMENT LIST/ MINIMUM EQUIPMENT LIST

This concept does not originate directly from the product type-certification standards, but from operational standards such as **JAR-OPS 1/EU OPS 1** (Commercial Air Transport—Aeroplanes), **JAR-OPS 3** (Commercial Air Transport—Helicopters),[39] and **FAR 91**.

5.3.1. The master minimum equipment list

The master minimum equipment list (MMEL) is a master list (approved by the authority) appropriate to an aircraft type that determines those instruments, items of equipment or function that, while maintaining the level of safety intended in the applicable standards, may temporarily be inoperative either due to the inherent redundancy of the design, and/or due to specified operational and maintenance procedures, conditions, and limitations, and in accordance with the applicable procedures for continued airworthiness.

This implies that all systems related to the airworthiness of the aircraft and not included in the list are automatically required to be operative, whereas nonsafety-related equipment, such as galley equipment and passenger convenience items, do not need to be listed.

The MMEL covers the type of operations for which the aircraft is certificated.

Certain MMEL items need to be supported by operational and maintenance procedures, which have to be identified to the authority during the MMEL approval process.

---

[39] JAR-OPS 2 (General Aviation) has not been issued.

The creation of the master list, as is obvious, is strictly related to the safety assessment criteria discussed in Chapter 4, and therefore it must be **prepared by the** TCH.

5.3.2. The minimum equipment list

The minimum equipment list (MEL) is a list that provides for the operation of aircraft, under specified conditions, with particular instruments, items of equipment or functions inoperative at the commencement of the flight. This list is **prepared by the operator** for his own aircraft taking into account the relevant operational and maintenance conditions, in accordance with a procedure approved by the authority.

The MEL is based (without being less restrictive) on the relevant MMEL approved by the authority.

FAR 91 gives criteria for instruments and equipment that may not be included in an MEL.

Criteria are also provided for operations conducted (under FAR 91) with inoperative instruments and equipment and without an approved MEL.

In any case (according to FAR 21.197), if an aircraft with inoperative instruments or equipment is considered capable of safe flight for *particular purposes* (e.g., delivering or exporting the aircraft, production flight testing new production aircraft, etc.), it can be operated under a **special flight permit**.[40]

5.4. TYPE CERTIFICATION OF IMPORTED PRODUCTS

The certification of an imported product is normally carried out through the assessment of the type certification performed in the exporting state, made by the authority of the importing state. The aim of this assessment is to ensure that the imported product meets a level of safety equivalent to that provided by the applicable laws, regulations, and requirements that would be effective for a similar product in the importing state. The result of this assessment is the type-certificate **validation**.

The TCH and the exporting authority are then ready to negotiate individually with the different importing authorities. The matter could also be further complicated by different requirements in different states.

This was simplified in Europe when the JAA Member States adopted the same JARs.

Furthermore, the JAA joint certifications and validations (and now the EASA certifications and validations), leading to the issue of a common type certificate, have further simplified the matter. Hence, the national authorities, to issue an airworthiness certificate, have only to check the compliance of single aircraft with the national operational requirements.[41]

---

[40] A permit to fly can be issued under EASA Part 21—Subpart P.
[41] See Chapter 4, Paragraph 4.3.48.

To simplify the TC validation processes, **bilateral agreements**[42] have been made between states; these agreements are based on a high degree of mutual confidence in the technical competence and regulatory capacity of the exporting authority for performing aircraft certification functions within the scope of the agreement. A bilateral agreement is not a trade agreement, but a technical agreement providing that "the importing state shall give the same validity to the certification made by the competent aeronautical authority of the exporting state as if the certification had been made by its (the importing country's) own competent aeronautical authority in accordance with its own applicable laws, regulations, and requirements."[43] Nevertheless, because these laws, regulations, and requirements could be different, the agreement permits the importing state to prescribe **additional technical conditions**, "which the importing state finds necessary to ensure that the product meets a level of safety equivalent to that provided by its applicable laws, regulations, and requirements that would be effective for a similar product produced in the importing state."[44]

Subpart N of JAR 21 prescribes the procedural requirements for certification of imported products, parts, and appliances in a JAA Member State, and approval of major changes under STC procedures when such changes are designed by a person that is not the TC holder and is located in a non-JAA country. As far as the United States is concerned, similar procedures are contained in FAR 21 Paragraphs 24 and 29, and Subpart N. Moreover, the FAA AC 21–23B provides ample advisory material on this subject.

5.4.1. EASA type certification

As regards the EASA, Subpart N is "not applicable." Nevertheless, guidance criteria are expressed in the EASA Internal Working Procedure "Type Certification Procedure" document, in which two cases are considered:

(1) **Type Certification under a bilateral (recognition) agreement with the State of Design.**

In the case of a formal recognition agreement between the Community and a third country in accordance with Article 12 of Regulation (EC) No. 216/2008, this agreement including the associated implementing procedures may supplement, change, or supersede the normal EASA certification procedures.

In this case, the EASA certification may be called **validation** and it is assumed that the imported product shall meet, with the same level of confidence, a level of safety equivalent to that required for a comparable product designed and manufactured within an EASA Member State.

NOTE: As long as the Community has not concluded own bilateral (recognition) agreements, according to Article 12 of the Basic Regulation, existing bilateral

---

[42] For example, the FAA Bilateral Aviation Safety Agreements and inherent IPAs, which are to replace the old Bilateral Airworthiness Agreements.

[43] FAA AC 21-23 B.

[44] See note 43.

(recognition) agreements—including their Implementation Procedures of Airworthiness (IPA)—between EU Member States and third countries, may be used for the validation of non-EU Type Certificates. This includes Type validation principles/post type validation principles (TVP/PTVP) as agreed with the FAA.

(2) Type Certification under a working arrangement with the State of Design.

In the case of a working arrangement between EASA and the competent authority of a third country, in accordance with Article 27 of the Basic Regulation, the normal EASA certification procedures shall apply.

However, based on the working arrangement, EASA may use the foreign certification system, which has demonstrated the same level of independent checking function, to find compliance with the EASA certification basis.

5.4.1.1. ACCEPTANCE OF PMA PARTS

There is a decision of July 2007 of the EASA Executive Director on the Acceptance of Parts Designed in the United States under the PMA System of the FAA. A synthesis is reported here.

Whereas:

- The Basic Regulation requires the Agency to issue certificates for the approval of the design of parts and appliance and of their installation into products subject to that Regulation.
- The Basic Regulation recognizes the possibility, in the absence of an agreement concluded by the Community, for the Agency to issue certificates in application of existing agreements between Member States and a third country.
- Several Member States have concluded bilateral agreements with the United States covering the reciprocal acceptance of certification findings, in particular the approval, under certain conditions, of PMA parts.
- When the conditions specified in the above-mentioned agreements are met, the Agency is bound to issue a certificate approving the design of those parts; it is more efficient to approve in advance the design of all those parts that meet the conditions specified by all the agreements and therefore limit direct Agency involvement in cases deserving specific attention.

Decision for the Approval of the design of certain PMA parts

An approval is hereby issued by the Agency to an organization under the regulatory oversight of the FAA for a part designed under their PMA system, provided that

- The PMA part is not a "critical component."

    The statement *"This PMA part is not a critical component"* should be written in Block 13 of the FAA Form 8130-3 or
- The PMA part conforms to design data obtained under a licensing agreement from the holder of the FAA design approval according to FAR 21.303(c)(4) of the FAA. The statement *"Produced under licensing*

agreement from the FAA design approval holder" should be written in Block 13 of FAA Form 8130-3 or

- The PMA holder can show that the part has received an explicit approval by means of a design change or STC from the Agency or, when this approval was granted before 28 September 2003, from any of the National Aviation Authorities of the Member States of the European Union. The reference to this authorization should be written in Block 13 of the FAA Form 8130-3.

5.4.2. FAA type certification

With reference to the above-mentioned AC 21-23B, we quote as a matter of interest some points relating to the FAA's technical involvement in the validation of imported (in the United States) products and inherent changes. This involvement, of which it is important to be aware because it is also related to European exports to the United States, consists of the following:

(1) To provide for the FAA familiarity with the general design, performance, and operational characteristics of the product, for the purpose of establishing the US certification basis to the extent necessary, and for the FAA to meet its post-certification responsibilities after the product enters service on the US registry.

(2) To establish the US type-certification basis and the means of compliance for the product under application by determining the US airworthiness and environmental standards that would be applied to a similar product if it were to be produced in the United States.

(3) To understand the airworthiness certification system (including the airworthiness and environmental standards, policies, and certification practices) applied by the exporting authority in their domestic certification of the product; this will include an understanding of the level of the exporting authority's involvement with prototype conformity inspections, tests, and flight programs.

(4) To compare the airworthiness and environmental standards, policies, and practices applied by the exporting authority in their domestic certification with the US type-certification basis or design requirements and certification policies and practices.

(5) To define and explain any additional technical conditions that should be met for FAA certification to provide for equivalency with the applicable US airworthiness and environmental standards.

(6) To maintain sufficient liaison and technical dialog with the exporting authority to ensure that technical questions and issues that might affect US certification of the product are identified and resolved between the FAA and the exporting authority as early as possible.

(7) To provide for effective management of the certification project and for the most cost-effective utilization of FAA resources on the project.

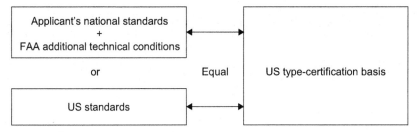

FIGURE 5.4 US type-certification basis

5.4.2.1. CERTIFICATION BASIS[45]

The applicable **US airworthiness standards** are those in effect on the date of application for the US TC,[46] while the applicable **US environmental standards** are those in effect on the date of the US type certification.[47]

Another way to define the FAA certification basis is the **addition of technical conditions (ATCs)** to the certification basis of the exporting authority. These ATCs take into account the following:

(1) Differences in the basic airworthiness and environmental standards of the United States and exporting state.

(2) Noncompliance with the exporting authority airworthiness or environmental standards because of exemptions[48] or equivalent safety findings[49] granted by the exporting authority.

(3) Special conditions issued by the FAA because of novel or unusual design features of the product that are not required in an equivalent manner by the exporting authority.

(4) Mandatory airworthiness actions (i.e., Airworthiness Directives) directed by the exporting authority to correct unsafe conditions experienced during operation before application for FAA approval.

(5) Optional conditions identified by the FAA to assist any eventual US operator to comply with current US operational or maintenance requirements.

Figure 5.4 illustrates how the US type-certification basis can be determined.

In the second option, if the findings of compliance are applicable to the US Standards, the FAA will not develop any ATCs.

The identification and discussion of the problems arising from the above-mentioned issues are reported in the Issue Papers (similar to the

---

[45] See Chapter 6, Paragraph 6.2.5 and 6.5.7.

[46] Unless the bilateral agreement states otherwise, the date of application to the exporting authority could be accepted.

[47] This is a very strict requirement because these standards could be amended at the last moment.

[48] "Exemption" means formal acceptance by the authority of noncompliance to a specific requirement.

[49] Equivalent safety finding: any airworthiness provisions not complied with, compensated for by factors that provide an equivalent level of safety.

JAA CRIs[50]) and the exporting authority is required to have a substantial involvement.

As mentioned above, also describing AC 21-23B, the TC validation is a rather complex process, sometimes more difficult than the original type-certification itself. Of course, the process can be simplified when two authorities have acquired sufficient experience as regards importing/exporting, but above all if the personnel are well aware of the philosophy of bilateral agreements.

Some problems arose in the early years of FAA/JAA validations, with complaints about manufacturing on both sides of the Atlantic, because it seemed that the validation teams had no clear vision of what they had to (or had not to) verify. This situation forced the JAA and FAA (and later the Canadian Authority as well) to discuss the matter and eventually come up with the "JAA/FAA Validation Procedures." This is a manual that explains the validation process, fixes organizational procedures, and above all the nature of the validation team's examinations and the team's relationship with the exporting authority. Furthermore, training courses for validation team members are organized in both Europe and the United States.

5.5. TRANSFER OF A TYPE CERTIFICATE

It is sometimes necessary to transfer a TC from one TC holder to another for various reasons: the sale or the bankruptcy of an enterprise, the sale of a certificate type design, and so on.

Procedural requirements for this transfer are prescribed by JAR/FAR 21.47 and EASA Part 21A.47. According to the JAA and EASA, the new TCH must be able to:

(1) Undertake the responsibility of a TCH as defined in Paragraph 21.44/ 21A.44.

(2) Demonstrate the ability to qualify for an appropriate DOA or have the authority's agreement for an alternative procedure.[51]

The approved design organization of the new TCH is a prerequisite for the transfer of the type certificate according to JAR 21 and EASA Part 21.

The FAA's approach is different. According to FAR 21.47:

A type certificate may be transferred to or made available to third persons by licensing agreements. Each grantor shall, within 30 days after the transfer of a certificate or execution or termination of a licensing agreement, notify in writing the appropriate Aircraft Certification Office. The notification must state the name and address of the transferee or licensee, date of the transaction, and in the case of a licensing agreement, the extent of authority granted the licensee.

---

[50] See Chapter 6, "Certification review items" Paragraph 6.2.5.3.
[51] See the "Design organization" section in this chapter.

This could be interpreted as the possibility of transferring the TC independent of the new TCH's organization. Actually, there is a possibility that the TC relates to aircraft no longer in existence, and in such a case, the presence of a design organization is irrelevant. Of course, the Administrator will intervene when the new TCH begins to produce aircraft according to the transferred TC, or it assumes the responsibility of the continuing airworthiness of a series of aircraft type certificated according to the same TC.

Significantly, the conditions for a correct management of the type design have to be maintained or recreated, both for production and continued airworthiness. Nevertheless, it is necessary to clarify that the transfer is also possible if the new TCH does not have a production organization. In this case, the TCH's responsibilities must be limited to the management of continuous airworthiness, which is a vital function for the already certificated and operational aircraft of the type covered by the TC.

The authority could therefore have to deal with various situations. We will consider some examples:

(1) The TC holder is an enterprise with a production organization, which is sold and changes its corporate name, but without substantial organizational changes. This case is the easiest to solve because, apart from some inevitable red tape (an FAA production certificate or non-EASA organization approval are not transferable), everything continues unchanged.

(2) The TC is transferred to a different enterprise intending to continue with the production (or take it up again). In this case, the authority, besides the assessments relating to the responsibilities of a TC holder (Paragraph 21.44), must also deal with the production organization[52] according to Subpart F or G of JAR/FAR/21/EASA Part 21. If the new enterprise does not have POA to expand with the new production, it is likely to begin with JAR/FAR 21/EASA Part 21 Subpart F procedures. The task of the authority is now a complex one, because conformity with the certificated type design must be ensured in a completely new environment that could also benefit from better means of production, but may require some type design changes, which have to be approved. It is certain that the authority, for the certification of the first aircraft produced, will not be satisfied by the normal verifications prescribed for the series aircraft; repetition of ground tests carried out for the aircraft type certification and a spot check of the certification flight tests could be required. All this is to ensure that the new series of aircraft are not inferior to the previous ones from a safety point of view.

(3) The TC is transferred to an enterprise that does not have its own means of production. The case has already been considered at the beginning of the chapter where we mentioned that the task of the enterprise is limited to

---

[52] See Chapter 7, "The production organization."

the management of continuous airworthiness. In the next point, we will see how important this TC transfer is, even if the aircraft production is interrupted—in a temporary or definitive manner.

(4) The TC holder disappears or is no longer able to cope with his or her responsibilities. This is not unusual, especially for small aeronautical enterprises, and serious problems could arise for the relevant aircraft that remain, so to speak, orphans. In this case and generally speaking, two scenarios are possible:

(a) The authority replaces the TC holder as far as responsibility for continued airworthiness is concerned. This is likely to happen for small aircraft of the General Aviation that normally request a lesser engagement for this task. This also allows the obligations toward the national authorities of the states that have imported aircraft of the concerned type to be maintained.

(b) The authority does not intend to (or cannot) assume the TC holder's responsibilities. In such a case, the type certificate could be suspended, pending the application for a new TCH or, in the worst case, the type certificate could be revoked. It is evident that the suspension or revocation of the TC will have similar consequences regarding the certificate of airworthiness issued for the aircraft of the type concerned if still operating.

NOTE: According to EASA, an aircraft becomes orphan when:

(1) The legal person holding the TC has ceased to exist. The TC automatically becomes invalid by law because there is no one to be in compliance with the TC holders responsibilities [21A.51(a)(1) and 21A.44] or

(2) The TC holder no longer complies with his regulatory obligations. A typical case is when the TC holder loses his DOA, or fails to comply with 21.A.14 before 28 September 2005. This makes the TC invalid [21A.51(a)(1)] or

(3) The TC holder has surrendered the TC. This also makes the TC invalid [21A.51 (a)(2)].

Under the current Part 21, orphan aircraft cannot be issued a Certificate of Airworthiness, which requires that a TC holder takes responsibility for the continued oversight of the design. They can therefore only continue to be operated if they hold a restricted certificate of airworthiness or a permit to fly. These documents can only be issued on the basis of a design approved by the Agency

5.6. INSTRUCTIONS FOR CONTINUED AIRWORTHINESS

Flight safety begins with the design of the aircraft. This means not only that the structures, systems, flight performance, flight qualities, and so on must comply with the applicable requirements, but they also need to provide instructions for maintenance of the aircraft and for repairs during its operational life.

JAR/FAR 21/EASA Part 21 use different wording, but have the same meaning. They require the following.

5.6.1. Provision of Instructions for Continued Airworthiness

The holder of a design approval, including either the type certificate or STC for an aircraft, aircraft engine, or propeller, shall furnish at least one set of complete Instructions for Continued Airworthiness, prepared in accordance with the applicable requirements, to the owner of each type of aircraft, aircraft engine, or propeller on its delivery, or on issuance of the first standard airworthiness certificate for the affected aircraft, whichever occurs later, and thereafter make those instructions available to any other person required by the regulation to comply with any of the terms of these instructions. In addition, changes to the instructions for continued airworthiness shall be made available to any person required by this regulation to comply with any of those instructions.

The above-mentioned applicable requirements are the relevant certification standard, FAR/JAR/CS-23, -25, -27, -29, -33, -35, and so on.

For instance, FAR 23 has the following requirement (**23.1529 Instructions for Continued Airworthiness**)[53]: "The applicant must prepare Instructions for Continued Airworthiness in accordance with **Appendix G** to this part that are acceptable to the Administrator. The instructions may be incomplete at type certification if a program exists to ensure their completion prior to delivery of the first airplane or issuance of a standard certificate of airworthiness, whichever occurs later."

To complete the example, we report an *extract* of Appendix G. This appendix specifies requirements for the preparation of Instructions for Continued Airworthiness as required by Paragraph 23.1529.

General. The Instructions for Continued Airworthiness for each airplane must include the Instructions for Continued Airworthiness for each engine and propeller (hereafter designated products), for each appliance required by this chapter, and any required information relating to the interface of those appliances and products with the airplane. If Instructions for Continued Airworthiness are not supplied by the manufacturer of an appliance or product installed in the airplane, the Instructions for Continued Airworthiness for the airplane must include the information essential to the continued airworthiness of the airplane.

Format. The Instructions for Continued Airworthiness must be in the form of a manual or manuals as appropriate for the quantity of data to be provided.

Content. The contents of the manual or manuals must be prepared in the English language. The Instructions for Continued Airworthiness must contain the following manuals or sections, as appropriate, and information:

(**1**) *Airplane maintenance manual or section*

(**a**) Introduction information that includes an explanation of the airplane's features and data to the extent necessary for maintenance or preventive maintenance.

---

[53] The other aircraft standards have the same number for the corresponding paragraph. JAR/CS-23 has equivalent requirements.

(**b**) A description of the airplane and its systems and installations, including its engines, propellers, and appliances.

(**c**) Basic control and operation information describing how the airplane components and systems are controlled and how they operate, including any special procedures and limitations that apply.

(**d**) Servicing information that covers details regarding servicing points, capacities of tanks, reservoirs, types of fluids to be used, pressures applicable to the various systems, location of access panels for inspection and servicing, locations of lubrication points, lubricants to be used, equipment required for servicing, two instructions and limitations, mooring, jacking, and leveling information.

(**2**) *Maintenance instructions*

(**a**) Scheduling information for each part of the airplane and its engines, APUs, propellers, accessories, instruments, and equipment that provides the recommended periods at which they should be cleaned, inspected, adjusted, tested, and lubricated, and the degree of inspection, the applicable wear tolerances, and work recommended at these periods. The recommended overhaul periods and necessary cross-reference to the Airworthiness Limitations section of the manual must also be included. In addition, the applicant must include an inspection program that includes the frequency and extent of the inspections necessary to provide for the continued airworthiness of the airplane.

(**b**) Troubleshooting information describing probable malfunctions, how to recognize those malfunctions, and the remedial action for those malfunctions.

(**c**) Information describing the order and the method of removing and replacing products and parts with any necessary precautions to be taken.

(**d**) Other general procedural instructions including procedures for system testing during ground running, symmetry checks, weighing and determining the center of gravity, lifting and shoring, and storage limitations.

(**e**) Diagrams of structural access plates and information needed to gain access for inspections when access plates are not provided.

(**f**) Details for the application of special inspection techniques including radiographic and ultrasonic testing where such processes are specified.

(**g**) Information needed to apply protective treatments to the structure after inspection.

(**h**) All data relative to structural fasteners such as identification, discard recommendations, and torque values.

(**i**) A list of special tools needed.

(**3**) *Airworthiness Limitations section*

The Instructions for Continued Airworthiness must contain a section titled Airworthiness Limitations *that is segregated and clearly distinguishable* from the rest of the document. This section must set forth each mandatory replacement time, structural inspection interval, and related structural inspection procedure required for type certification. If the Instructions for Continued Airworthiness consist of multiple documents, the section required by this paragraph must be included in the principal manual.

5.7. REPAIRS

5.7.1. Introduction

An aircraft is subject to damages that have to be repaired. A "repair" means elimination of damage and/or restoration to an airworthy condition of a product, part, or appliance.

Elimination of damage by replacement of parts or appliances without the necessity for design activity does not require authority approval (under subpart M of JAR 21/EASA Part 21).

Because a repair normally involves a change of configuration, it is considered as a change to the type design and consequently must be approved.

There are types of damage that can be anticipated, so that the repair of this damage can be studied in advance. Manual and other Instructions for Continued Airworthiness (such as Manufacturer Structural Repair Manual) are provided by the TCH for the aircraft operators and contain useful information for the development and approval of repairs.

When these data are explicitly identified and approved, they may be used by the operators without further approval to cope with anticipated in-service problems arising from normal usage provided that they are used strictly for the purpose for which they have been developed.

Of course, damage that cannot be anticipated has to be approved on a case-by-case basis.

5.7.2. Subpart M of JAR 21/EASA Part 21

Subpart M (Repairs) of JAR 21/EASA Part 21 prescribes procedural requirements for the approval of repairs made on products, parts, and appliances. A summary of these requirements is given below.

5.7.2.1. CLASSIFICATION OF REPAIRS

A repair can be "major" or "minor" and the classification must be made in accordance with the criteria applicable for a change in type design (see "Changes in type design" section in this chapter).

According to, in particular, EASA GM 21A.435, a repair is classified as major if it needs extensive static, fatigue, and damage tolerance strength justification and/or testing, or if it needs unusual methods, techniques, or practices.

Furthermore, repairs requiring reassessment and reevaluation of the original certification substantiation data to ensure that the aircraft still complies with all the relevant requirements are considered as major repairs.

Repairs whose effects are considered to be minor and require minimal or no assessment of the original certification substantiation data to ensure that the aircraft still complies with all the relevant requirements are considered as minor.

5.7.2.2. DEMONSTRATION OF CAPABILITY

An applicant for major repair design approval shall demonstrate its capability by holding a DOA issued by the Agency.

By way of derogation, as an alternative procedure to demonstrate its capability, an applicant may seek the Agency's agreement for the use of procedures compatible with the requirements of Subpart M.

5.7.2.3. REPAIR DESIGN

The applicant shall

(1) Show compliance with the type-certification basis and environmental protection requirements incorporated in the type certificate or STC, as applicable, plus any amendments to those requirements or special conditions the Agency find necessary to establish a level of safety equal to that established by the type certification basis.

(2) Submit all necessary substantiation data, when requested by the Agency.

(3) Declare compliance with the above requirement.

(4) Where the applicant is not the TC or STC holder, compliance with the TC basis may be done through the use of its own resources or through an arrangement with the TC or STC holder as applicable.

5.7.2.4. ISSUE OF A REPAIR DESIGN APPROVAL

When it has been declared and shown that the repair design meets the applicable conditions, it shall be approved

(1) By the Agency or

(2) By an appropriately approved organization that is also the TC or STC holder under a procedure agreed with the Agency or

(3) For minor repairs only, by an appropriately approved design organization, under a procedure agreed with the Agency.

5.7.2.5. REPAIR EMBODIMENT

The embodiment of a repair shall be made by an appropriately approved maintenance organization or by a production organization appropriately approved in accordance with the privileges of Subpart G of Part 21.[54]

5.7.2.6. INSTRUCTIONS FOR CONTINUED AIRWORTHINESS

A holder of the repair approval shall furnish at least one complete set of those changes to the Instructions for Continued Airworthiness that result from the design of repair, comprising descriptive data and accomplishment instructions prepared in accordance with the applicable requirements, to each operator of aircraft incorporating the repair.

Repair manuals are provided by the TCH for the aircraft operators and contain useful information for the development and approval of repairs.

---

[54] See Chapter 7, "The production organization" Paragraph 7.2.2.2.

When these data are explicitly identified and approved, they may be used by the operators without further approval to cope with anticipated in-service problems arising from normal usage provided that they are used strictly for the purpose for which they have been developed.

Of course, damage that cannot be anticipated has to be approved case-by-case.

Figures 5.5 and 5.6, extracted from EASA AMC&GM for Part 21, although appearing complicated at first sight, give a clear idea of the repair process approval for products for which the state of design is an EU Member State and when the state of design is not a Member State.

5.7.2.7. GENERAL REMARKS

We may question why a major repair needs an Instruction for Continued Airworthiness to be added to the instruction of the relevant product.

The answer is that major repairs can change the existing maintenance practices or inspection intervals. For example, major structural repairs may need more inspection. Repairs on static engine components could even influence the life limits of critical rotating parts. The person holding the inspection authorization or authority to approve the return to service is responsible for determining whether any changes are necessary to the existing product Instructions for Continued Airworthiness resulting from the major repair.

5.7.3. FAA repairs

FAR 21 does not have a subpart dedicated to repairs.

FAR 1 defines a *major alteration* as an alteration not listed in the aircraft, aircraft engine, or propeller specifications that might appreciably affect weight, balance, structural strength, performance, power plant operation, flight characteristics, or other qualities affecting airworthiness or that is not done according to accepted practices or cannot be done by elementary operations.

FAR 1 defines a *major repair* as a repair that, if improperly done, might appreciably affect weight, balance, structural strength, performance, power plant operation, flight characteristics, or other qualities affecting airworthiness, or that it is not done according to accepted practices or cannot be done through elementary operations.

A *minor repair* is a repair other than a major repair.

FAR 43 (Maintenance, Preventive Maintenance, Rebuilding, and Alteration) prescribes rules governing the maintenance, preventive maintenance, rebuilding, and alteration of any aircraft having a US airworthiness certificate, foreign-registered civil aircraft used in common carriage or carriage of mail under the provisions of FAR 121 or 135, and airframe, aircraft engines, propellers, appliances, and component parts of such aircraft.

We will report an excerpt of Appendix A to FAR 43: major alterations, major repairs, and preventive maintenance.

(1) Airframe major repairs. Repairs to the following parts of an airframe and repairs of the following types, involving the strengthening, reinforcing,

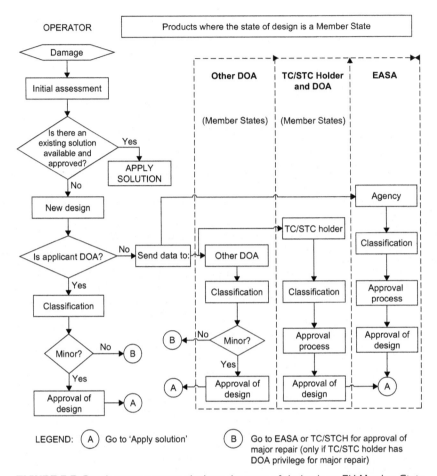

FIGURE 5.5 Repair process approval where the state of design is an EU Member State

splicing, and manufacturing of primary structural members or their replacement, when replacement is by fabrication such as riveting or welding, are airframe major repairs. (i) Box beams; (ii) monocoque or semimonocoque wings or control surfaces; (iii) wing stringers or chord members; (iv) Spars; (v) Spar flanges; (vi) members of truss-type beams; (vii) thin sheet webs of beams; (viii) keel and chine members of boat hulls or floats; (ix) corrugated sheet compression members that act as flange material of wings or tail surfaces; (x) wing main ribs and compression members, (xi) wing or tail surface brace struts; (xii) engine mounts; (xiii) fuselage longerons; (xiv) members of the side truss, horizontal truss, or bulkheads; (xv) main seat support braces and brackets; (xvi) landing gear brace struts; (xvii) axles; (xviii) wheels; (xix) skis and ski pedestals; (xx) parts of the control system such as control columns, pedals, shafts, brackets, or horns; (xxi) repairs involving the substitution of material; (xxii) the repair of damaged areas in metal- or plywood-stressed covering exceeding 6 inches

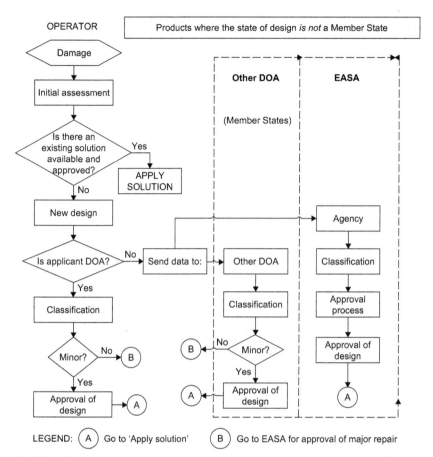

FIGURE 5.6 Repair process approval where the state of design is not an EU Member State

in any direction; (xxiii) the repair of portions of skin sheets by making additional seams; (xxiv) the splicing of skin sheets; (xxv) the repair of three or more adjacent wing or control surface ribs or the leading edge of wings and control surfaces, between such adjacent ribs; (xxvi) repair of fabric covering involving an area greater than that required to repair two adjacent ribs; (xxvii) replacement of fabric on fabric-covered parts such as wings, fuselages, stabilizers, and control surfaces; and (xxviii) repairing, including rebottoming, of removable or integral fuel tanks and oil tanks.

(2) **Power plant major repairs.** Repairs of the following parts of an engine and repairs of the following types are power plant major repairs. (i) Separation or disassembly of a crankcase or crankshaft of a reciprocating engine equipped with an integral supercharger; (ii) separation or disassembly of a crankcase or crankshaft of a reciprocating engine equipped with other than spur-type propeller reduction gearing; (iii) special repairs to structural engine parts by welding, plating, metalizing, or other methods.

(3) Propeller major repairs. Repairs of the following types to a propeller are propeller major repairs. (i) Any repairs to, or straightening of, steel blades; (ii) repairing or machining of steel hubs; (iii) shortening of blades; (iv) retipping of wood propellers; (v) replacement of outer laminations on fixed pitch wood propellers; (vi) repairing elongated bolt holes in the hub of fixed pitch wood propellers; (vii) inlay work on wood blades; (viii) repairs to composition blades; (ix) replacement of tip fabric; (x) replacement of plastic covering; (xi) repair of propeller governors; (xii) overhaul of controllable pitch propellers; (xiii) repairs to deep dents, cuts, scars, nicks, and so on, and straightening of aluminum blades; and (xiv) the repair or replacement of internal elements of blades.

(4) Appliance major repairs. Repairs of the following types to appliances are appliance major repairs. (i) Calibration and repair of instruments; (ii) calibration of radio equipment; (iii) rewinding the field coil of an electrical accessory; (iv) complete disassembly of complex hydraulic power valves; and (v) overhaul of pressure type carburetors, and pressure type fuel, oil and hydraulic pumps.

FAR 145 (Repair Stations) prescribes the requirements for issuing repair station certificates and associated ratings to facilities for the maintenance and alteration of airframes, power plants, propellers, or appliances, and prescribes the general operating rules for the holders of those certificates and ratings.

We can conclude that the FAA prescribes the rules for *repairs* in the same context as the rules for *alteration* and, more generally, in the frame of *maintenance*, an issue that is discussed in Chapter 9.

Chapter six

The Type-Certification Process

6.1. JAA JOINT CERTIFICATIONS AND NATIONAL CERTIFICATIONS

This section, written at the end of 2000, mainly aims to provide an historical perspective because the EASA has since introduced new procedures.

It is, however, interesting to see how the JAA has operated over many years, and this is presented in Sections 6.1, 6.1.1, and 6.1.2.

In the previous chapters, we described how the JAA performed joint type certifications to simplify the exchange of aeronautical products among the Member States. Not being a legal authority, the JAA could not issue a type certificate, but just a "recommendation" at the end of the process, allowing each Member State to issue a TC without further verification. Hence, the national authorities could issue certificates of airworthiness on this basis for single products, after the assessment of compliance with operational rules in force in the relevant states.

Joint certifications were essentially performed according to two procedures.

6.1.1. JAA multinational procedure

This was for type certification of products of the "higher" range. Without listing them all, we can mention large aircraft, commuters, turbine engines, and so on.[1] Teams of specialists (e.g., in structures, flight tests, and systems) from different national authorities were put together for the certification of such products. In summary, the national authorities proposed some specialists for each certification; these specialists were evaluated on the basis of their experience (a curriculum vitae was required), and then the JAA assessed and approved the composition of the team. A Program Manager, with the main task of coordinating the team's work, was appointed in a similar way.

6.1.2. JAA local procedure

This was for type certification of products of the "lower" range, such as very light aeroplanes, sailplanes, and powered sailplanes, some JAR 23

---

[1] A precise list of these products is documented.

Airworthiness: An Introduction to Aircraft Certification.

single-engine aeroplanes, and so on.[2] Type certification of this kind of product involved, in general, a smaller team and less-complex management. Therefore, the certification process was assigned to a national authority (possibly to the applicant's national authority), which had to be acknowledged by the JAA as the **Primary Certification Authority (PCA)**. This happened after an assessment of the national authority's suitability, performed by a JAA commission. The PCA's job was monitored by the JAA Certification Division, which issued the usual recommendation for national type certifications, at the end of the type-certification process.

It is nevertheless worth considering that an applicant was not bound to perform a joint certification. In consideration of the legal status of the JAA, national certification was still possible. In this case, of course, the validity of the type certificate was limited to the state of the national authority, so that the TC had to be validated by each country importing the product (the same procedure existing before the joint certifications).

There are several examples of applicants who chose the national certifications. At first sight, this could seem to be a lack of awareness, but it was in reality a technical and economic choice. JAA certifications, especially if multinational, were necessarily more complex than national certifications and took more time. Furthermore, they were unbalanced from the authorities' tariffs point of view, because there was no "joint" charging system. This means, for example, that a certification team that was predominantly English was much more expensive than a team that was predominantly French. All that considered, if an enterprise had no immediate interest in the European market, because it was looking at the national and (possibly) the US market (an FAA validation was due for both joint and national certifications), the choice of a national certification could be judged more convenient.

However, we were in a period of transition toward the establishment of the EASA, with the issue of new rules, and the abolition of national certifications.[3]

6.2. THE EASA TYPE-CERTIFICATION PROCESS

Although the basic philosophical concepts of type-certification procedures are generally the same for EASA, and FAA, there are some peculiarities in the type-certification process that necessitate a separate description of the FAA process.

We will deal with this subject also keeping in mind that the same basic concepts are applicable for changes to TC, Supplemental type-certificate approvals, and European Technical Standard Order authorizations.[4]

---

[2] Refer Note 1.
[3] This prediction has become reality.
[4] See Chapter 5.

The following information is based on EASA "Type-Certification Procedure (C.P008-02)," which is the applicable document in Europe at present.

6.2.1. Application

Applications for an EASA Type Certificate (EASA Form 30) shall be sent to the Applications and Procurement Services Department and made in accordance with Paragraph 21A.15 of EASA Part 21, Section A.

6.2.2. Attribution of technical investigation tasks

After eligibility has been fully assessed and the application has been accepted by the Agency, the responsible Certification Manager (CM) will decide whether the technical investigation will be further processed internally or the technical investigation should be allocated to an NAA.[5]

When the application is further processed internally, the technical investigation is performed by the EASA Products Certification Department, using EASA staff and/or NAA seconded staff. Individual NAA staff, selected by the EASA Products Certification Department, may be invited to participate in a specific technical task under the direct technical management of the Agency, when there is a framework service contract concerning the provision of services in place between the Agency and the NAA seconding staff.

The Agency may only allocate the technical investigation task to the NAAs of EASA Member States that have been accredited for this task and when there is an appropriate legal arrangement in place between that NAA and the Agency.

All certification tasks, irrespective of whether they are performed internally or allocated to an NAA shall be executed following the provisions of this EASA Type-Certification Procedure.

For the product type-certification task, the following cases can occur:

6.2.2.1. EU PRODUCTS

(A) In case the Agency does not perform a product type-certification task itself, this task shall be allocated to the NAA of the State of Design.

(B) Should it not be possible to allocate the certification task to the NAA of the State of Design for reasons that may include inadequate scope of accreditation, inability, or unwillingness to provide services in accordance with the Agency's procedures or within the allocated timeframe, internalization of the certification task may be reconsidered or the certification task may be allocated instead to an alternative NAA that is suitably accredited. The alternative NAA shall be selected in relation with its particular experience and competence for this task, that is, the NAA has built up the necessary

---

[5] National Aviation Authority.

expertise in a particular field, domain, or category of product, and it is accredited to perform tasks in such area(s).

6.2.2.2. NON-EU PRODUCTS

For imported products designed by foreign organizations, if the Agency cannot perform the task by itself, NAAs of EU Member States may be allocated type-certification tasks in accordance with the criteria listed in Section 6.2.2.1 (B) EU Products.

6.2.3. Certification team

6.2.3.1. GENERAL

The investigation process for type certification of an aviation product is performed by a team of experts, led by a PCM. The PCM is accountable to the responsible EASA CM.

6.2.3.2. DETERMINATION OF THE CERTIFICATION TEAM

To establish an appropriate certification team, the applicant, if deemed necessary by the responsible EASA CM, will organize an initial briefing for general familiarization with the project. This briefing will take place at a convenient and cost-effective location. The attendance at the initial briefing will normally include at least a representative of the EASA Certification Directorate and the PCM if already appointed.

Following the general familiarization, the responsible EASA Products CM together with the Experts Head of Department, the Certification Environmental Protection Section Manager, and the appointed PCM will select the members of the certification team from EASA and/or NAA staff from NAAs with which EASA has appropriate contractual arrangements.

The composition and size of certification teams can vary and are dependent on the product that needs to be type-certificated. When the extent of the investigation does not justify the need for a team, one person may perform the investigation.

A certification team for a Propeller may consist of only one specialist, whereas a new Large Transport Aircraft for example may likely need more experts covering the disciplines of:

Flight (including Performance), Human factors, Structures, Hydromechanical systems, Electrical systems, Avionic systems, Power plant, Transmissions, Cabin safety, Environmental Control systems Electronic Controls and Software, Noise and Environmental protection.

For the certification of derivatives or major changes or major repairs, the certification team involved in the initial certification of the concerned product and its continuing airworthiness should be used as much as possible, without prejudice to adjust team size to the nature and complexity of the project.

Trainees may be involved in the investigation, at no direct cost to the applicant.

6.2.4. Type certification of EU Products

The EASA type-certification process can generally be divided into the following phases:

- Phase I—Technical familiarization and establishment of the type-certification basis
- Phase II—Agreement of the certification program
- Phase III—Compliance determinations
- Phase IV—Final report and issue of a type certificate.

NOTE: The following description of the type-certification's phases comes from practical and tested certification procedures and technicalities, and it does not necessarily report the content of the above mentioned EASA Type-Certification Procedure.

6.2.4.1. PHASE I—TECHNICAL FAMILIARIZATION AND ESTABLISHMENT OF THE TYPE-CERTIFICATION BASIS

The objective of this phase is to provide technical information about the project to the team specialists to enable the definition of and agreement on the initial EASA type-certification basis.

The type-certification basis is generally given by the applicable CS that is effective on the date of application, plus special conditions[6] if deemed necessary. All these special conditions are not necessarily issued at the beginning of the certification process, because they could also be the result of better design knowledge during the certification process. In Chapter 5, we described how the establishment of the certification basis could become somewhat complex in cases such as changes to TCs and TC validations.

6.2.4.2. PHASE II—AGREEMENT OF THE CERTIFICATION PROGRAM

The objective of this phase is the definition of and agreement on the proposed means of compliance (MoC) with each paragraph of the certification basis and the identification of the team involvement.

Full use of applicant Design Organization Approval (DOA) privileges should be made when defining the certification team involvement,[7] in particular for the agreement on the compliance document to be accepted without further verification.

To be more specific in this matter, we define some technicalities associated with this phase:

(1) Terms of reference (TORs). A list of all paragraphs and subparagraphs of the relevant certification basis is normally produced by the authority's PCM, with the identification of the specialists responsible for compliance with the same requirements. There could be different specialists who are responsible for the same paragraph (e.g., a specialist on systems, one on

---

[6] See Chapter 4, "Special conditions."
[7] See Chapter 5, "Design Organization Approval."

structures, and one on the flight manual). Each has to do their own part of the job, and coordination among them will be provided to ensure that the whole paragraph be complied with.

(2) **MoC definition.** The MoCs are the categorization of the means used to demonstrate compliance with the requirements. A requirement can be complied with, for example, by a flight test, a static test, and/or a substantiation report. These MoCs are defined in the JAA procedures, and some examples are as follows:

MC2: Calculation/analysis. Reports for the evaluation of loads, strength, performance, flying qualities, or other characteristics.

MC3: Safety assessment. Documents describing safety analysis philosophy and methods, safety evaluation plans (software), system safety assessment, zonal safety assessment, and others.

MC6: Flight tests. Reports of flight tests written in the "Flight Test Program" and performed by a flight test crew.

MC7: Inspections. Conformity inspections to verify that materials, parts, processes, and fabrication procedures conform to the type design. Aircraft inspection to verify the compliance with the requirement, which cannot be determined adequately from evaluation of technical data only.

The MoC definition is a very important phase of the certification process because it lays the foundations of the job to be carried out. For this reason, the authority's team and the applicant must agree with it and in sufficient detail to ensure good mutual understanding.

(3) **The compliance checklist (CCL).** A record of compliance with every applicable certification requirement must be produced by the applicant. This record, based on the above-mentioned MoCs, must refer to the documents necessary to demonstrate compliance with the applicable requirements and would take the form of **compliance record sheets (CRSs)**. As the demonstration of compliance progresses, when a single paragraph is "closed" the CRSs will be entered in the **CCL**, containing all references able to single out the compliance demonstrations that have been carried out (identification of the relevant report, its title and edition, page number, reference to other documents).

The CCL is a key document in type-certification; it actually allows tracing back to the compliance documents, even from many years in the past. It is therefore fundamental in the post-TC phase for approval of changes, in cases that are contested due to incidents/accidents or to other reasons.

6.2.4.3. PHASE III—COMPLIANCE DETERMINATION

The objective of this phase is the demonstration of compliance with the certification basis and the acceptance of the compliance demonstrations.

Having established and agreed the MoCs, the applicant must provide the authority with tests and calculations demonstrating compliance with the certification basis, normally by means of documents and reports. The reports must make precise references to the inherent requirements, not only quoting the paragraph but also which is the MoC concerned, where different MoCs have to be complied with.

In the CCL, each document mentioned must contain a statement by the applicant declaring (total or partial) compliance with the applicable requirements.

We will now describe some implications of this crucial phase.

(1) Tests on prototypes and test articles. We have previously mentioned that demonstrations of compliance often require tests to be carried out not only on the prototype(s) but also on single parts of the aircraft.[8]

Something apparently obvious, but that should be clearly kept in mind, is that the prototype or single part to be tested must be *representative* of the type design. To this end, it is required that for any certification test the applicant submits in advance a statement of conformity to the type design or, in the presence of deviations, a statement that such deviations are not influential on the test to be performed.

To give a clear example, the assessment of the stall characteristics of an aeroplane with a cabin configuration with deviation from the type design, or with an inefficient system that has nothing to do with the flight controls, will not be influenced by these anomalies. It should be different for an unapplied change referring to flaps or control surfaces.

Therefore, it is of paramount importance to establish the correct configuration control of prototypes and test articles during the certification process. It is also necessary to check what would be the effect of a type-design change on tests already performed and on documentation already produced. An integration of the above-mentioned tests and documentation could be necessary or, in the worst case, a need for them to be rewritten.

(2) The certification review item (CRI). The certification review item is a document recording each step leading to the closure of a subject in particular cases such as the following:

(a) To record the process followed to define the content of the type-certification basis (CRI A-1)

(b) To develop and administer special conditions

(c) To administer new policies, for example, unusual MoC/interpretations

(d) To administer exemptions[9] or equivalent safety findings[10]

(e) To deal with subjects involving controversial discussions between the team and the applicant.

The authority's PCM, in the "conclusion" statement of his or her report, will document the decision on how to resolve an issue when this has been reached (sometimes with the concourse of the authority at higher level).

(3) The action item (AI). The purpose of an AI is to administer the progress of an item not requiring a CRI but requiring special attention of the applicant or the team. An AI may be opened for the following cases:

---

[8] For example, a drop test of a landing gear unit, a static test of a flap, an aileron, and so on.

[9] See Chapter 5, "Certification basis."

[10] See Chapter 5, "Certification basis."

(a) To review the suitability of compliance demonstration of selected subjects

(b) To follow-up a closed CRI, when necessary[11]

(c) To administer matters interfacing certification and flight operations

(d) Any other case, as deemed necessary.

The AI will define the characteristics to be checked, the relevant requirements, the interpretations to be used, the actions, the responsibilities, and the basis for conclusions, as necessary.

6.2.4.4. PHASE IV—FINAL REPORT AND ISSUE OF A TYPE CERTIFICATE

The objective of this phase is the establishment of a project's final report recording details of the type investigation and, based on approval of the final report by the responsible CM, the issue of the EASA type certificate.

(1) **Statement of compliance.** On completion of the certification program, the applicant shall provide a declaration of compliance that the type design of the product to be type-certificated complies with the type-certification basis.

The team members issue a statement of satisfaction to the PCM with the applicant's compliance declaration of the disciplines involved.

On acceptance of all necessary statements of satisfaction by the EASA certification team, the PCM shall issue a compliance statement to the responsible EASA CM confirming that the type design of the product complies with the type-certification basis.

(2) **Final certification report.** The PCM, in conjunction with the team, shall produce and present to the responsible EASA CM, a report that will record the type design on which the type-investigation process is based, the significant subjects investigated the details of that investigation, the CRIs that have been discussed, the process followed, and the conclusions regarding compliance with the type-certification basis.[12]

If there are some open actions, the so-called **post-TC items**, a list of the same has to be issued, making sure that this is not a mere excuse to postpone some demonstrations of compliance that are necessary for the TC issue.

(3) **Type certificate.** After approval of the final report, the responsible EASA CM shall take the necessary steps inside EASA for the issue of the type certificate.

A type-certificate data sheet (TCDS)[13] will form part of the EASA type certificate.

---

[11] The CRI is "closed" when a *decision* about the actions to be carried out has been reached; the *realization* of these actions represents a further phase.

[12] The Internal Working Procedure TCP in its appendix defines working procedures addressing the content and presentation of such a final report.

[13] TCDS: document attached to the TC, containing the product's main characteristics, the certification basis, the type-certification date, and so on.

We will now describe this final phase in more detail.

Normally, a final **type-certificate board meeting**[14] is held

(1) To ratify:

(a) Closure of AIs

(b) Completion of certification review items

(c) Approval of CRSs/CCL

(d) Authority's flight test results.

(2) To approve:

(a) Aircraft flight manual and airworthiness limitation section

(b) Certification maintenance requirements

(c) Type-design definition

(d) Post-TC items

(e) Draft of TCDS.

(3) To endorse: the applicant's and the team's statement of compliance.

6.2.4.5. THE AUTHORITY'S INVOLVEMENT

In Chapter 5, we mentioned that the authority's intervention can be modulated on the basis of the DOA privileges, if the applicant has achieved DOA.

Of course, the authority has to be particularly involved in those phases of the certification process we can define as "preliminary phases": familiarization, certification basis definition, and CRS approval. It must also deal with the administration of certification review items. Nevertheless, the authority has the choice of arranging with the applicant which reports should be checked and which tests should be witnessed.

In the case of flight tests, the authority usually employs its own flight personnel and carries out a flight test program based on the applicant's flight test reports.

It goes without saying that it is impossible to establish definite rules about the authority's intervention, because this is influenced by various factors such as the design complexity and, above all, the design organization experience demonstrated in previous type certifications.

If the applicant does not have DOA because the object of the application does not require a DOA, the DOA privileges being absent, the authority is not allowed to delegate anything, and in principle, it is involved in each report and each test.

Also, in this case, the nature of the design and the applicant's experience are very important. This means that it is up to the authority team's professionalism to decide whether their checks are sufficient or whether they should "revise all the calculations."

---

[14] The TCB meetings are official meetings attended by the team, the design organization, and some authority representatives responsible for the type certification. Normally, these meetings open and close the type-certification process, with some intermediate meetings assessing the state of the certification process.

6.2.4.6. POST-TC ACTIVITIES

After the TC issue, the same certification team is usually involved in the following activities:

(1) Changes in the type design made by the type-certificate holder (TCH)

(2) Changes in the type design made by someone other than the TCH

(3) Continued airworthiness actions, including approval of service bulletins[15] and issuance of Airworthiness Directives[16]

(4) Approval of repairs.

6.3. THE FAA TYPE-CERTIFICATION PROCESS

6.3.1. Introduction

In dealing with the applicant design organization, we have found that FAR 21 does not mention a formal approval.

To understand the FAA type certification, we encounter a fundamental peculiarity of the FAA's organization: **delegation**.

The Federal Aviation Act of 1958 was the original statute that allowed the FAA to delegate activities to authorized private individuals employed by aircraft manufacturers. Although paid by the manufacturers, these **designees** act as surrogate for the FAA in examining aircraft design, production quality, and airworthiness. The FAA is responsible for overseeing the designees' activities and determining whether the designs meet the FAA's requirements for safety.

It is important to note that, according to the Code of Federal Regulations, where the regulations make reference to the "Administrator," this also includes any person authorized by the Administrator to exercise or perform that specific power, duty, or function.

Private individuals have been examining, testing, and inspecting aircraft as part of the FAA's regulatory system for aviation safety since at least 1927. The FAA's Act of 1958 gives the current legislative authority to appoint a wide variety of designees to issue certificates.

The functional roles and responsibilities for designees are set forth in FAA Orders 8110.37 D for **Designated Engineering Representatives (DER)** and 8100.8 C for **Designated Manufacturing Inspection Representatives (DMIR), Designated Airworthiness Representatives (DAR)**, and **Organizational Designated Airworthiness Representatives (ODAR)**.

The FAA relies on both individual and organizational delegations in the certification process. Delegation is used to the maximum practicable extent with appropriate oversight safeguards as defined in the FAA's delegation management process policies.

---

[15] Documents issued by the TCH containing instructions for corrective actions (changes, inspections, etc.), improvements, and so on.

[16] Documents issued by the authority making mandatory particular actions (changes, inspections, etc.). See Chapter 9.

The FAA and the applicant agree to manage all designee activity within the regulations and policy regarding designee appointment, procedures, and oversight. It is essential that the FAA and the public have confidence in the integrity of the designee system and that it functions properly. Both the FAA and applicant agree to foster an environment where open communication between the designees and applicant's management and between the designees and their FAA counterparts is a standard practice. That environment should encourage the designees, within the scope of their delegation, to openly communicate certification items with the FAA which is necessary to maintain confidence in the designee system. The applicant agrees to create a working environment in which designees can make judgments on compliance and conformity findings free from undue pressure and with the support and knowledge of the FAA. It should be clearly understood by FAA personnel and designees that their objective is to find compliance with the regulations and not to dictate design.

6.3.2. Designated Engineering Representatives

The DER may approve engineering technical data within the limits of his or her authority and, when authorized by the ACO,[17] may witness FAA compliance tests and perform compliance inspections. DERs will follow the procedures of FAA Order 8110.4 C, "Type-Certification." The specific role, authorized area, and responsibility of the DER will be established by agreement between the ACO and the DER.

6.3.2.1. COMPANY DERs

An individual may be appointed to act as company DER for his or her employer and may only approve, or recommend approval to the FAA, technical data for the company. Company DERs may perform their FAA functions at different administrative levels, as agreed on between the FAA and the company. In some cases, a DER may personally evaluate and approve technical data. In other cases, a DER may ensure, through the company management system, the proper evaluation of technical data by other persons; then the DER will approve data by certifying that the data complies with the applicable regulations.

6.3.2.2. CONSULTANT DER

An individual may be appointed to act as an independent (self-employed) consultant DER to approve, or recommend approval of, technical data to the FAA for a client.

6.3.2.3. THE DER's DESIGNATION

DERs are experts acting within well-defined limits of their appointment. The list includes

---

[17] Aircraft Certification Office.

(1) Structural DERs,

(2) Power plant DERs,

(3) System and equipment DERs,

(4) Radio DERs,

(5) Engine DERs,

(6) Propeller DERs,

(7) Flight analyst DERs,

(8) Flight test pilot DERs, and

(9) Acoustical DERs.

Order 8110.37 D specifies the items of competence of each DER.

It is also worth mentioning some "special" delegations/authorizations, which are appointments not specifically listed in the above-mentioned items of competence. The following are examples of special delegations.

6.3.2.4. ADMINISTRATIVE/MANAGEMENT DERs

A qualified person may be appointed as an administrative coordinator or as a manager of an applicant's certification program. These designations free the FAA from having to carry out the normal project administration, technical coordination, and guidance usually associated with a certification program.

(1) Administrative DER. Usually a company DER acts as a focal point for FAA coordination activity, including organizing technical DER activity, correspondence, scheduled meetings, conformity inspections, and FAA participation in official tests.

(2) Management DER. Usually a consultant DER, performs FAA certification management duties similar to the FAA program manager. This includes organizing the certification program, directing, overseeing, and managing the task of technical assessment and finding of compliance. The DER assures that all technical data required to show compliance is reviewed and approved by the appropriate DER, except in those areas reserved by the FAA for approval.

6.3.3. Guidance material for the type-certification process

Fundamental guidance material for the type-certification process is provided by:

(1) Order 8110.4 C, Type-Certification prescribes the responsibilities and procedures for FAA aircraft certification personnel responsible for certification process of civil aircraft, engines, and propellers.

(2) "The FAA and Industry Guide to Product Certification" (CPI Guide) containing a description of the purpose and vision of the improved certification process. It also includes an overview of the phases of product certification, including the process flow and detailed descriptions of the key players' roles. This guide describes how to plan, manage, and document an effective, efficient product certification process, and working relationship between the FAA and an applicant. The guide can be used for type-certification, Supplemental type-certification, significant amendments to

TC or STC, production approval, and other design approvals, including PMA and TSO authorization. The guide is used as a supplement to existing FAA guidance.

A more in-depth understanding of this subject can be gained by consultation of these documents, training, and on-job training; we will describe the main issues of the FAA type-certification process on the basis of these two documents.

6.4. THE CPI GUIDE

We begin with a summary of the **CPI Guide**, starting from the description of two documents that are the basis of the type-certification process.

6.4.1. Partnership for Safety Plan

The Partnership for Safety Plan (PSP) is a written "umbrella" agreement between the FAA and the applicant that defines *generic procedures* for product certification, establishes the general expectation or operating norms, and identifies deliverables.[18] The PSP also defines the general discipline and methodology to be used in planning and administering certification projects, and it includes project schedule milestone development, generic delegation procedures, conformity procedures, communications protocol, an issue resolution process, and the generic operating norms for developing metrics for project evaluation.

Appendix I of the CPI Guide provides instruction for producing the PSP.

6.4.2. Project-Specific Certification Plan

The Project-Specific Certification Plan (PSCP) applies the agreed principles of the PSP to a *specific certification project*. Each project will have a PSCP designed to be used as a project management tool, providing milestones, performance measures, and information unique to a certification project. The PSCP captures procedures based on the generic methodologies of the PSP and applies them to a specific project.

Figure 6.1 is a diagrammatic representation of the relationship between the PSP and PSCPs.

6.4.3. Phases of type-certification

There are five certification phases. They range from early project concept and initiation through post-certification activities. The five phases are illustrated in Fig. 6.2.

We will only cover the definitions of the five phases. The CPI Guide contains detailed descriptions of each phase, including the phase's definition, tasks, required information, deliverables, and criteria for success.

---

[18] Deliverables: prerequisites for subsequent phases to be completed before entering a new phase.

FIGURE 6.1 Relationship between the Partnership for Safety Plan (PSP) and Project-Specific Certification Plans (PSCPs)

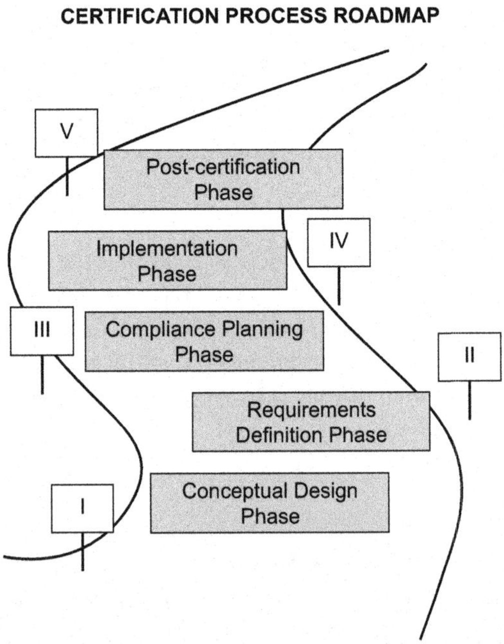

CERTIFICATION PROCESS ROADMAP

FIGURE 6.2 "Roadmap" of the certification process

Furthermore, each table is followed by a phase evaluation checklist as a tool for project evaluation during the appropriate phase.

The FAA and applicant Project Managers (PMs) should jointly prepare a phase evaluation checklist at the end of each phase of a product certification. These forms should be continuously evaluated by the applicant/FAA team for immediate improvement of the process.

6.4.3.1. PHASE I—CONCEPTUAL DESIGN

This phase is initiated when the applicant begins a design concept for a product that may lead to a viable certification project. The intent is to ensure early, value added, joint involvement with the expectation of covering critical areas and the related regulatory issues, and to begin formulating a preliminary PSCP. This is an opportunity to apply the PSP principles to develop a mutual understanding of potential new projects.

6.4.3.2. PHASE II—REQUIREMENT DEFINITION

Efforts in this phase clarify the product definition and the associated risks and conclude with a mutual commitment to move forward with product certification. Specific regulatory requirements and methods of compliance or critical issues are formulated. A more formal PSCP is developed.

6.4.3.3. PHASE III—COMPLIANCE PLANNING

During this phase, a PSCP is completed. The plan is a tool to which the responsible parties commit and use to manage the product certification project.

6.4.3.4. PHASE IV—IMPLEMENTATION

During this phase, the applicant and FAA work closely in managing, refining, and achieving their agreed PSCP to ensure that all agreed upon product-specific certification requirements are met.

6.4.3.5. PHASE V—POST-CERTIFICATION

During this phase, closeout activities provide the foundation for continued airworthiness activities and certificate management for the remainder of the product's life cycle.

6.4.4. The "key players" of the type-certification process

Figure 6.3 gives a breakdown of the people involved in all phases of the type-certification process and descriptions of their roles.

We will provide only a brief description of the key players; the CPI Guide contains detailed information on their responsibilities, accountability, communication, and so on.

(1) FAA and applicant's management—provide leadership and resources. The applicant and the FAA work to establish a PSP to reach a clear common understanding of their respective responsibilities for the design and production definition and the certification requirements. The respective managements provide leadership and resources to product certification teams through the PMs to accomplish the project and resolve issues. The management has ultimate responsibility through the product certification team for the quality of compliance finding work, standard application of regulatory compliance policy and procedures, and the timely, efficient completion of the product certification projects.

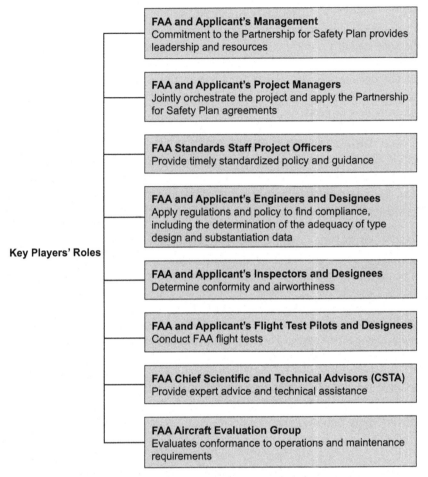

Key Players' Roles

FAA and Applicant's Management
Commitment to the Partnership for Safety Plan provides
leadership and resources

FAA and Applicant's Project Managers
Jointly orchestrate the project and apply the Partnership
for Safety Plan agreements

FAA Standards Staff Project Officers
Provide timely standardized policy and guidance

FAA and Applicant's Engineers and Designees
Apply regulations and policy to find compliance,
including the determination of the adequacy of type
design and substantiation data

FAA and Applicant's Inspectors and Designees
Determine conformity and airworthiness

FAA and Applicant's Flight Test Pilots and Designees
Conduct FAA flight tests

FAA Chief Scientific and Technical Advisors (CSTA)
Provide expert advice and technical assistance

FAA Aircraft Evaluation Group
Evaluates conformance to operations and maintenance
requirements

Note: Appendix III describes key players' roles and responsibilities as they apply to
avionics approvals

FIGURE 6.3 The "key players" involved in the type-certification process

(2) **FAA and applicant's PMs**—orchestrate the project and get the job done. The
FAA, designees, and applicant's PMs are the principal focal points for the
project. They coordinate and direct the certification team's effort and ensure
that things are kept moving to achieve the product certification objectives.

(3) **FAA Standard Staff Project Officer**—coordinates the directorate interaction. The Standard Staff Project Officer provides the certification team with
clear and timely regulatory and policy guidance specific to the project. He
or she is the focal point within the responsible project directorate for that
policy and for engaging other appropriate directorate staff on installation
issues across FAR Parts, for example, engines, propellers, APUs.

(4) **FAA engineers and/or designees**—apply regulations and policy to find
compliance. The engineers as assigned for appropriate disciplines are the principal contacts for the applicant. Their activity is always in coordination with

the FAA PM and follows the agreed PSCP for guiding the certification process, communication guidelines, and how rules and policy will be applied. The engineers and designees understand the technical details of the project, application of applicable rules and policy, and are responsible for the majority of the compliance findings associated with the project. They also evaluate sufficiency of the type design and substantiation data with the discretion to review any of the data therein, such as critical material process specifications.

(5) **FAA inspectors and/or designees**—determine conformance and airworthiness. The FAA aviation safety inspectors provide consultation and advice on production processes proposed in the design. They conduct and oversee, through designees, a variety of conformity inspections, evaluations of aircraft airworthiness, and issue airworthiness certificates or other approvals. They conduct progressive evaluation of the manufacturer's quality and production systems for eventual production approvals. The inspector is made aware of conformance issues on critical parts that cannot be determined solely from type-design data. This would then require focused process control, inspection, or evaluation within the production quality system.

(6) **FAA flight test pilots and/or designees**—Conduct product certification flight tests. The flight test pilots provide technical advice to the team on aircraft configuration, operation, flight testing, and instrumentation needed for compliance determinations. They conduct FAA flight tests and other appropriate evaluations, find compliance to flight test requirements, and provide guidance to the applicant on preparing the flight manual and related operational procedures.

(7) **FAA Chief Scientific and Technical Advisor (CSTA)**—provides expert advice and technical assistance. The CSTA provides professional technical guidance, advice, and assistance in their discipline. They are a direct link to an extensive professional network in the R&D community, professional and academic organizations, industry, other government, and national and international experts in their discipline.

(8) **FAA Aircraft Evaluation Group (AEG)**—evaluates conformance to operations and maintenance requirements. The FAA AEG provides a link to applicable Flight Standards technical services. This lends an aircraft operational and maintenance perspective to the type-design assessment, thereby allowing FAA engineering and their designees to determine appropriate compliance requirements in those areas. The AEG carries knowledge of the product and how it was type-certificated to the aircraft Maintenance Review Board (MRB), Flight Operations Evaluation Board (FOEB), and Flight Standardization Board (FSB) activities.

6.5. FAA ORDER 8110.4 C, TYPE-CERTIFICATION

The CPI Guide, as we have seen, is an operative document that should be used by the FAA and applicants together to fulfil their respective roles and expedite certification of products focusing on safety significant issues.

FAA Order 8110.4 C is essentially orientated to prescribe the responsibility and procedures for FAA aircraft certification personnel for the certification of civil products under FAR 21.

We will now give just an idea of the content of this valuable document to better understand how the responsibilities are distributed and to better clarify certain aspects of the type-certification process.

Figure 6.4 provides a typical summary of the type-certification process.

6.5.1. Application for TC, amended TC, STC, and PC[19]

Information is provided for submission of application for the various certifications, including the FAA forms to be used, the documents to be enclosed, the applicable paragraphs of FAR 21, and so on.

6.5.2. Establishment of TC project

(1) **General.** An applicant submits a TC, amended TC, or STC application to the geographically responsible ACO.[20]

(2) **Certification Project Notification (CPN).** The ACO is responsible for assigning a project number, a PM, and notifying the accountable directorate of each project completing the CPN with information on the project. On the basis of the importance of the project, National Resources Specialists (NRS) and the AEG are requested. The accountable directorate assigns a project officer for significant projects.

The PM and the Project Officer are the focal points for the ACO and the accountable directorate, respectively.

(3) **Assignment and duties of the PM.** The PM is responsible for planning, reviewing, evaluating, and coordinating all aspects of a certification project in accordance with the Certification Program Plan (CPP), which is a fundamental document in the certification process (it will be discussed later). The PM is responsible for initiating this CPP and coordinating with the project officer and the Certificate Management ACO. The PM also coordinates with the appropriate manager(s) in the selection of other team members.

(4) **Project team.** A project team is established for all projects that require significant involvement by technical personnel and normally consist of the following:

(a) A PM

(b) Engineers or technical specialists

(c) Pilots and/or flight test engineers

(d) Manufacturing inspectors

(e) Operations and/or airworthiness inspectors from the AEG

(f) A Project Officer and other staff at the discretion of the accountable directorate.

---

[19] Product certificate, see Chapter 7.
[20] For FAA organization, see Chapter 3.

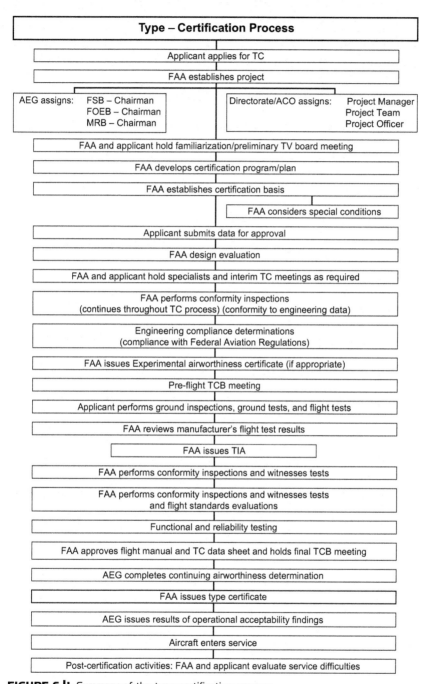

FIGURE 6.4 Summary of the type-certification process

6.5.3. Type-Certification Board

(1) General. A Type-Certification Board (TCB) is established for all aircraft and engine projects in which complete type certification is involved. For instance, TCBs are not always required for STC projects.

The purposes of a TCB are to acquaint the applicant and the FAA with the certification project, resolve significant problems, establish milestones and schedules for the overall accomplishment of the type-certification program, review the applicant's certification plan, review the proposed basis for certification, and assure that all outstanding certification issues are resolved.

(2) TCB members. The FAA members are as follows:

(a) The ACO Manager

(b) The PM

(c) The managers, supervisors, or senior personnel from the appropriate engineering disciplines, flight test, manufacturing inspection, and AEG.

In addition, there is a list of participants, other than TCB members, who may be invited to participate on an advisory basis—for example, Washington Headquarters, NRS (National Resources Specialists), additional AEG personnel, and so on, applicants and their representatives.

(3) TCB meetings. The following TCB meetings are normally organized:

(a) Familiarization TCB meeting

(b) Preliminary TCB meeting

(c) Interim TCB meeting

(d) Preflight TCB meeting

(e) Final TCB meeting.

Depending on the type and/or the size of the project, all the TCB meetings may not be necessary.

The ACO Manager or his/her representative serves as Chairman. (Order 8110.4 C provides details on each of the above-mentioned TCB meetings.)

6.5.4. Certification Program Plan

The CPP defines the working relationship between the accountable directorate and the geographic ACO or within an accountable directorate during a specific TC project. The CPP is the principal program coordination tool and is updated throughout the program by the PM, as required.

An applicant's certification plan may take the place of the CPP if it includes all information that would be addressed in the CPP and is coordinated with the Project Officer.

6.5.5. Issue paper

An issue paper provides means for the identification and resolution of significant technical, regulatory, and administrative issues that occur during a certification process. Issue papers are primarily intended to provide an overview of significant issues, a means to determine the status of issues, and a post-certification summary statement on how issues were resolved.

6.5.6. Issue book

The PM assembles issue papers and publishes them in the form of an issue book, which is distributed to the TCB members, project team members, applicant, and accountable aircraft certification directorate.

6.5.7. Type-certification basis

The proposed certification basis is established by the FAA at the beginning of a TC program. The applicant is advised of all aspects at the beginning of the program, including operational requirements.

Once the certification basis has been established and agreed by the FAA and applicant, new policy will not be introduced unless an unsafe condition is found to exist in a product that has a design feature affected by that policy.

(1) Special class of aircraft. Special class of aircraft includes airships, gliders, motor gliders, very light airplanes, and other nonconventional aircraft for which airworthiness standards have not been issued under FAR 21. The procedures necessary to establish and receive approval for the certification basis are provided by the relevant ACs (Order 8110.4 C provides the list).

(2) Changes. Order 8110.4 C provides instruction for the establishment of a certification basis for the different cases we have described in Chapter 5.

(3) Additional requirements. Additional requirement are the following:

  (a) *Special conditions.* Starting from the definition, Order 8110.4 C provides instructions and guidance for the issue of special conditions.

  (b) *Equivalent level of safety findings.* These are made when literal compliance with a certification regulation cannot be shown and compensating factors exist that can be shown to provide an equivalent level of safety. They are normally proposed by the applicant to the ACO and submitted to the directorate.

  (c) *Exemptions.* In a type-certification program, any interested person may petition the FAA for a temporary or permanent exemption from an FAR. The petition for exemption is made to the accountable directorate through the ACO and processed according to Order 8110.4 C information.

  (d) Applicable requirement of FAR 34 and FAR 36 for environmental protection.

6.5.8. Type-certification program

In this paragraph, Order 8110.4 C provides a great amount of information and instructions for the applicant's submission to the FAA of the type design, test reports, and computations necessary to show that the product to be certificated meets the applicable type-certification basis. In particular, the content of the applicant test plan is defined.

Information/instructions are also provided to the FAA about the use to be made of data submitted by the applicant, for the witnessing of tests, conformity inspections, notifications of noncompliance, and so on.

6.5.9. Type-inspection authorization

The type-inspection authorization (TIA), prepared by the ACO, is issued to authorize official conformity airworthiness inspections, and ground and flight tests necessary to fulfil certain certification requirements. Order 8110.4 C provides information/instructions for TIA issuance.

6.5.10. Operational and airworthiness evaluations

AEGs (Aircraft Evaluation Group)[21] are responsible for the operational and maintenance aspects of the aircraft type-certification process and, once the aircraft enters service, are the coordination point for activities involving Flight Standards.

The AEGs advise manufacturers of pertinent operational and maintenance requirements during the design and certification process.

The AEGs have the primary responsibility for evaluation of aircraft and its systems for operational suitability and continued airworthiness.

Each directorate AEG is responsible for those AEG functions dealing with the TC product for which its directorate has responsibility.

The AEG makes recommendations to FAA field offices regarding operations specifications, training and maintenance program, and airmen qualification through management of several FAA boards, such as the FSB (Flight Standardization Board), FOEB (Flight Operating Evaluation Board), and MRB (Maintenance Review Board).

6.5.11. Flight manual

The ACO responsible for the project approves flight manuals including revisions and supplements.

The flight manual should not be approved until:

(1) The FAA project flight test pilot and/or flight test engineer, the AEG operation specialist, and appropriate FAA engineers concur with the operational limitations and normal and emergency procedures.

(2) The FAA flight test engineer recommends approval of the performance section of the flight manual.

(3) The AEG has reviewed and coordinated information in the flight manual.

6.5.12. Type certificates

The certifying ACO issues a type certificate when an applicant completes the requirements of the applicable FAA Regulations for the product. Order 8110.4 C provides guidance for the preparation of the applicable FAA form.

---

[21] This is a Flight Standard group colocated with each directorate and is responsible for determining operational acceptability and continued airworthiness requirements for newly certified products.

6.5.13. The type-certificate data sheet

The TCDS, which is part of the TC, provides a concise definition of the configuration of a type-certificated product. Therefore, a standard format for the TCDS is necessary to allow information about a specific product to be easily found. The Order also provides guidance for the preparation of this document.

NOTE: FAA Order 8110.4 C contains a multitude of other information that we will not discuss here because it is outside the scope of this book, which is not a "certification manual" but a means to enable technical people to understand the principles of airworthiness.

The Order also contains the following useful list of ACs related to type certification.

(1) AC 20-135, Power plant Installation and Propulsion System Component Fire Test Methods, Standards and Criteria

(2) AC 21.17-1, Type Certification—Airships

(3) AC 21.17-2, Type Certification—Fixed Wing Gliders

(4) AC 21.17-3, Type Certification of Very Light Airplanes

(5) AC 21-23, Airworthiness Certification of Civil Aircraft, Engines, Propellers, and Related Products Imported into the United States

(6) AC 21-24, Extending a Production Certificate to a Facility Located in a Bilateral Airworthiness Agreement Country

(7) AC 21-40, Application Guide for Obtaining a Supplemental Type Certificate

(8) AC 23-8, Flight Test Guide for Certification of Part 23 Airplanes

(9) AC 25-7, Flight Test Guide for Certification of Transport Category Airplanes

(10) AC 25-19, Certification Maintenance Requirements

(11) AC 25.571-1, Damage Tolerance and Fatigue Evaluation of Structure

(12) AC 27-1, Certification of Normal Category Rotorcraft

(13) AC 29-2, Certification of Transport Category Rotorcraft

(14) AC 33-2, Aircraft Engine Type-Certification Handbook

(15) AC 36-4, Noise Certification Handbook

(16) AC 121-22, MRB

6.6. CONSTRUCTION OF PROTOTYPES AND TEST ARTICLES

Type certification consists mainly of type-design approval. A TC is actually valid even if, for whatever reason, there are no more aircraft of that type. Nevertheless, it is not possible to carry out a type certification "on paper." One or more prototypes[22] and test articles have to be built.

---

[22] In the type certification of sailplanes and light aeroplanes, for economic reasons, only one prototype is often built.

The applicant's design organization could be part of an enterprise that has the means of mass production and even the POA,[23] otherwise it could be an independent organization working with an enterprise having these capabilities. In the first case, the DO has two options:

(1) To perform the prototype construction inside the production organization of the enterprise.

(2) To perform the prototype construction inside an **experimental department**.

In Case 1, the DO has the advantages emanating from well-organized production, because the authority has already given approval. Hence, when the product is type-certificated it will be ready for mass production. The disadvantage of such an arrangement comes from the necessity of being subject to rules that, especially in the case of large companies, are rather complex. For instance, the acquisition of a changed part can be subject to a long delay. If we consider that, during type certification, changes are frequently required, it is clear why the second option is preferred in many cases.

Inside an experimental department, the DO technicians are in close contact with the prototype material, making the introduction of changes simpler and the activities easier, so time can be saved. Because aeronautical production is at stake, the department must respect the quality assurance rules, with its own controllers and procedures. A possible guide for departmental organization can be found in Subpart F of JAR/FAR21/EASA Part 21, "Production without Production Organization Approval," even if this relates to already certificated products and parts.

In the case of an applicant being an independent DO, working with an enterprise with production facilities, the above-mentioned remarks are still valid in principle, and the choice between the two solutions could also depend on the size of the enterprise.

Having defined some principles, it is not possible to establish fixed rules, because there could be many different situations and suitable choices. A large enterprise would prefer to arrive at the end of the type certification with an industrialized product ready for mass production. A small enterprise could have built a hand-crafted product, could have it certificated this way, and could be considered for future industrialization (changes in the type design would be necessary) if mass production is possible.

For example, one of the best known experimental departments is Lockheed Martin's **Skunk Works**, which began under the direction of the legendary Kelly Johnson, the designer of the P-38 "Lightning" (just one of his many celebrated aircraft). In 1943, he was commissioned to design the first American jet fighter and to build a prototype in only 180 days. For reasons of secrecy, Kelly Johnson rented a big circus tent and set up shop next to a noxious plastic factory, whose stench kept the curious at bay. One day one of the engineers went to work

---

[23] See Chapter 7, "Production Organization Approval."

wearing a gas mask as a gag, and another employee picked up a ringing phone and announced, "Skonk Works." That was a fashionable expression at the time, originating from All Capp's cartoons and referring to special juice made by a dead skunk. The expression became popular and, changed to "Skunk Works" for editorial reasons, became the registered name of the department. The P-80 "Shooting Star" was built in only 143 days, 37 days ahead of schedule; it is thought that it was probably the smell that spurred Kelly's workers to build the aeroplane in such a short time!

Among the most celebrated designs of Skunk Works are the F-104 "Star-fighter," the U-2 spy plane capable of flying at 70,000 ft (in the 1950s), and the SR-71 "Blackbird," capable of flying at Mach 3 and at an altitude of more than 80,000 ft (in the 1960s).

In the 1980s, the creation of the stealth aeroplane F-117A was destined to begin a new era in the design of military aircraft, exploiting some theoretical principles discovered by Russian scientists, but never before put in practice in the Soviet Union.

One of Kelly Johnson's basic rules was that "engineers must always work within a stone's throw of the airplane being built."

Skunk Works represents a brains trust where nothing is impossible, having carried out previous testing and demonstrations.

Chapter seven

Production of Products, Parts, and Appliances

After the prototype phase of a product has led to the type certification, what typically follows is its mass production. In the previous chapters, we explained how this mass production could be performed by an individual who is not the type-certificate holder. In any case, the TCH is required to collaborate with the production organization to ensure:

(1) Satisfactory coordination of design and production.
(2) Proper support for the continuing airworthiness of the product.

7.1. THE JAA/EASA PRODUCTION ORGANIZATION

JAR 21/EASA Part 21 provides two options as regards the production organization:

(1) **Production Organization Approval (POA)** according to Subpart G.
(2) **Production without POA** according to Subpart F.

In the first case, the approval bears similarity with the DOA.[1] As for the DOA, the approval aims to highlight the responsibility of the organization, allowing the authority to perform less fiscal, but more efficient, control.

For example, it has always been the norm for the authorities to survey the construction of every single aircraft and to also carry out flight tests to issue a certificate of airworthiness. The holder of a POA, on the basis of the POA privileges, may obtain a certificate of airworthiness on presentation of a statement of conformity, with no further showing.

All this obviously requires that the authority obtain a deep knowledge of the organization, performing careful checks to ensure the continuous validity of the organization approval.

The second case (Subpart F of JAR21/EASA Part 21) is applicable to manufacturing organizations for which a production approval under Subpart G would be inappropriate—for example, because production is limited to a number of units, or because production is initiated under this Subpart F in

---

[1] See Chapter 5, "Design Organization Approval." **163**

Airworthiness: An Introduction to Aircraft Certification.

advance of issue of a POA under Subpart G. Such an organization does not have the privileges of POA; this means that it will also be exposed to closer authority supervision for the issue of the final certification. We have previously noted how something similar can happen for design organization without a DOA.

We will now describe these two types of production organization in more detail.

7.1.1. Production Organization Approval

As we have just mentioned, if an applicant for POA is not the TCH, he or she must have an appropriate cooperation agreement with the TCH.

Production is intended to relate to products (aircraft, engines, and propellers), parts, and appliances (JTSO/ETSO articles, JPA parts, and other parts)[2] and changes in type-design certificated as Supplemental type certificates (STCs).[3]

Among others, the most important characteristics of a POA are:

(1) A **Quality System** to enable the organization to ensure that each product, part, or appliance produced by the organization, or by its partner or supplied from or subcontracted to outside parties, conforms to the applicable design data and is in condition for safe operation. This structure (the tasks are defined in detail in Appendix B of JAR 21 and AMC and GM of EASA Part 21) provides the organization with all control procedures and, among others, the following are most important:

(a) Manufacturing processes

(b) Verification of incoming materials

(c) Vendor and subcontractor assessment, audit, and control

(d) Nonconforming item control

(e) Personnel competence and qualification

(f) Inspection and testing, including production flight tests

(g) Airworthiness coordination with the TCH

(h) Internal quality audits and resulting corrective action.

This structure provides all involved personnel with written information to allocate their relevant responsibilities.

The Quality System must include an **Independent Quality Assurance Function** to monitor compliance with, and adequacy of, the documented procedures of the Quality System. "Independent" is related to the lines of reporting, authority, and access within the organization, and assumes an ability to work without technical reliance on the monitored functions.

The aim is to have the organization able to produce, in conformity with the applicable design, products, parts, and appliances in condition for **safe operation**.[4] To ensure the above, the Quality Assurance Function has to

---

[2] See Chapter 5, "Parts and appliances approval."
[3] See Chapter 5, "The Supplemental type certificate (STC)."
[4] It is worth emphasizing that the safety goal is intrinsic to the organization and independent of the authority's control.

perform planned, continuing, and systematic evaluations or audits of factors that affect the conformity and safe operation.

(2) The **Organization**. The following individuals have to be appointed:

(a) A manager **accountable** to the authority. He or she should be responsible for the organization's activities on these matters and has the corporate authority for ensuring that all production work is carried out to the required standards. The authority may be delegated in writing to another manager of the organization.

(b) A manager or group of managers with responsibilities and tasks clearly defined, reporting (directly or indirectly) to the manager accountable. One of these managers, normally known as the **Quality Manager**, is responsible for monitoring the organization's compliance with Subpart G of JAR 21/EASA Part 21; he or she should have a direct link with the manager accountable.

(c) Staff at all levels with appropriate authority to be able to fulfil their allocated responsibilities, with full and effective coordination within the part of the production organization dealing with airworthiness matters.

(d) **Certifying staff.** Defined as those employees who are authorized to sign final documents (e.g., statements of conformity, JAA Form One/EASA Form 1[5]).

(e) **Privileges.** As in DOA, privileges exist in POA to release the organization from strict authority control. Then, the organization may

(i) Obtain, in the case of complete aircraft and upon presentation of a Statement of Conformity (EASA Form 52), an aircraft certificate of airworthiness and a noise certificate without further showing.

(ii) Issue, in case of other products, parts, and appliances, authorized release certificates (JAA Form One/EASA Form 1)[6] without further showing.

---

[5] JAA Form 1/EASA Form 1 (Authorized Release Certificate) identifies the **conformity** or **airworthiness** and eligibility status of products/parts/appliances/components/assemblies (referred to as "part" or "parts") after manufacture or to release maintenance work carried out under the authority's approval. There are two types of certificate:

1. JAA Form 1/EASA Form 1 **for airworthiness purposes**, related to parts that fully conform to an approved design standard, and then qualified for installation and operation.
2. JAA Form 1/EASA Form 1 **for conformity**, related to parts that conform to designs and data that are not yet approved. For example, a landing gear unit undergoing certification dynamic tests could match a design, but it will not necessarily be in compliance with the applicable certification standards. Furthermore, even if the tests are successful, the part could be damaged by tests and then eventually be no more airworthy.

NOTE: Some JAA forms such as JAA Form 1 may continue to be used by non-EU member countries.

NOTE: Appendix I of the EASA Part 21 provides instructions related to the use of EASA Form 1 for manufacturing purposes.

[6] See Note 5.

    (iii) Maintain a new aircraft produced by the organization and issue a certificate of release to service (EASA Form 53) in respect of that maintenance.

    (iv) Under procedures agreed with its competent authority for production, for an aircraft it has produced and when the production organization itself is controlling under its POA the configuration of the aircraft and is attesting conformity with the design conditions approved for the flight, issue a permit to fly in accordance with 21A.711(c) including approval of the flight conditions in accordance with 21A.710(b).[7]

(3) Exposition. The organization must supply a **Production Organization Exposition** (POE) (Part 21A.143), a document similar to the DOA Handbook we have previously mentioned. The document provides a general description of the organization and its scope of work, titles, and names of managers with their duties and responsibilities, a list of certifying staff, a description of the Quality System, inherent procedures, and so on.

The Competent Authority requires the POE to be an accurate definition and description of the production organization. The document does not require approval in itself, but it will be considered as such by virtue of the approval of the organization.

NOTE: There are plenty of information for the "Subpart G—POA for products, parts, and appliances" in the EASA AMC and GM to Part 21.

General remarks. We can make the same considerations for POA that we made for DOA in Chapter 5. Also, in POA, there is a true leap of quality, leading to a condition of self-control for the organization, with advantages for safety and authority efficiency.

7.1.2. The EASA POA

The objective of this working procedure—a short summary is provided here—is to enable the Agency to process foreign Part-21 Subpart G approval applications and allocate internal/external resources as necessary to carry out the organization audit and finally issuance of a POA certificate following a satisfactory recommendation.

This procedure describes how EASA will internally handle the approval of production organizations located outside the territory of the Member States or on specific request from a Member state the approval of a production organization located inside the territory of that Member State.[8]

---

[7] This privilege comes from the amendment of EASA Part 21 of April 2007. For details on permit to fly, see Paragraph 8.4.3 of Chapter 8 and Notes 17 and 18 of Chapter 5.

[8] As mentioned in the "EASA certification" section in Chapter 3, EU product organizations are normally approved by the local competent authority.

7.1.2.1. ACCEPTANCE OF APPLICATION

Applications for an EASA POA shall be sent to the EASA Manager of Applications Certification (MAC) and made in accordance with Part 21 and its AMC and GM.

When the application is made by an organization located within an EU country, the application for POA to the Agency needs to be supported by a statement of the Competent Authority of the applicant that this Authority is requesting the Agency to handle the application.

7.1.2.2. ALLOCATION OF TECHNICAL INVESTIGATION TASKS

After eligibility has been fully assessed and once principle acceptance is given, the MAC will check with the responsible EASA Production Organizations Manager (POM), whether the application shall be further processed internally or the technical investigation should be allocated to an external party.

In cases where the technical investigation shall be performed internally, the POM will establish an appropriate EASA certification team using EASA staff and/or NAA staff under appropriate contractual arrangements.

In cases where the technical investigation shall be allocated to an external party that will handle the technical investigation on behalf of EASA, the selected external party may only be, by the time being, an NAA which is appropriately accredited and has appropriate contractual arrangements with EASA. In this case, a Production Oversight Coordinator (POC) of this Designated Authority (DA) will be appointed.

In the case that the technical investigation shall be performed internally, the POM will act as EASA POC.

7.1.2.3. DETERMINATION OF THE POA TEAM

The POC will nominate a team leader or members to carry out the investigation process. The composition and size of the basic investigation team can consist of only the team leader but may vary and is dependent on the features of the Organization.

Trainees may participate in investigation teams at no direct cost to the applicant.

7.1.2.4. INVESTIGATION FOR INITIAL ORGANIZATION APPROVAL

The investigation process will be performed according to Section B of Part 21 and its associated AMC/GM and the EASA procedure.

When the full investigation for compliance of the applicant with Part 21 has been satisfactorily determined, the POC shall carry out a quality review of the pertinent documentation.

The POC/DA shall verify that the continued surveillance plan covers all elements required by 21B.235.

The POC/DA shall notify the POM of any major delays, serious problems or rejection of key staff members of the applicant during the investigation process.

7.1.2.5. ISSUING THE ORGANIZATION APPROVAL CERTIFICATE

The POC/DA shall forward to the POM the proposal for the EASA approval certificate, and the current accepted continued surveillance plan.

When satisfied with the above recommendation package, the POM shall prepare and sign the EASA approval certificate.

7.1.3. Production without POA

We previously mentioned cases where Subpart F of JAR 21/EASA Part 21 applies, which we can now summarize:

(1) The authority considers production approval under Subpart G inappropriate.

(2) Production is initiated under Subpart F in advance of issue of a POA under Subpart G.

Applicants may apply showing conformity of individual products, parts, or appliances under Subpart F, if they hold or have applied for an approval covering the design of that product, part, or appliance, or (as for POA) have ensured satisfactory coordination between production and design, through an appropriate arrangement with the applicant for, or holder of, an approval of such a design.

In the EASA AMC and GM for PART 21, AMC No. 1 and No. 2 to Paragraph 21A.122 explain what is a suitable "arrangement"; furthermore, an important number of AMC or GM help the applicant *"for demonstrating the conformity with the applicable design data of a product, part and appliance that is intended to be manufactured without a production organization approval under Subpart G."*

7.1.3.1. THE ORGANIZATION

Without going into details that can be found in Subpart F and inherent advisory material, the following are required:

(1) A Production Inspection System.

(2) An **Organization Manual** that describes the production inspection system required, ensuring that each product, part, or appliance conforms to the applicable design data and is in condition for safe operation. This means that procedures must be established, for example, for control of incoming materials (and bought or subcontracted parts), processes, manufacturing techniques, design changes (including material substitutions), and so on. Furthermore, it must contain a general description of the organization.

In this organization, we find the same basic concepts defined by POA. The Product Inspection System is the equivalent of the POA Quality System. The Organization Manual contains items bearing similarity with those provided by the POA Exposition. (In any case, we do not believe that an aircraft built under Subpart F could be less safe than one built under Subpart G.)

What is then the difference between the two types of organization?

The true difference is the presence, in the POA, of the Independent Quality Assurance System, which, through Quality System monitoring, has the responsibility of making the organization truly reliable, independent of authority intervention.

In the production organization without POA, this monitoring task pertains to the authority, which has to perform control quite different compared with that performed with POA.

Because in many cases (but not always), Subpart F relates to small organizations and simple products, the procedures can be conveniently simplified. It is therefore clear why POA privileges are not granted to these organizations.

In conclusion, with or without POA, the right balance must be found to ensure that the production responds to the safety concepts acquired in the type certifications and approvals of products, parts, and appliances.

7.2. PRODUCTION UNDER FAR 21

FAR 21 also provides two alternatives for production:
(**1**) A production certificate, under Subpart G.
(**2**) Production under type certificate only, under Subpart F.

7.2.1. The production certificate

7.2.1.1. APPLICABILITY

According to Subpart G of FAR 21, a type-certificate holder or private individuals holding the right to benefit from that type certificate under a licensing agreement, or a Supplemental type-certificate holder, may apply for a production certificate for the product concerned.

7.2.1.2. PRIVILEGES

A PC holder has the privileges specified in FAR 21.163. In addition, a PC holder is eligible to have a qualified employee(s) designated as Designated Manufacturing Inspection Representative (DMIR). The PC holder may also be authorized to represent the Administrator as an Organizational Designated Airworthiness Representative (ODAR). Among the above-mentioned privileges, the PC holder can

(**a**) Obtain an aircraft airworthiness certificate without further showing, except that the Administrator may inspect the aircraft for conformity with the type design.

(**b**) In the case of other products, obtain approval for installation on type-certificated aircraft.

As we have seen for the JAA/EASA POA, the privileges tend to release the manufacturer from strict Administrator control.

To obtain such privileges, manufacturers must show that they have established and can maintain a **Quality Control System** for any product, for which they request a production certificate, so that each article will meet the design provisions of the pertinent certificate.

7.2.1.3. QUALITY CONTROL SYSTEM[9]

Paragraph 21.143 prescribes a list of data to be submitted to the Administrator, describing the inspection and test procedures necessary to ensure that each article produced conforms to the type design and is in condition for safe operation. In particular, what is required is

(a) A statement describing assigned responsibilities and delegated authority of the quality control organization, together with a chart indicating the functional relationship of the quality control organization to management and to other organizational components, and indicating the chain of authority and responsibility within the quality control organization.

(b) A description of inspection procedures for raw materials, purchased items, and parts and assemblies produced by manufacturers' suppliers, including methods used to ensure acceptable quality of parts and assemblies that cannot be completely inspected for conformity and quality when delivered to the prime manufacturer's plant.

(c) A description of the methods used for production inspection of individual parts and complete assemblies, including the identification of any special manufacturing processes involved, the means used to control the processes, the final test procedure for the complete product, and, in the case of aircraft, a copy of the manufacturer's production flight test procedures and check-off list.

(d) An outline of the materials review system, including the procedure for recording review board decisions and disposing of rejected parts.

(e) An outline of a system for informing company inspectors of current changes in engineering drawings, specifications, and quality control procedures.

(f) A list or chart showing the location and type of inspection stations.

7.2.1.4. PROCESSING AN APPLICATION FOR A PC

The application, made on the relevant FAA form, is submitted to the manager of the competent Manufacturing Inspection Office in the directorate in which the applicant's principal manufacturing facility is located.

After a preliminary audit, a team is selected to make the suitable evaluations.

Of course, the FAA provides guidance documents for the development of this process, such as Order 8120.2E and AC 21-1.

---

[9] **Quality System.** A documented organizational structure containing responsibilities, procedures, processes, and resources that implement a management function to determine and enforce quality principles. A quality system encompasses quality assurance and quality control.

(1) **Quality Assurance.** A management system for programming and coordinating the quality maintenance and improvement efforts of the various groups in a design and/or manufacturing organization, so as to permit design and/or production in compliance with regulatory and customer requirements.

(2) **Quality Control.** Conducting and directing supervision of the quality tasks (inspection of product) to ensure the quality requirements of the product are achieved.

7.2.1.5. PERIODIC FAA PRODUCTION FLIGHT TESTS

FAA production flight tests will be conducted periodically at the PC holder's facility to ensure continued compliance with all parameters as specified in pertinent type-certificate data with respect to performance, flight characteristics, operation qualities, equipment operations, and so on.

7.2.1.6. PC HOLDER'S RESPONSIBILITY

The PC holder is responsible for maintaining the Quality Control System in conformity with the data and procedure approved for the PC, and/or determining that each completed product submitted for airworthiness certification or approval conforms to the TC or STC and is in condition for safe operation.

7.2.2. Production under type certificate only

7.2.2.1. APPLICABILITY

According to Subpart F of FAR 21.123, each manufacturer of a product being manufactured under a type certificate only shall

(a) Make each product available for inspection by the Administrator.
(b) Maintain at the place of manufacture the technical data and drawings necessary for the Administrator to determine whether the product and its parts conform to the type design.
(c) Except as otherwise authorized by the Aircraft Certification Directorate Manager for the geographic area in which the manufacturer is located, for products manufactured more than 6 months after the date of issue of the type certificate, establish and maintain an approved production inspection system (APIS) that insures that each product conforms to the type design and is in condition for safe operation.
(d) On the establishment of the APIS [as required by paragraph (c) of this section] submits a manual to the Administrator, which describes that system and the means for making the determinations required by Paragraph 21.125(b).

Detailed information about the application of Subpart F can be found in AC 21-6A and Order 8120.2E.

7.2.2.2. PRIVILEGES

A manufacturer of a product or part(s) in accordance with Subpart F of FAR 21 is not granted any privileges.

However, on establishment of an APIS, the APIS holder is eligible to have a qualified employee(s) as DMIR. The APIS holder may also be authorized to represent the Administrator as an ODAR.

To better understand the matter, a manufacturer who has been issued a type certificate is given 6 months under FAR 21.123(c) to establish and implement a production inspection system, unless the manufacturer has applied for a production certificate under FAR 21 Subpart G. During the 6-month period, each complete product or part thereof is subjected to FAA inspection prior to the issuance of airworthiness certificates. This procedure is normally time consuming and is likely to allow only a very slow production rate. Therefore, it is to the manufacturer's advantage to develop and implement an approvable production inspection system as quickly as possible. As the manufacturer's individual fabrication, assembly, and inspection operations are found to be in compliance with the regulations, they may be FAA approved on a progressive basis. When areas are found to be in compliance, the FAA may thereafter reduce its inspection and increase its reliance on the manufacturer's production inspection system. When the total production inspection system is found to be in compliance with the regulations, the established ACO will issue the letter of Approval of the Production Inspection System (APIS). Subsequent FAA inspections will be for the purpose of surveillance of the approved system to determine continued compliance.

7.2.2.3. PRODUCTION INSPECTION SYSTEM: MATERIAL REVIEW BOARD

An effective Material Review Board is of primary importance for an efficient Production Inspection System, because it controls the inspections, identification, rework, and use of damaged or nonconforming articles, including the isolation or scrapping of unusable articles.

An APIS is based on compliance with the inspection standards specified in FAR 21.125. The APIS holder is required to establish a Material Review Board (to include representatives from the inspection and engineering department). He or she is also required to have process specifications, Material Review Board records, test procedures, and flight check forms that are acceptable to the FAA. It would be advantageous to the TC applicant to develop these data concurrently with the manufacture, inspection, and testing of prototypes of the product.

7.2.2.4. TC HOLDER'S RESPONSIBILITY

Prior to the issuance of an APIS, a TC holder or licensee who makes a product is particularly responsible for complying with Paragraphs 21.123, 21.127 (Tests: aircraft), 21.128 (Tests: aircraft engines), 21.129 (Tests: propellers), and 21.130 (Statement of conformity), as appropriate for the particular product concerned.

7.2.2.5. STATEMENT OF CONFORMITY

On receipt of the statement of conformity (21.130, the FAA will inspect the completed product to determine that it conforms to the type design and is in condition for safe operation. If so, an airworthiness certificate will be issued for an aircraft, or an Airworthiness Approval Tag (FAA Form 8130-3) will be issued for an engine or propeller.

7.2.2.6. DESIGNATED MANUFACTURING INSPECTION REPRESENTATIVES

Following the approval of the production inspection system, the manufacturer may obtain the appointment of individuals to be employed as DMIR for the purpose of issuing airworthiness certificates and/or airworthiness approval tags.

Chapter eight

Certificates of Airworthiness

8.1. INTRODUCTION

In this chapter, we describe the basic requirements governing the certificates of airworthiness—airworthiness certificates for the FAA. However, we will not report these requirements in their entirety. Therefore, to find practical applications of these requirements, the reader will have to refer directly to JAR/FAR 21/EASA Part 21, other standards cited in the requirements, and finally, relevant advisory material.

To facilitate the import and export of aircraft, and to facilitate operations of aircraft in international air navigation, Article 33 of the ICAO places the burden on the State of Registry to recognize and render valid an airworthiness certificate issued by another Contracting State, subject to the condition that the airworthiness requirements under which such a certificate is issued or rendered valid are equal to or above the minimum standards of ICAO Annex 8.

Special certificates of airworthiness—with some exceptions, such as the Restricted certificate and other certificates that require type certification—can be defined in the same way as the EASA's permits to fly, that is, issued to aircraft that do not meet or have not been shown to meet, applicable certification specification but are capable of safe flight under defined conditions.

Recalling what was mentioned in Chapter 5, the type certificate is not an authorization for aircraft operation, which is obtained when a certificate of airworthiness is issued.

With regard to the duration, as a general rule, unless suspended or revoked sooner, or a termination date is otherwise established by the authority, a certificate of airworthiness is effective within any period specified therein, as long as maintenance is performed in accordance with the applicable requirements, and provided the aircraft remains in the same register. A certificate of airworthiness is invalid when the type certificate under which it is issued is suspended or revoked by the authority.

To be considered "airworthy" and eligible for issuance of an airworthiness certificate, a **type-certificated** aircraft must meet two conditions:

(a) The aircraft must conform to its TC. Conformity to type design is considered attained when the aircraft configuration and the components installed are consistent with the drawings, specifications, and other data that are part

173

of the TC, which includes any supplemental type certificate (STC) and field-approved alterations incorporated in the aircraft.

(b) The aircraft must be in a condition for safe operation. This refers to the condition of the aircraft relative to wear and deterioration, for example, skin corrosion, window delaminating/crazing, fluid leaks, and tire wear.

NOTE: If one or both these conditions are not met, the aircraft would be considered unairworthy.

8.2. GENERAL CLASSIFICATION

The general classification of the certificates is provided by JAR 21, EASA Part 21, and FAR 21.

8.2.1. JAR 21 (Amendment 5)[1] certificates of airworthiness

(1) Subpart H provides requirements for the issue of **Standard certificates of airworthiness**.

(2) Subpart L provides requirements for **Export Airworthiness Approval**.

8.2.2. EASA Part 21 certificates of airworthiness

NOTE: See Appendix 8.4.

Subpart H provides requirements for the following:

(1) Certificates of airworthiness issued to aircraft in accordance with Part 21.

(2) Restricted certificates of airworthiness.

(3) Permits to fly.

8.2.3. FAR 21 airworthiness certificates

NOTE: See Appendix 8.5.

(1) Subpart H provides requirements for the following:

  (a) Standard airworthiness certificates—airworthiness certificates issued for aircraft type-certificated in one of the normal, utility, acrobatic, commuter, or transport categories, and for manned free balloons, and for aircraft designated by the Administrator as *special classes of aircraft*.[2]

  (b) Special airworthiness certificates—Primary, Restricted, Limited, Light-Sport, and Provisional airworthiness certificates, special flight permits, and Experimental certificates.

(2) Subpart I provides requirements for **Provisional airworthiness certificates**.

(3) Subpart L provides requirements for **Export Airworthiness Approvals**.

---

[1] See Note 49 of Chapter 4 and Section 4.8.

[2] Special classes of aircraft: see "FAR 21 Standard airworthiness certificates" section in this chapter.

8.3. JAR 21 (AMENDMENT 5)[1] CERTIFICATES OF AIRWORTHINESS

8.3.1. Standard certificates of airworthiness

Standard certificates of airworthiness are issued for aircraft for which a type certificate has been issued in accordance with JAR 21.

8.3.1.1. ISSUE OF A CERTIFICATE

(1) Any owner (or the agent of the owner) may apply for a certificate of airworthiness.

(2) A Standard certificate of airworthiness can be issued for new or used aircraft without prejudice to other provisions of national laws, applicable in the absence of a comprehensive set of JAA rules, on presentation to the competent authority of the relevant documentation required by JAR 21.

(3) In particular, for used aircraft, historical records to establish the production, modifications, and maintenance standards of the aircraft must be submitted.

The wording "without prejudice … etc." is used to take into account the fact that the JARs do not cover issues such as environmental certification procedures and others that may interfere with the certification procedures of JAR 21. This wording should not be interpreted as having the potential for additional requirements deviating from JARs, but only for additional national administrative requirements for subjects that are not otherwise addressed by the JARs.

As explained in Chapter 3, succeeding a joint type certification, the type certificate was issued by the national authorities on the basis of the JAA's recommendations.

The certificate of airworthiness was issued by the national authorities for aircraft that conformed to a type design approved under a type certificate, and with the applicable national rules for operations and environmental protection.

8.3.2. Export airworthiness approval

8.3.2.1. TYPES OF APPROVAL

(1) Export Airworthiness Approval of complete aircraft issued in the form of **Export certificate of airworthiness**. Such certificates do not authorize the operation of the aircraft.

(2) Export Airworthiness Approval of other products, parts (except standard parts), or appliances issued in the form of an **Authorized Release Certificate (JAA Form 1)**, in accordance with applicable JARs.

8.3.2.2. APPLICATION FOR AN EXPORT CERTIFICATE OF AIRWORTHINESS

The manufacturer or owner (or its agent) of a new aircraft, or the owner (or its agent) of a used aircraft may apply for an Export certificate of airworthiness on presentation to the competent authority of the relevant documentation required by JAR 21.

In particular, for used aircraft, historical records to establish the production, modification, and maintenance standards of the aircraft must be submitted.

8.3.2.3. ISSUE OF EXPORT CERTIFICATE OF AIRWORTHINESS

The certificate is issued if the applicant shows that

(1) The aircraft conforms to the type design acceptable to the importing country.[3]

(2) New aircraft have been produced under Subpart F or G of JAR 21.[4]

(3) Used aircraft possess or qualify for a valid certificate of airworthiness issued by the exporting authority.

(4) The aircraft meets the additional requirements for import of the importing country.

(5) All documents prescribed by JAR 21 have been submitted.

8.3.2.4. EXPORT APPROVAL EXCEPTIONS

Export approvals can be issued for aircraft, parts, or appliances that do not meet all the requirements prescribed for the issue of an Export certificate of airworthiness or a JAA Form 1, if the importing authority provides a written statement of acceptability.

In these cases, the requirements that are not met and the difference in configuration, if any, between the product, part, or appliance to be exported and the related type-approved product, part, or appliance must be listed on the Export Airworthiness Approval as exceptions.

For example, it is possible to obtain Export certificates of airworthiness for damaged aircraft, or aircraft to be completed in the importing country, when the "status" of the aircraft is clearly defined.

8.4. EASA PART 21 CERTIFICATES OF AIRWORTHINESS

8.4.1. Certificates of airworthiness issued to aircraft in accordance with Part 21

8.4.1.1. APPLICABILITY

The certificates of airworthiness will be issued to aircraft that conform to a type certificate that has been issued in accordance with EASA Part 21.[5]

8.4.1.2. APPLICATION

Each application shall include the following:

(1) For a new aircraft, a statement of conformity issued by the manufacturer under POA privileges or validated by the competent authority,[6] a weight and balance report, and the flight manual.

---

[3] For many years, this concept has often been misleading. In fact, some authorities used to *require* a statement of conformity to **their own** type certificate for imported aircraft, and to *issue* a statement of conformity (also) to **their own** type certificate for exported aircraft, creating an unbalanced situation. The latest bilateral agreements have solved this problem.

[4] See Chapter 7.

[5] The definition is equivalent to the definition of the Standard certificate of airworthiness (JAR 21) or the Standard airworthiness certificate (FAR 21).

[6] The modalities of this statement are prescribed in Paragraph 21A.174.

(2) For used aircraft originating from a Member State, airworthiness review certificate (EASA Form 15a).

(3) For used aircraft originating from a non-Member State:

(a) A statement by the competent authority of the state where the aircraft is, or was registered, reflecting the airworthiness status of the aircraft on its register at the time of transfer.

(b) A weight and balance report.

(c) The flight manual.

(d) Historical records to establish the production, modification, and maintenance standards of the aircraft.

(e) A recommendation for the issuance of a certificate of airworthiness and an airworthiness review certificate.

8.4.1.3. ISSUE OF CERTIFICATE OF AIRWORTHINESS

The competent authority of the state of registry shall issue a certificate of airworthiness for

(1) *New aircraft*, on presentation of the documentation required by 21A.174(b) 2, when the aircraft conforms to an approved design and is in condition for safe operations. This may include inspection by the competent authority of the Member State of registry.

(2) *Used aircraft*, on presentation of the documentation required by 21A.174(b) 3 demonstrating that the aircraft conforms to a type design approved under a type certificate and any STC, change, or repair approved in accordance with EASA Part 21, and to applicable Airworthiness Directives, and the aircraft has been inspected in accordance with the applicable provision.

8.4.2. Restricted certificates of airworthiness

Restricted certificates of airworthiness shall be issued to aircraft, which conform to a type certificate that has been issued in accordance with EASA Part 21, or which has been shown to the Agency to comply with specific airworthiness specifications ensuring adequate safety.

8.4.2.1. DEFINITION OF A RESTRICTED TYPE CERTIFICATE

For an aircraft that does not meet the provisions of 21A.21(c),[7] the applicant shall be entitled to have a Restricted type certificate issued by the Agency after:

(1) Complying with the appropriate type certification basis established by the Agency ensuring adequate safety with regard to the intended use of the aircraft, and with the applicable environmental protection requirements.

(2) Expressly stating that it is prepared to comply with 21A.44.[8]

(3) Furthermore, the engine or propeller, or both, installed in the aircraft shall have a type certificate or have been shown to be in compliance with the airworthiness specifications ensuring adequate safety.

---

[7] "Standard" type certificate.
[8] Obligations of the holder.

8.4.2.2. APPLICATION
As per "Standard" certificates.

8.4.2.3. ISSUE OF RESTRICTED CERTIFICATE OF AIRWORTHINESS
The competent authority of the Member State of registry shall issue a Restricted certificate of airworthiness for:

(1) *New aircraft*, on presentation of the documentation required by 21A.174(b)2, demonstrating that the aircraft conforms to a design approved by the Agency under a Restricted type certificate or in accordance with specific airworthiness specifications and in conditions of safe operation.

(2) *Used aircraft*, on presentation of the documentation required by 21A.174(b) 3 demonstrating that the aircraft conforms to a design approved by the Agency under a Restricted type certificate or in accordance with certification specifications, and the applicable Airworthiness Directives have been complied with, and the aircraft has been inspected in accordance with the applicable provision.

8.4.3. Permits to fly

Permits to fly shall be issued in accordance with Subpart P to aircraft that do not meet, or have not been shown to meet, applicable certification specifications but are capable of safe flight under defined conditions and for the following purposes:

(1) development;

(2) showing compliance with regulations or certification specifications;

(3) design organizations or production organizations crew training;

(4) production flight testing of new production aircraft;

(5) flying aircraft under production between production facilities;

(6) flying the aircraft for customer acceptance;

(7) delivering or exporting the aircraft;

(8) flying the aircraft for Authority acceptance;

(9) market survey, including customer's crew training;

(10) exhibition and air show;

(11) flying the aircraft to a location where maintenance or airworthiness review are to be performed, or to a place of storage;

(12) flying an aircraft at a weight in excess of its maximum-certificated takeoff weight for flight beyond the normal range over water, or over land areas where adequate landing facilities or appropriate fuel is not available;

(13) record breaking, air racing, or similar competition;

(14) flying aircraft meeting the applicable airworthiness requirements before conformity to the environmental requirements has been found; and

(15) for noncommercial flying activity on individual noncomplex aircraft or types for which a certificate of airworthiness or restricted certificate of airworthiness is not appropriate.

8.4.3.1. APPLICATION

If the applicant has not been granted the privilege to issue a permit to fly,[9] an application for a permit to fly shall be made to the Competent Authority[10] in a form and manner established by that authority and shall include the following:

(1) the purpose(s) of the flight(s), in accordance with the purpose of the permit to fly;

(2) the ways in which the aircraft does not comply with the applicable airworthiness requirements;

(3) the flight conditions approved by the Agency or an appropriately approved design organization, under its privileges.

Flight conditions include the following:

(a) the configuration(s) for which the permit to fly is requested;

(b) any condition or restriction necessary for safe operation of the aircraft. If the applicant has not been granted the privilege to approve the flight conditions, an application for approval of the flight conditions shall be made

(1) when approval of the flight conditions is related to the safety of the design, to the Agency in a form and manner established by the Agency or

(2) when approval of the flight conditions is not related to the safety of the design, to the competent Authority in a form and manner established by that authority.

8.4.3.2. ISSUE OF PERMITS TO FLY

(a) The Competent Authority shall issue a permit to fly

(1) on presentation of the data required (see Section 8.4.3.1);

(2) when the flight conditions have been approved (see Section 8.4.3.1); and

(3) when the Competent Authority, through its own investigations, which may include inspections, or through procedures agreed with the applicant, is satisfied that the aircraft conforms before flight to the design of the configuration for which the permit to fly is requested.

(b) An appropriately approved design organization may issue a permit to fly (EASA Form 20b) under the privilege granted under 21A.263(c)(7), when the flight conditions have been approved (see Section 8.4.3.1).

(c) An appropriately approved production organization may issue a permit to fly (EASA Form 20b) under the privilege granted under 21A.163(e), when the flight conditions have been approved (see Section 8.4.3.1).

NOTE

(a) Any **change** that invalidates the flight conditions or associated substantiation established for the permit to fly shall be approved (see Section 8.4.3.1).

(b) A change affecting the content of the permit to fly requires the issuance of a new permit to fly.

---

[9] See Note 17 of Chapter 5.

[10] For the purpose of Subpart P, the "Competent Authority" shall be (a) the authority designated by the Member State of registry or (b) for unregistered aircraft, the authority designated by the Member State that prescribed the identification marks.

The Amendment of April 2007 of "AMC and GM to Part 21" provides plenty of information about the procedure for the issuing of a permit to fly.

Figures 8.1 and 8.2 give a comprehensive idea of the approval of flight conditions and the issue of permit to fly.

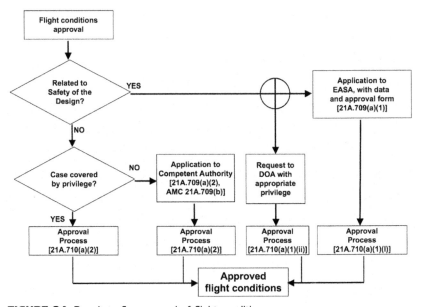

FIGURE 8.1 Permit to fly: approval of flight conditions

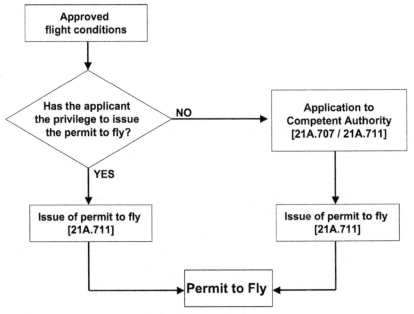

FIGURE 8.2 Issue of Permit to fly

8.4.3.3. EXAMPLES OF PERMIT TO FLY

A certificate of airworthiness or restricted category certificate of airworthiness may not be appropriate for an individual aircraft or aircraft type when it is not practicable to comply with the normal continued airworthiness requirements and the aircraft is to a design standard that is demonstrated to be capable of safe flight under defined conditions.

The EASA GM 21A.701(a) provides a list (not exhaustive) of cases in which the issuance of a permit to fly (instead of a certificate of airworthiness or restricted certificate of airworthiness) is appropriate.

(1) Development:
- Testing of new aircraft or modifications;
- Testing of new concepts of airframe, engine propeller, and equipment; and
- Testing of new operating techniques;

(2) Showing compliance with regulations or certification specifications:
- Certification flight testing for type certification, STCs, changes to type certificates or European Technical Standard Order authorization;

(3) Design organizations or production organizations crew training:
- Flights for training of crew that will perform design or production flight testing before the design approval and Certificate of Airworthiness (C of A) can be issued.

(4) Production flight testing of new production aircraft:
- For establishing conformity with the approved design, typically this would be the same program for a number of similar aircraft;

(5) Flying aircraft under production between production facilities:
- Green aircraft ferry to follow on final production.

(6) Flying the aircraft for customer acceptance:
- Before the aircraft is sold and/or registered.

(7) Delivering or exporting the aircraft:
- Before the aircraft is registered in the State where the C of A will be issued.

(8) Flying the aircraft for Authority acceptance:
- In the case of inspection flight test by the authority before the C of A is issued.

(9) Market survey, including customer's crew training:
- Flights for the purpose of conducting market survey, sales demonstrations, and customer crew training with non-type-certificated aircraft or aircraft for which conformity has not yet been established or for nonregistered a/c and before the Certificate of Airworthiness is issued.

(10) Exhibition and air show:
- Flying the aircraft to an exhibition or show and participating in the exhibition or show before the design approval is issued or before conformity with the approved design has been shown.

(11) Flying the aircraft to a location where maintenance or airworthiness review are to be performed, or to a place of storage:
- Ferry flights in cases where maintenance is not performed in accordance with approved programs, where an AD has not been complied with where certain equipment outside the Minimum Equipment List (MEL) is

unserviceable or when the aircraft has sustained damage beyond the applicable limits.

(12) **Flying an aircraft at a weight in excess of its maximum-certificated takeoff weight for flight beyond the normal range over water, or over land areas where adequate landing facilities or appropriate fuel is not available:**
   − Oversees ferry flights with additional fuel capacity.

(13) **Record breaking, air racing, or similar competition:**
   − Training flight and positioning flight for this purpose are included.

(14) **Flying aircraft meeting the applicable airworthiness requirements before conformity to the environmental requirements has been found:**
   − Flying an aircraft that has been shown to comply with all applicable airworthiness requirements but not with environmental requirements.

(15) **For noncommercial flying activity on individual noncomplex aircraft or types for which a certificate of airworthiness or restricted certificate of airworthiness is not appropriate.**
   − For aircraft that cannot practically meet all applicable airworthiness requirements, such as certain aircraft without TC-holder ("generically termed orphan aircraft") or aircraft that have been under national systems of Permit to Fly and have not been shown to meet all applicable requirements. The option of a permit to fly for such an aircraft should only be used if a certificate of airworthiness or restricted certificate of airworthiness cannot be issued due to conditions that are outside the direct control of the aircraft owner, such as the absence of properly certified spare parts.

NOTE: The above listing is of cases when a permit to fly MAY be issued; it does not mean that in the described cases a permit to fly MUST be issued. If other legal means are available to allow the intended flight(s), they can also be used.

8.4.4. General remarks on EASA certificates of airworthiness

We previously mentioned that the issue of EASA certificates of airworthiness is made by the competent authority of the state of registry.

Referring again to the EASA classification of the certificates of airworthiness, the permits to fly have replaced all Special certificates or other airworthiness certificates issued for aircraft that did not hold a certificate of airworthiness or restricted certificate of airworthiness issued under EASA Part 21.

The Special certificates already issued by the Member States will be gradually reissued as permits to fly. The amendment of Part 21 of 30 March 2007 (as Annex to Commission Regulation EC No. 1702/2003) has filled a gap in the current regulation. As a matter of fact, the Member States did not have common rules for the issue of special certificates or similar documents. Today, the publication of a new Subpart P "Permit to fly" and relevant amendments of other Subparts, settle these common rules, as summarized in Section 8.4.3.

The Member States can adapt the different Permits to fly to local requirements; they can issue internal circulars, but always in compliance with Part

21 and with the direct or indirect (through DOA) involvement of the Agency for approval of the flight conditions related to the safety of design.

8.5. FAR 21 AIRWORTHINESS CERTIFICATES

8.5.1. Standard airworthiness certificates

8.5.1.1. APPLICABILITY

Standard airworthiness certificates are airworthiness certificates issued for aircraft type-certificated in one of the normal, utility, acrobatic, commuter, or transport categories; for manned free balloons; and for aircraft designated by the Administrator as special classes of aircraft.[11]

8.5.1.2. ISSUE OF STANDARD AIRWORTHINESS CERTIFICATES

(1) *New aircraft manufactured under a production certificate (PC).* An applicant for a Standard airworthiness certificate for a new aircraft manufactured under a PC is entitled to a Standard airworthiness certificate without further showing, except that the Administrator may inspect the aircraft to determine conformity to the type design and condition for safe operation.

(2) *New aircraft manufactured under type certificate only.* An applicant for a Standard airworthiness certificate for a new aircraft manufactured under a type certificate only is entitled to a Standard airworthiness certificate on presentation, by the holder or licensee of the type certificate, of the statement of conformity prescribed in Paragraph 21.130 if the Administrator finds after inspection that the aircraft conforms to the type design and is in condition for safe operation.

(3) *Import aircraft.* An applicant for a Standard airworthiness certificate for an import aircraft type-certificated in accordance with Paragraph 21.29[12] is entitled to an airworthiness certificate if the country in which the aircraft was manufactured certifies, and the Administrator finds, that the aircraft conforms to the type design and is in condition for safe operation.

Paragraph 21.183 also prescribes requirements for noise, passenger emergency exits, fuel venting, and exhaust emissions, when applicable.

8.5.2. Special airworthiness certificates

At first sight, we could question why the FAA has so many types of Special airworthiness certificates.

The answer, as already mentioned, is the necessity to solve the multitude of different problems linked with day-to-day operation, on the basis of different written rules established for the different cases.

---

[11] Special classes of aircraft include gliders (sailplanes for the Europeans) and powered gliders, airships, and other kinds of aircraft, which would be eligible for a Standard airworthiness certificate, but for which no FAA airworthiness standards have yet been established.

[12] 21.29 Issue of type certificate for import products manufactured in a foreign country with which the United States has an agreement for the acceptance of these products for export and import.

It is essential to recall what was written at the beginning of this chapter about the content of this panoramic exposition of the rules governing the certificates of airworthiness; readers will not find a copy of FAR 21 paragraphs, but only the basics of these requirements are used to explain their meaning. Therefore, readers are advised that, for practical applications of the requirements, they will have to refer to FAR 21, other FARs cited by the requirements, and finally, advisory material (ACs, Orders, etc.).

It is worth quoting in particular the FAA Order 8130-2F establishing procedures for accomplishing original and recurrent airworthiness certification of aircraft and related products.

It is also important to know that most national authorities have had national regulations for aircraft certifications for a long time, based on the FAA's regulations, so that many of their certificates of airworthiness bear similarity with the corresponding FAA certificates. This implies that some considerations at the end of the description of the single certificate are not only applicable to US aircraft but can also assume more general value.

Another practical consideration: when a certificate of airworthiness for a type-certificated aircraft is issued, safety is assured by the conformity of the single aircraft to the relevant type certificate and applicable operational requirements.

Because most of the Special certificates are not based on the existence of a type certificate, the issue of such a certificate of airworthiness should assure a "sufficient level of safety"; this statement involves evaluations not only of the physical conditions of the single aircraft but also of its design.

These types of evaluations normally require experience, skill, and common sense.

NOTE: According to FAR 21.187 *"Issue of multiple airworthiness certification,"* an applicant for an airworthiness certificate in the restricted category, and in one or more other categories except primary category, is entitled to the certificate, if

(1) He shows compliance with the requirements for each category, when the aircraft is in the configuration for that category; and

(2) He shows that the aircraft can be converted from one category to another by removing or adding equipment by simple mechanical means.

8.5.2.1. SPECIAL AIRWORTHINESS CERTIFICATE FOR PRIMARY CATEGORY AIRCRAFT

(1) Definition of Primary category aircraft

An aircraft type-certificated in the Primary category:

(a) Is unpowered or it is an airplane powered by a single, naturally aspirated engine with a 61-knot or less V_{so} stall speed as defined in Paragraph 23.49, or a rotorcraft with a 6 lb/ft^2 main rotor disk loading limitation, under sea level standard day conditions.

(b) Weighs not more than 2700 lb or, for seaplanes, not more than 3375 lb.

(c) Has a maximum seating capacity of not more than four persons, including the pilot.

(d) Has an unpressurized cabin.

An applicant may include a special inspection and preventive maintenance program as a part of the aircraft's type design or Supplemental type design.

(2) Issue of an airworthiness certificate

(a) *New Primary category aircraft manufactured under a PC.* An applicant for an original, Special airworthiness certificate—Primary category for a new aircraft manufactured under a PC, including aircraft assembled by another person from a kit provided by the holder of the PC and under the supervision and quality control of that holder, is entitled to a Special airworthiness certificate without further documentation required, except that the Administrator may inspect the aircraft to determine conformity to the type design and condition for safe operation.

(b) *Imported aircraft.* An applicant for a Special airworthiness certificate—Primary category for an imported aircraft type-certificated under Paragraph 21.29[13] is entitled to a Special airworthiness certificate if the civil airworthiness authority of the country in which the aircraft was manufactured certifies, and the Administrator finds, after inspection, that the aircraft conforms to an approved type design that meets the applicable criteria.

(3) General remarks

Aircraft in this category are of simple design and intended exclusively for recreation and personal use. Although these aircraft may be available for rental and flight instruction under certain conditions, the transport of persons or property for hire is prohibited.

One benefit of the certification in this category is the possibility for the pilot/owner to perform preventive maintenance beyond what is already allowed under Appendix A of FAR 43. Of course, there are rules that the pilot/owner has to follow to be considered properly qualified.

FAR 21.184(c) allows an applicant to exchange a standard airworthiness certificate for a special airworthiness certificate in the primary category. The conversion will be made through the normal STC process. The only benefit for making a conversion is so the pilot/owner may perform preventive maintenance beyond what already is allowed under Appendix A to part 43.

FAR 21.17(f)(l)[14] sets the Designation of Applicable Regulations.

The intent is to provide a means whereby private industry can develop airworthiness design standards for primary category aircraft and submit them to the FAA for approval. These private industries include, but are not limited to, associations such as the Experimental Aircraft Association (EAA), consensus standards developing groups such as the Society of Automotive Engineers, manufacturers, aircraft designers, and individuals.

---

[13] 21.29 Issue of type certificate: import products.

[14] For primary category aircraft, the requirements are the applicable airworthiness requirements contained in FAR 23, 27, 31, 33, and 35, or such other airworthiness criteria as the Administrator may find appropriate and applicable to the specific design and intended use and provide a level of safety acceptable to the Administrator.

8.5.2.2. SPECIAL AIRWORTHINESS CERTIFICATES FOR RESTRICTED CATEGORY AIRCRAFT

(1) Definition of Restricted category aircraft

An aircraft type-certificated in the Restricted category for special purposes:

(a) Meets the airworthiness requirements of an aircraft category *except those requirements that the Administrator finds inappropriate* for the special purpose for which the aircraft is to be used.

(b) Is of a type that has been manufactured in accordance with the requirements of, and accepted for use by, an Armed Force of the United States and has been later modified for a special purpose.

(c) "Special purpose operations" include the following:

  (i) Agricultural (spraying, dusting and seeding, and livestock and predatory animal control)

  (ii) Forest and wildlife conservation

  (iii) Aerial survey (photography, mapping, and oil and mineral exploration)

  (iv) Patrol (pipelines, power lines, and canals)

  (v) Weather control (cloud seeding)

  (vi) Aerial advertise (skywriting, banner towing, airborne signs, and public address systems)

  (vii) Any other operation specified by the Administrator.

(2) Issue of an airworthiness certificate

(a) *Aircraft manufactured under a PC or type certificate only.* An applicant for the original issue of a Restricted category airworthiness certificate for an aircraft type-certificated in the Restricted category, that was not previously type certificated in any other category, must comply with the appropriate provisions of Paragraph 21.183.[15]

(b) *Other aircraft.* An applicant for a Restricted category airworthiness certificate for an aircraft type-certificated in the Restricted category, that was either a surplus aircraft of the Armed Forces or previously type certificated in another category, is entitled to an airworthiness certificate if the aircraft has been inspected by the Administrator and found to be in a good state of preservation and repair and in a condition for safe operation.

(c) *Imported aircraft.* An applicant for the original issue of a Restricted category airworthiness certificate for an imported aircraft type certificated in the Restricted category only in accordance with FAR 21.29 is entitled to an airworthiness certificate if the country in which the aircraft was manufactured certifies, and the Administrator finds that the aircraft conforms to the type design and is in a condition for safe operation.

---

[15] 21.183 Issue of standard airworthiness certificates for normal, utility, acrobatic, commuter, and transport category aircraft, manned free balloons, and special classes of aircraft.

Paragraph 21.185 also prescribes requirements for noise, venting, and exhaust emissions, as applicable.

(3) General remarks

To better understand the meaning of this special aircraft category, we will take one example of many.

An aeroplane, already type certificated according to FAR 23, is provided with an agricultural spraying installation. The certificate may tolerate an increase in the maximum takeoff weight with consequent reduction in the rate of climb (also under the minimum allowed by FAR 23) due to the higher weight and also due to the drag increase caused by the external installation. Obviously, it should be demonstrated that the aircraft's flight qualities are still acceptable: a quick drain to cope with emergencies may be installed; airspace restrictions could be enforced; and so on. In other words, all the appropriate checks shall be carried out and limitations prescribed to take into account deviations from essential requirements for airworthiness depicted in the basic regulations.

8.5.2.3. SPECIAL AIRWORTHINESS CERTIFICATE FOR LIMITED CATEGORY AIRCRAFT

(1) Definition of Limited category aircraft

A limited category Special airworthiness certificate is issued to operate surplus military aircraft that have been converted to civilian use under the following conditions:

(a) The aircraft has a Limited type certificate.[16]

(b) The aircraft conforms to its type certificate.

(c) The FAA has determined that the aircraft is safe to operate.

(d) Operation may not include carrying passengers or cargo for hire. The FAA may prescribe additional limitations as necessary for safe operation.

(2) Issue of an airworthiness certificate

An applicant for an airworthiness certificate for an aircraft in the Limited category is entitled to the certificate when

(a) He or she can show that the aircraft has been previously issued a Limited category type certificate and that the aircraft conforms to that type certificate.

(b) The Administrator finds, after inspection (including a flight check by the applicant), that the aircraft is in a good state of preservation and repair and is in a condition for safe operation.

(c) The Administrator prescribes limitations and conditions necessary for safe operation.

---

[16] FAA Order 8130.2, "Airworthiness Certification of Aircraft and Related Products," contains a list of aircraft models that have been issued Limited category type certificates.

8.5.2.4. SPECIAL AIRWORTHINESS CERTIFICATE FOR A LIGHT-SPORT CATEGORY AIRCRAFT

(1) Definition

A **Light-Sport aircraft (LSA)** is an aircraft, other than a helicopter or powered lift that, since its original certification, has continued to meet the following criteria:

(a) A maximum takeoff weight of not more than
- (i) 660 lb (300 kg) for lighter-than-air aircraft;
- (ii) 1320 lb (600 kg) for aircraft not intended for operation on water; and
- (iii) 1430 lb (650 kg) for an aircraft intended for operation on water.

(b) A maximum airspeed in level flight with maximum continuous power (V_H) of not more than 120 knots CAS (Calibrated Air Speed) under standard atmospheric conditions at sea level.

(c) A maximum never-exceed speed (V_{NE}) of not more than 120 knots CAS for a glider.

(d) A maximum stalling speed or minimum steady flight speed without the use of lift-enhancing devices (V_{S1}) of not more than 45 knots CAS at the aircraft's maximum certificated takeoff weight and most critical center of gravity.

(e) A maximum seating capacity of no more than two persons, including the pilot.

(f) A single, reciprocating engine, if powered.

(g) A fixed or ground-adjustable propeller if a powered aircraft other than a powered glider.

(h) A fixed or autofeathering propeller system if a powered glider.

(i) A fixed-pitch, semirigid, teetering, two-blade rotor system, if a gyroplane.

(j) A nonpressurized cabin, if equipped with a cabin.

(k) Fixed landing gear, except for an aircraft intended for operation on water or a glider.

(l) Fixed or repositionable landing gear, or a hull, for an aircraft intended for operation on water.

(m) Fixed or retractable landing gear for a glider.

(2) Issue of an airworthiness certificate

(a) *Eligibility.* To be eligible for a Special airworthiness certificate in the Light-Sport category, an applicant must provide the FAA with
- (i) The aircraft's operating instructions;
- (ii) The aircraft's maintenance and inspection procedures;
- (iii) The manufacturer's statement of compliance as described in Paragraph 21.190(c); and
- (iv) The aircraft's flight-training supplement.

The aircraft must not have been previously issued a Standard, Primary, Restricted, Limited, or Provisional airworthiness certificate, or an equivalent airworthiness certificate issued by a foreign civil aviation authority. The aircraft must be inspected by the FAA and found to be in a condition for safe operation.

(**b**) *Manufacturer's statement of compliance for Light-Sport category aircraft.* The content of the manufacturer's statement to be provided is described in Paragraph 21.190(c). In particular, the document shall state the compliance with the provisions of the **consensus standard**.

Consensus standard means, for the purpose of certificating LSA, an industry-developed consensus standard that applies to aircraft design, production, and airworthiness. It includes, but is not limited to, standards for aircraft design and performance, required equipment, manufacturer quality assurance systems, production acceptance test procedures, operating instructions, maintenance and inspection procedures, identification and recording of major repairs and major alterations, and continued airworthiness.

(**c**) *LSA manufactured outside the United States.* For aircraft manufactured outside the United States to be eligible for a Special airworthiness certificate in the Light-Sport category, an applicant must meet the requirements of eligibility and provide to the FAA evidence that:

    (**i**) The aircraft was manufactured in a country with which the United States has a Bilateral Airworthiness Agreement concerning airplanes or Bilateral Aviation Safety Agreement with associated Implementation Procedures for Airworthiness concerning airplanes, or an equivalent airworthiness agreement.

    (**ii**) The aircraft is eligible for an airworthiness certificate, flight authorization, or other similar certification in its country of manufacture.

(**3**) **General remarks**

The recent institution of this new category of aircraft in the United States, with special certification, could represent a revolution in the general aviation arena.

The boom of general aviation in the United States has been exhausted for a long time, mainly for economic reasons caused by many factors.

To create a revival in the field of the sport and recreational aviation, the FAA, after several years of study and discussions, issued the new rules for the **LSA** on 1 September 2004, relating to the certification of such aircraft and also to the licenses to operate them.

These rules, recommended for some time by the EAA (Experimental Aircraft Association), aim to make it possible to fly a variety of machines characterized by a low cost of production and operation and with pilot's licenses obtainable in simplified manner. Significantly, it is possible to credit sport pilot flight time toward more advanced pilot ratings.

According to an FAA summary:

The intended effect of this action is to provide for the manufacture of safe and economical certificated aircraft that exceed the limits currently allowed by ultralight regulation, and to allow operation of these aircraft by certificated pilots for sport and recreation, to carry a passenger, and to conduct flight training and towing in a safe manner.

A sport pilot may exercise flight privileges in one or more of the following aircraft categories:

(a) Aeroplane (single-engine only)
(b) Sailplane
(c) Lighter-than-air craft (airship or balloon)
(d) Rotorcraft (gyroplane only)
(e) Powered parachute (PPC)
(f) Weight-shift control aircraft (e.g., trikes).

We mentioned the limitations stated by the rules in the definition.

The certification of these aircraft excludes a type certification. The FAA issues a Special airworthiness certificate for a Light-Sport category aircraft on the basis of a manufacturer's statement of compliance to an above-mentioned **consensus standard**. In particular, the airworthiness standard mentioned in the definition of consensus standard could be a new one or a standard already accepted by the FAA.

The statement of conformity to a consensus standard, which is accepted (but not approved) by the FAA, actually involves an autocertification. In any case, the FAA must be allowed by the manufacturer to have unrestricted access to its facilities, and perform a final inspection for the issue of the airworthiness certificate.

Other attractive privileges are the possibility of obtaining an **Experimental airworthiness certificate** for the operation of LSA, if the aircraft was assembled from an eligible kit without the supervision and quality system of the manufacturer. In this case, the assemblage is without the burden of the 51 percent imposed to the amateur-built aircraft.[17]

The aircraft can be used only for the purpose of sport and recreation and for flight training.

The continued airworthiness of LSA-issued Experimental certificates would follow the experience and precedent that has been established for the continued airworthiness of experimental amateur-built aircraft. The aircraft owner would be responsible for ensuring the continued airworthiness of the aircraft.

The FAA has instituted a new repairman certificate called "Repairman— LSA." There are two ratings for this certificate: *Inspection* and *Maintenance*. To earn this certificate, an applicant must be at least 18 years old; speak, read, and understand English; complete the amount of training appropriate for the rating; and must be a US citizen or permanent legal resident.

The big FAA initiatives normally lead to fallout in the rest of the world.

For example, in 2006, the LSA category was introduced in Australia by the Civil Aviation Safety Authority, with very little differences from the FAA LSA category.

As expected, the introduction of the LSA category in the United States has been very successful.

---

[17] As explained in the "Operating amateur-built aircraft" section in this chapter, the main portion of the aircraft must be fabricated and assembled by persons who undertook the construction.

According to a recent FAA oversight, after the first special airworthiness certificate issued in April 2005, more than 90 different manufacturers have produced "ready-to-fly" aeroplanes, PPCs, and Weight-Shift-Control (WSC, i.e., "trikes") that have been registered as S-LSA (Special Light Sport Aircraft).

More than 900 such aircraft have now received S-LSA airworthiness certificates and there are reports of a very low accident rate, which would indicate a good behavior of the LSA consensus standard and the inherent FAA regulations.

For the consensus standard process, ASTM International was selected as the standards development organization and the FAA has been working within their process for the development, approval, and revision of the standards.

8.5.2.5. EXPERIMENTAL CERTIFICATES OF AIRWORTHINESS
(1) Definition

The Experimental certificates of airworthiness are issued for aircraft that are not type certificated and for type-certificated aircraft that embody nonapproved changes or likely to exceed the approved limitations.

There are various types of Experimental certificates of airworthiness issued for different purposes. We now list these certificates and then explain their utilization from a quite general point of view: detailed information can be found in the Order 8130-2F:

(a) Research and development

(b) Showing compliance with regulations

(c) Crew training

(d) Exhibition

(e) Air racing

(f) Market surveys

(g) Operating amateur-built aircraft

(h) Operating primary kit-built aircraft

(i) Operating an LSA.

(2) Issue of an experimental certificate

The requirements for issuing experimental certificates are contained in Sections 21.191, 21.193, and 21.195.

Section 91.319 prescribes operating limitations that are applicable to all aircraft having experimental certificates. In addition, the Administrator may prescribe other limitations as may be considered necessary under Section 91.319(e).

(a) *Research and development*

Testing new aircraft design concepts, new aircraft equipment, new aircraft installations, and new aircraft operating techniques or new uses for aircraft.

To better understand what we are referring to, we can take the example of a person who wants to test a new type of engine (even an engine of new conception) installed on a type-certificated aircraft and, at least in the short term, who is not interested in achieving a type certificate (or STC).

Flights carried out with such aircraft must not have consequences from a type-certification point of view. The authority's intervention in such a case should be limited to general information about the activities to be performed by the applicant to establish some limitations that must essentially be of operative nature (i.e., areas over which the experiments will be conducted and how to reach such areas).[18] The authority will not perform flight tests for the issue of a certificate of airworthiness.

(b) *Showing compliance with regulations*

This pertains to conducting flight tests and other operations to show compliance with the airworthiness regulations, including flights to show compliance for issuance of type and STCs, flights to substantiate major design changes, and flights to show compliance with the function and reliability requirements of the regulations.

In this case, the authority's involvement is quite different because the flight tests to be performed are inherent to the type certification. It is important to be aware of the aircraft's configuration and the state of demonstration of compliance already carried out.[19] The flight envelope cannot be frozen because flight tests are carried out to gradually enlarge the same. Hence, the applicant must agree with the authority about the criteria necessary to fix the limitations for each flight test and for gradual enlargement of the flight envelope.

(c) *Crew training*

Regarding training of the applicant's flight crews, the certificate of airworthiness is normally issued during the type-certification process to train the applicant's crews for type-certification or mass-production test flights.

In this case also, the aircraft is involved in a type-certification process. Then the remarks made in Subsection (b) are still valid, with the exception of the authorized flight envelope, which should be well defined and explored.

(d) *Exhibition*

This refers to exhibition of the aircraft's flight capabilities, performance, or unusual characteristics at air shows; motion pictures, television, and similar productions; and the maintenance of exhibition flight proficiency, including (for persons exhibiting aircraft) flying to and from such air shows and productions.

We will consider two cases:

(i) *Aircraft with type certification in process.* This case can be seen as an extension of the certificate of airworthiness for crew training. Sometimes, fortunately not often, an authorization is

---

[18] Normally, the applicant has to produce a program of the experimentation and the number of flights he reckons as necessary.

[19] For example, static tests, system and equipment assessments, and so on.

requested to perform maneuvers that should not be allowed even with a Standard certificate of airworthiness. The authority might allow such maneuvers (that must be well identified) if supported by serious justifications (structural analysis, flight tests, etc.).

(ii) *Other aircraft.* This case refers to "non-type certificated aircraft" for which it is possible to express a judgment about a sufficient safety level for operations limited to those described in the certificate of airworthiness. This case is also interesting because of the possibility it offers in restoring historical or ex-military aircraft.

NOTE: It is worth remembering that aircraft certificated for exhibition are not allowed for indiscriminate tourist use, but only for the operations permitted by the certificate of airworthiness.

(e) *Air racing*

This refers to participation in air races, including (for such participants) practicing for such air races and flying to and from racing events. The description in Subsection (d) is applicable, inclusive of the final note.

(f) *Market surveys*

Use of aircraft for purposes of conducting market surveys, sales demonstrations, and customer crew training includes

(i) A manufacturer of aircraft within the United States may apply for an Experimental certificate for an aircraft that is to be used for market surveys, sales demonstrations, or customer crew training.

(ii) A manufacturer of aircraft engines who has altered a type-certificated aircraft by installing different engines, manufactured by them within the United States, may apply for an Experimental certificate for that aircraft to be used for market surveys, sales demonstrations, or customer crew training if the basic aircraft, before alteration, was type certificated in one of the normal, acrobatic, commuter, or transport categories.

(iii) A private individual who has altered the design of a type-certificated aircraft may apply for an experimental certificate for the altered aircraft to be used for market surveys, sales demonstrations, or customer crew training if the basic aircraft, before alteration, was type certificated in one of the normal, utility, acrobatic, or transport categories.

(g) *Operating amateur-built aircraft*

This refers to operating an aircraft, the major portion of which has been fabricated and assembled by persons who undertook the construction project solely for their own education or recreation.

The determination of major portion will be made by evaluating the amount of work accomplished by the amateur builder(s) against the total amount of work necessary to complete the aircraft, excluding standard procured items.

NOTE: The major portion of the aircraft is defined as more than 50 percent of the fabrication and assembly tasks, commonly referred to as the *51-percent rule*.[20,22]

For this type of aircraft, the demonstration of compliance to airworthiness standards is not required. Furthermore, the certification of the applicant for design or production organization is not required.

Amateur-built aircraft are eligible for an experimental airworthiness certificate when the applicant presents satisfactory evidence of the following:

(a) The aircraft was fabricated and assembled by an individual or group of individuals.

(b) The project was undertaken for educational or recreational purposes.

(c) The FAA finds that the aircraft complies with acceptable aeronautical standards and practices.

NOTE: Aircraft that are manufactured and assembled as a business for sale to other persons are not considered amateur-built aircraft.

The authority (or delegate organization) control of amateur-built aircraft is quite different from the control performed in other cases. The aim of these controls is to ascertain the technical skill of the applicant for building the aircraft, a sufficient qualitative level of construction and assembly, and flight behavior that obviously must not be perilous.

The authority does not have the responsibility of guaranteeing to third persons (e.g., customers) the airworthiness of the aircraft; hence, formalities such as material certificates of origin and standardized procedures can be avoided. It is important to investigate the means by which the applicant is able to guarantee himself or herself (he or she is going to operate the aircraft) about the adequacy of materials and parts, technical processes, and checks. All these establish a peculiar relationship between the authority and the applicant and implie a great sensibility and experience of the professional controlling the construction, whose experience is sometimes integrated with the applicant's experience.[21]

We will consider two categories of amateur-built aircraft:

(i) Aircraft already certificated somewhere as amateur-built aircraft

(ii) Aircraft of a new design.

The first is a "relaxed" case, because knowing that a certain type of aircraft is already flying (sometimes tens or even hundreds of units)

---

[20] See AC 20-27G.

[21] It is worth mentioning that the authority does not have the task of teaching how to build an aircraft. Amateur builder associations, normally of a national nature, provide a valuable advisory activity.

allows the limitations of controls to a good realization of the design according to the drawings and instructions provided by the design holder, who sometimes supplies a kit of parts and materials.[22]

In the case of a new design, even if compliance to an airworthiness standard is not required, a design made by one or more competent persons should be presented. The authority does not require the design documentation, but it should be informed about the design criteria, the tests to be performed, and the standards taken as reference (not necessarily the type-certification standards required for similar aircraft).

Analogous remarks are valid for major changes in aircraft described in the former case.

Amateur-built aircraft must be provided with a flight manual and instructions for continued airworthiness. The applicant is responsible for the maintenance of the aircraft, which could be directly performed by him or her, if capable, or by maintenance organizations.

The AC 20-27G (30 September 2009) provides specific information and guidance to amateur aircraft builders on certificating and operating an amateur-built aircraft; what to do and know before building an amateur-built aircraft; designing and constructing an amateur-built aircraft; fabricating and assembling an amateur-built aircraft; registering an amateur-built aircraft; identifying and marking an amateur-built aircraft; applying for certification of an amateur-built aircraft; FAA inspection of an amateur-built aircraft; issuing an airworthiness certificate for an amateur-built aircraft; flight testing an amateur-built aircraft; and operating an amateur-built aircraft after flight testing.

(h) *Operating primary kit-built aircraft*

This refers to operation of a Primary category aircraft that meets the criteria of Paragraph 21.24(a)(1)[23] that was assembled by a person from a kit manufactured by the holder of a PC for that kit, without the supervision and quality control of the PC holder.

(i) *Operating LSA*

Operating an LSA that has been assembled:

(i) From an aircraft kit for which the applicant can provide the information required by Paragraph 21.193(e)[24] and

(ii) In accordance with manufacturer's assembly instructions that meet an applicable consensus standard or

(iii) It has been previously issued a Special airworthiness certificate in the Light-Sport category.

---

[22] In these cases, the authority checks that the prefabricated parts are no more than 50 percent (in terms of working hours) of the total. This (not always easy) evaluation has to be performed before the beginning of the construction.

[23] For a Primary category aircraft type certificate.

[24] Requirements for LSA assembled from a kit.

8.5.3. Special flight permits

8.5.3.1. DEFINITION

A special flight permit may be issued for an aircraft that may not currently meet applicable airworthiness requirements but is capable of safe flight, for the following purposes:

(1) Flying the aircraft to a base where repairs, alterations, or maintenance are to be performed, or to a point of storage.

(2) Delivering or exporting the aircraft.

(3) Production flight testing new production aircraft.

(4) Evacuating aircraft from areas of impending danger.

(5) Conducting customer demonstration flights in new production aircraft that have satisfactorily completed production flight tests.

A special flight permit may also be issued to authorize the operation of an aircraft at a *weight in excess* of its maximum certificated takeoff weight for flight beyond the normal range over water or over land areas where adequate landing facilities or appropriate fuel is not available. The excess weight that may be authorized is limited to the additional fuel, fuel-carrying facilities, and navigation equipment necessary for the flight.

On application, a special flight permit with a continuing authorization may be issued for aircraft that may not meet applicable airworthiness requirements but are capable of safe flight for the purpose of flying aircraft to a base where maintenance or alterations are to be performed.

8.5.3.2. ISSUE OF SPECIAL FLIGHT PERMITS

To issue a special flight permit, the authority will gather all the necessary information for the purpose of prescribing operating limitations and may make, or require, the applicant to make appropriate inspections or tests necessary for safety.

Order 8130.2F provides information for application and issuance; aircraft inspections; special operating limitations; and special flight permit for operation of overweight aircraft, production flight testing, conducting customer demonstration flights, and so on.

8.5.4. Provisional airworthiness certificates

8.5.4.1. DEFINITION OF PROVISIONAL AIRWORTHINESS CERTIFICATE[25]

A Special airworthiness certificate in the Provisional category is issued to conduct *special purpose* operations of aircraft with **Provisional type certificates**. The duration of this airworthiness certificate is limited to the duration of the provisional type certificate.

---

[25] FAR 21 Subpart I prescribes procedural requirements for the issue of provisional airworthiness certificates.

The special purpose operations for which provisionally certificated aircraft may be operated are included in FAR 91.317 (operating limitations) as follow:

Unless otherwise authorized by the Administrator, no person may operate a provisionally certificated civil aircraft except

(1) In direct conjunction with the type or supplemental type certification of that aircraft; (2) For training flight crews, including simulated air carrier operations; (3) Demonstration flight by the manufacturer for prospective purchasers; (4) Market surveys by the manufacturer; (5) Flight checking of instruments, accessories, and equipment that do not affect the basic airworthiness of the aircraft; or (6) Service testing of the aircraft.

8.5.4.2. PROVISIONAL TYPE CERTIFICATE

Two classes of Provisional type certificates may be issued. Class I certificates may be issued for all categories and have a duration of 24 months. Class II certificates are issued for Transport category aircraft only and have a duration of 12 months.

FAR 21 Subpart C prescribes procedural requirements for the issue of provisional type certificates, amendments to provisional type certificates, and provisional amendments to type certificates; and it rules governing the holders of those certificates.

In particular, FAR 21 Subpart C prescribes requirements based on the compliance of the aircraft with certain applicable paragraphs of operational standards such as FAR 91 and 121.

As reported in FAR 21:

(1) Any manufacturer of aircraft within the United States who is a United States citizen may apply for Class I or II Provisional type certificates, for amendments to Provisional type certificates they hold and for provisional amendments to type certificates they hold.

(2) Any manufacturer of aircraft in a foreign country with which the United States has an agreement for the acceptance of those aircraft for export and import may apply for a Class II Provisional type certificate, for amendments to Provisional type certificates they hold, and for provisional amendments to type certificates they hold.

(3) An aircraft engine manufacturer who is a US citizen and has altered a type-certificated aircraft by installing different type-certificated aircraft engines manufactured by him within the United States may apply for a Class I Provisional type certificate for the aircraft and for amendments to Class I Provisional type certificates he holds, if the basic aircraft, before alteration, was type certificated in one of the normal, utility, acrobatic, commuter, or transport categories.

8.5.4.3. GENERAL REMARKS

In Section 8.5.4.1, we indicated a list of special purpose operations for which a provisional type certificate and a provisional airworthiness certificate are issued. Generally speaking, the provisional type certificate is issued in advance of a (nonprovisional) type certificate during the type-certification process, when the applicant shows compliance with the relevant FAR 21 Subpart C requirements.

Typically, the provisional type-certificated prototype aircraft is not in conformity with the type design when the type certificate is issued. Nevertheless, it is worth considering that the associated provisional airworthiness certificate does not expire, unless previously surrendered, superseded, revoked, or otherwise terminated, for the duration of the corresponding provisional type certificate.

8.5.5. Export airworthiness approvals

8.5.5.1. APPLICABILITY

FAR 21, **Subpart L** contains procedural requirements for issuing **export airworthiness approvals** and the rules governing the holders of those approvals. The AC No. 21-44—a short summary is provided here—describes an acceptable means to comply with these requirements.

8.5.5.2. TYPES OF EXPORT AIRWORTHINESS APPROVALS

The FAA issues export airworthiness approvals for aircraft, aircraft engines, propellers, and articles. The requirements are described below.

(a) Export Airworthiness Approvals for Aircraft. FAA Form 8130-4, Export Certificate of Airworthiness (C of A), is used to issue an export airworthiness approval for an aircraft. The C of A represents a certifying statement from the FAA that a given aircraft

 (1) conforms to its FAA type design or properly altered condition and

 (2) is in a condition for safe operation at the time of examination and issuance of the certificate.

When required by the importing country or jurisdiction, the export C of A also includes a supplemental statement attesting to the aircraft's conformity to the importing country's type design.

An export C of A is not an authorization to operate the aircraft.

(b) Export Airworthiness Approvals for Engines, Propellers, and Articles. FAA Form 8130-3, Authorized Release Certificate, is used for issuing export airworthiness approvals to aircraft engines, propellers, and articles. The authorized release certificate is a certifying statement from the FAA that a given aircraft engine, propeller, or article

 (1) conforms to its FAA design approval or properly altered condition and

 (2) is in a condition for safe operation at the time of examination and issuance of the certificate.

8.5.5.3. ISSUANCE OF FAA FORMS 8130-4 AND 8130-3 FOR PRODUCTS OR ARTICLES LOCATED IN ANOTHER COUNTRY

Forms 8130-4 and 8130-3 may be issued for any product or article located in another country as long as the FAA finds no undue burden in administering the applicable requirements.

8.5.5.4. APPLICATION

Any person may apply for an export airworthiness approval. Each applicant must apply in a form and manner prescribed by the FAA.

8.5.5.5. REQUIREMENTS TO BE MET BEFORE THE FAA ISSUES AN EXPORT C OF A FOR A NEW OR USED AIRCRAFT

(a) **New or Used Aircraft Manufactured Under FAR 21, Subparts F or G.**
   The FAA issues an export C of A for a new or used aircraft manufactured under FAR 21, Subpart F, Production Under Type Certificate
   (TC)[26], or Subpart G, PCs, if the aircraft meets the airworthiness
   requirements of FAR 21, Subpart H, Airworthiness Certificates. Such
   aircraft are eligible for either a standard airworthiness certificate or
   a special airworthiness certificate in either the primary or the restricted
   category.

(b) **New or Used Aircraft Not Manufactured Under Part 21, Subparts
   F or G.** The FAA will also issue an export C of A for a new or used
   aircraft that was not manufactured under Subpart F or G. In this case,
   the aircraft already has a valid standard airworthiness certificate or
   a valid special airworthiness certificate in either the primary or
   restricted category (issued in accordance with the requirements of
   Subpart H). Examples of aircraft not manufactured under Subpart F
   or G include import aircraft that have been issued an FAA type
   design in accordance with §21.29 (Issue of Type Certificate: Import
   Products), and aircraft that have been constructed using spare and
   surplus parts.

NOTE: A product does not need to meet a requirement specified in Section **8.5.5** as applicable, if acceptable to the importing country and the importing country indicates that acceptability on the basis of a written statement; the requirements that are not met and the differences in configuration, if any, between the product to be exported and the related type-certificated product are listed as exceptions on the Export Airworthiness Approval.

8.5.5.6. EXPORTING AN AIRCRAFT TO A COUNTRY OR JURISDICTION THAT DOES NOT HAVE A BILATERAL AGREEMENT WITH THE UNITED STATES

When exporting an aircraft to a country or jurisdiction that does not have a bilateral agreement with the United States, and no definitive special import requirements have been formally submitted to the FAA, an FAA export C of A is not necessary.

However, the FAA will permit the issuance of Form 8130-4 for export of all eligible aircraft when these aircraft conform to their FAA-approved design or properly altered condition and are in a condition for safe operation.

---

[26] See section 7.2.2, Chapter 7

8.5.5.7. REQUIREMENTS TO BE MET BEFORE THE FAA ISSUES AN EXPORT AIRWORTHINESS APPROVAL FOR A NEW OR USED AIRCRAFT ENGINE, PROPELLER, OR ARTICLE

(a) New Aircraft Engines, Propellers, or Articles. The FAA or its designee may issue an export airworthiness approval, Form 8130-3, to export a new aircraft engine, propeller, or article that is manufactured under Part 21. The aircraft engine, propeller, or article is required to conform to its approved design and be in a condition for safe operation.

(b) Used Aircraft Engines, Propellers, or Articles. Any person (e.g., distributor, operator, private owner) may obtain from the FAA or its designee an export airworthiness approval for a used aircraft engine, propeller, or article. The used aircraft engine, propeller, or article is required to conform to its FAA-approved design and be in a condition for safe operation. This includes a statement from the applicant that used an aircraft engine, a propeller, or an article that has been properly maintained in accordance with Part 43.

NOTE: Articles, new or used aircraft engines, and propellers do not need to meet a requirement specified in Section 8.5.5.7 as applicable, if acceptable to the importing country and the importing country accepts a deviation from that requirement. Form 8130-3 will list, as an exception, each difference between the aircraft engine, propeller, or article and its approved design.

8.5.5.8. EXPORTING AN AIRCRAFT ENGINE, PROPELLER, OR ARTICLE TO A COUNTRY OR JURISDICTION THAT DOES NOT HAVE A BILATERAL AGREEMENT WITH THE UNITED STATES

When exporting a new or used aircraft engine, propeller, or article to a country or jurisdiction that does not have a bilateral agreement with the United States, and no definitive special import requirements have been formally submitted to the FAA, a Form 8130-3, with certain exceptions, is not issued.

8.5.5.9. GENERAL REMARKS ON EXPORT AIRWORTHINESS APPROVALS

In Chapter 5, the section "Type certification of imported products" relates to the validation of a type certificate by the authority of the importing country.

The Export certificate of airworthiness does not authorize flight operations; as previously mentioned, it is essentially a statement of conformity to the type certificate of the **importing** country, including additional requirements for import and a list of possible nonconformities accepted by the authority of the importing country.

Therefore, it is possible to issue an Export Airworthiness Approval for "nonairworthy" aircraft as well.

The Order 8130.21G dated 26 October 2009 describes the procedures for completion and use of the FAA Form 8130-3, Airworthiness Approval Tag. The order describes the procedures for completion and use of the FAA Authorized Release Certificate, FAA Form 8130-3, and Airworthiness Approval Tag.

The order describes the use of the form for domestic airworthiness approval, conformity inspections, and prepositioning; airworthiness approval of new products and articles; and so on...

8.6. ADDITIONAL AIRWORTHINESS REQUIREMENTS FOR OPERATION

8.6.1. Introduction

The operational life of an aircraft begins with the issue of a certificate of airworthiness or equivalent document, as has been shown in this chapter.

We have described that such a certificate can be issued either because the aircraft has been found to comply with a type certificate or, having not met (or have not been shown to meet) applicable certification specifications, it has been found to be capable of safe flight under defined conditions.

Because the same aircraft can be used in different kinds of operations, besides the basic certification requirements the aircraft also has to satisfy the requirements issued by the authority for each particular kind of operation.

For example, a single-engine FAR 23 airplane can be operated for personal use or for compensation or hire (aerotaxi, aerial working, etc.), but also according to different flight rules (VFR, IFR, etc.). Depending on the particular type of operation allowed, additional airworthiness requirements, which influence the airplane's configuration, shall be complied with (equipment, instruments, etc.).

To better illustrate the above remarks, Fig. 8.3 presents a simplified summary of the certification of an aircraft from design to operation.

Starting from the airworthiness and environmental standards (1), through the type-certification process (2), a type certificate is issued (3). To obtain a certificate of airworthiness (6), it is necessary to take into consideration the additional requirements for operation (4) and carry out a demonstration of compliance for the relevant kinds of operation to be authorized (5) (if not already incorporated in the type certificate).

Figure 8.3 also considers the cases of aircraft that, having not met (or having not been shown to meet) applicable certification specifications (according to Subpart H of FAR 21/EASA Part 21) (7), have been found to be capable of safe flight under defined conditions (8); they must be demonstrated to comply with additional airworthiness requirements for operations, if applicable (5), to obtain a certificate of airworthiness or a permit to fly (9).

It is correct to recognize that Fig. 8.3 is a schematic simplification, because generally aircraft are type certificated also in consideration of operating rules, then with operational requirements already incorporated in the type-certification basis.

In any case, to obtain a certificate of airworthiness, the aircraft must be eligible by make, model, and serial number, using TCDS, aircraft specifications, and/or applicable aircraft listing. Then, the inspection records and technical data must reflect that the aircraft conforms to the type design, that all

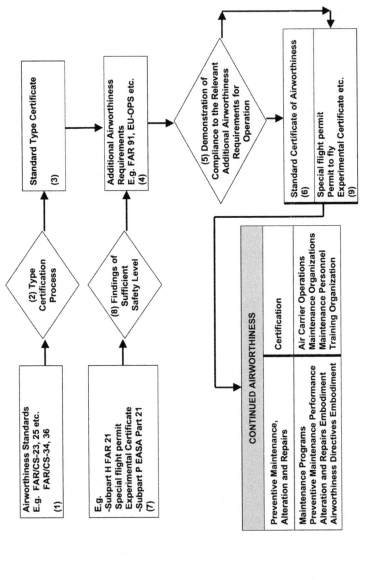

FIGURE 8.3 Summary of the certification of an aircraft from design to operation

required inspections and tests have been satisfactorily completed, and that the records are complete and reflect no unapproved design changes.

8.6.2. Operational standards

Operational standards prescribe requirements for the operation of aircraft, including prescriptions for the certification of operators, and in particular their organization, procedures, manuals, crew employment and training, equipment, aircraft adequacy and maintenance, transport of dangerous goods, and protection against acts of unlawful interference.

These operational standards, already listed in Chapter 4, are complex documents, which we will attempt to summarize. We will also mention their "applicability" and *partially* report some significant paragraphs or titles—significant for the scope of this book—related to additional airworthiness requirements.

We will consider:

(1) The FAA standards FAR 91, 121, 125, 129, 135, and 137.
(2) The/JAA standards JAR-OPS 1, JAR-OPS 3; the EASA OPS 1 and CS-AWO.

NOTE: Appendices 8.7 and 8.8/8.9 give a summary of the applicability of the above-mentioned standards for the additional airworthiness requirements.

8.6.2.1. DEFINITIONS (FROM FAR 1 AND FAR 119)

We will list some definitions to better understand the content of the above-mentioned operational standards.

Air carrier means a person who undertakes directly by lease, or other arrangement, to engage in air transportation.

Air commerce means interstate, overseas, or foreign air commerce or the transportation of mail by aircraft or any operation or navigation of aircraft within the limits of any Federal airway or any operation or navigation of aircraft that directly affects, or that may endanger safety in, interstate, overseas, or foreign air commerce.

Category II operation, with respect to the operation of aircraft, means a straight-in Instrument Landing System (ILS) approach to the runway of an airport under a Category II ILS instrument approach procedure issued by the Administrator or other appropriate authority.

Category III operation, with respect to the operation of aircraft, means an ILS approach to, and landing on, the runway of an airport using a Category III ILS instrument approach procedure issued by the Administrator or other appropriate authority.

Class:

(1) As used with respect to the certification, ratings, privileges, and limitations of airmen, means a classification of aircraft within a category having similar operating characteristics. Examples include single engine; multiengine; land; water; gyroplane; helicopter; airship; and free balloon; and
(2) As used with respect to the certification of aircraft, means a broad grouping of aircraft having similar characteristics of propulsion, flight, or landing.

Examples include airplane; rotorcraft; glider; balloon; landplane; and seaplane.

Commercial operator means a person who, for compensation or hire, engages in the carriage by aircraft in air commerce of persons or property, other than as an air carrier or foreign air carrier or under the authority of FAR 375. Where it is doubtful that an operation is for "compensation or hire," the test applied is whether the carriage by air is merely incidental to the person's other business or is, in itself, a major enterprise for profit.

Common carriage[27] means any operation for compensation or hire in which an operator holds itself out (by advertising or any other means), as willing to furnish transportation for any member of the public who seeks the services that the operator is providing.

Noncommon carriage means an aircraft operation for compensation or hire that does not involve holding out to others.[27]

8.6.2.1.1. Kind of operation

Kind of operation means one of the various operations a certificate holder is authorized to conduct, as specified in its operations specifications, that is, domestic, flag, supplemental, commuter, or on-demand operations.

Commuter operation means any scheduled operation conducted by any person operating one of the following types of aircraft with a frequency of operations of at least five round trips per week on at least one route between two or more points according to the published flight schedules: (1) airplanes, other than turbojet-powered airplanes, having a maximum passenger-seat configuration of nine seats or less, excluding each crew member seat, and a maximum payload capacity of 7500 lb or less or (2) rotorcraft.

Domestic operation means any scheduled operation conducted by any person operating any airplane described in Paragraph (1) of this definition at locations described in Paragraph (2) of this definition:
(1) Airplanes—(i) turbojet-powered airplanes; (ii) airplanes having a passenger-seat configuration of more than nine passenger seats, excluding each crew-member seat; or (iii) airplanes having a payload capacity of more than 7500 lb.
(2) Locations—(i) between any points within the 48 contiguous States of the United States or the District of Columbia; or (ii) operations solely within the 48 contiguous States of the United States or the District of Columbia; or (iii) operations entirely within any State, territory, or possession of the United States.

Flag operation means any scheduled operation conducted by any person operating any airplane described in Paragraph (1) of this definition at the locations described in Paragraph (2) of this definition:
(1) Airplanes—(i) turbojet-powered airplanes; (ii) airplanes having a passenger-seat configuration of more than nine passenger seats, excluding

---

[27] *Noncommon carriage*: see also in 8.6.2.1 Definitions, *When "common carriage" is not involved*. These aircraft operations often require an accurate evaluation to avoid "pitfall" that could result in illegal common carriage operations.

each crew member seat; or (iii) airplanes having a payload capacity of more than 7500 lb.

(2) Locations—(i) between any point within the State of Alaska; or (ii) between any point within the 48 contiguous States of the United States or the District of Columbia and any point outside the 48 contiguous States of the United States and the District of Columbia; (iii) between any point outside the United States and another point outside the United States.

On-demand operation means any operation for compensation or hire that is one of the following:

(1) Passenger-carrying operations conducted as a public charter ... that are any of the following types of operations: (i) common carriage operations conducted with airplanes, including turbojet-powered airplanes, having a passenger-seat configuration of 30 seats or fewer, excluding each crew member seat, and a payload capacity of 7500 lb or less; (ii) noncommon or private carriage operations conducted with airplanes having a passenger-seat configuration of less than 20 seats, excluding each crew member seat, and a payload capacity of less than 6000 lb; or (iii) any rotorcraft operation.

(2) Scheduled passenger-carrying operations conducted with one of the following types of aircraft with a frequency of operations of less than five round trips per week on at least one route between two or more points according to the published flight schedules: (i) airplanes, other than turbojet-powered airplanes, having a maximum passenger-seat configuration of nine seats or less, excluding each crew member seat, and a maximum payload capacity of 7500 lb or less; or (ii) rotorcraft.

(3) All-cargo operations conducted with airplanes having a payload capacity of 7500 lb or less, or with rotorcraft.

Supplemental operation means any common carriage operation for compensation or hire conducted with any airplane described in Paragraph (1) of this definition that is a type of operation described in Paragraph (2) of this definition:

(1) Airplanes: (i) airplanes having a passenger-seat configuration of more than 30 seats, excluding each crew member seat; (ii) airplanes having a payload capacity of more than 7500 lb; or (iii) each propeller-powered airplane having a passenger-seat configuration of more than nine and less than 31 seats, excluding each crew member seat, that is also used in domestic or flag operations and that is so listed in the operations specifications as required by Paragraph 119.49(a)(4) for those operations; or (iv) each turbojet-powered airplane having a passenger seat configuration of one or more and less than 31 seats, excluding each crew member seat, that is also used in domestic or flag operations and that is so listed in the operations specifications as required by Paragraph 119.49(a)(4) for those operations.

(2) Types of operation: (i) operations for which the departure time, departure location, and arrival location are specifically negotiated with the customer or the customer's representative; (ii) all-cargo operations; or (iii) passenger-carrying public charter operations conducted under Part 380 of this title.

8.6.2.1.2. Operations

Foreign air transportation means the carriage by aircraft of persons or property as a common carrier for compensation or hire, or the carriage of mail by aircraft, in commerce between a place in the United States and any place outside of the United States, whether that commerce moves wholly by aircraft or partly by aircraft and partly by other forms of transportation.

Interstate air transportation means the carriage by aircraft of persons or property as a common carrier for compensation or hire, or the carriage of mail by aircraft in commerce: between a place in a State or the District of Columbia and another place in another State; between places in the same State through the airspace over any place outside that State; or between places in the same possession of the United States.

Intrastate air transportation means the carriage of persons or property as a common carrier for compensation or hire, by turbojet-powered aircraft capable of carrying 30 or more persons, wholly within the same State of the United States.

Overseas air transportation means the carriage by aircraft of persons or property as a common carrier for compensation or hire, or the carriage of mail by aircraft, in commerce: between a place in a State or the District of Columbia and a place in a possession of the United States; or between a place in a possession of the United States and a place in another possession of the United States; whether that commerce moves wholly by aircraft or partly by aircraft and partly by other forms of transportation.

Scheduled operation means any common carriage passenger-carrying operation for compensation or hire conducted by an air carrier or commercial operator for which the certificate holder or its representative offers in advance the departure location, departure time, and arrival location. It does not include any passenger-carrying operation that is conducted as a public charter operation ....

When "common carriage is not involved" or "operations not involving common carriage" means any of the following:

(1) Noncommon carriage. (2) Operations in which persons or cargo are transported without compensation or hire. (3) Operations not involving the transportation of persons or cargo. (4) Private carriage.

Wet lease means any leasing arrangement whereby a person agrees to provide an entire aircraft and at least one crew member ......

8.7. FAA OPERATIONAL STANDARDS (ADDITIONAL AIRWORTHINESS REQUIREMENTS)

NOTE: See Appendix 8.7.

NOTE: To give an idea of the content of the operational standards, we quote the most noteworthy articles of these standards, often only partially or referring to the titles. This is done for practical reasons and for reference; however, we are not suggesting that this could replace the good practice of reading the original texts in full.

8.7.1. FAR 91. General operating and flight rules

SUBPART A: GENERAL

91.1. Applicability

(a) Except as provided in Paragraphs (b) and (c) of this section and FAR 91.701 and 91.703,[28] this part prescribes rules governing the operation of aircraft (**other than** moored balloons, kites, unmanned rockets, and unmanned free balloons, which are governed by FAR 101, and ultralight vehicles operated in accordance with FAR 103) within the United States, including the waters within three nautical miles of the US coast.

(b) Each person operating an aircraft in the airspace overlying the waters between three and 12 nautical miles from the coast of the United States must comply with FAR 91.1 to 91.21…

SUBPART C: EQUIPMENT, INSTRUMENT, AND CERTIFICATE REQUIREMENTS

91.203. Civil aircraft: certifications required

(a) Except as provided in FAR 91.715,[29] no person may operate a civil aircraft unless it has within it the following: (1) an appropriate and current airworthiness certificate….

(c) No person may operate an aircraft with a fuel tank installed within the passenger compartment or a baggage compartment unless the installation was accomplished pursuant to FAR 43, and a copy of FAA Form 337 authorizing that installation is on board the aircraft.

(d) No person may operate a civil airplane (domestic or foreign) in or out of an airport in the United States unless it complies with the fuel venting and exhaust emissions requirements of FAR 34.

91.205. Powered civil aircraft with Standard category
US airworthiness certificates: instrument and equipment
requirements

(a) *General.* Except as provided in Paragraphs (c)(3) and (e) of this section, no person may operate a powered civil aircraft with a Standard category US airworthiness certificate in any operation described in Paragraphs (b) to (f) of this section unless that aircraft contains the instruments and equipment specified in those paragraphs (or FAA-approved equivalents) for that type of operation, and those instruments and items of equipment are in operable condition.

---

[28] Paragraphs 701 and 703 belong to Subpart H of FAR 91, which applies to the operations of civil aircraft of US registry outside the United States and the operations of foreign civil aircraft within the United States.

[29] 91.715 Special flight authorizations for foreign civil aircraft. Foreign civil aircraft may be operated without airworthiness certificates required under Paragraph 91.203 if a special flight authorization for that operation is issued under this section.

(b) *Visual flight rules (VFRs) (day).* For VFR flight during the day, the following instruments and equipment are required:

(1) Airspeed indicator, (2) altimeter, (3) magnetic direction indicator, (4) tachometer for each engine, (5) oil pressure gauge for each engine using pressure system, (6) temperature gauge for each liquid-cooled engine, (7) oil temperature gauge for each air-cooled engine ....

(c) *VFRs (night).* For VFR flight during the night, the following instruments and equipment are required:

(1) Instruments and equipment specified in Paragraph (b) of this section, (2) approved position lights, (3) an approved aviation red or aviation white anticollision light system on all US-registered civil aircraft, (4) If the aircraft is operated for hire, one electric landing light ....

(d) *Instrument flight rules (IFR).* For IFR flight, the following instruments and equipment are required:

(1) Instruments and equipment specified in Paragraph (b) of this section, and, for night flight, instruments and equipment specified in Paragraph (c) of this section. (2) Two-way radio communications system and navigational equipment appropriate to the ground facilities to be used. (3) Gyroscopic rate-of-turn indicator, except on the following aircraft:....

(f) *Category II operations.*[30] The requirements for Category II operations are the instruments and equipment specified in: (1) Paragraph (d) of this section; and (2) Appendix A to this FAR.

(g) *Category III operations.*[31] The instruments and equipment required for Category III operations are specified in Paragraph (d) of this section.

(i) *Exclusions.* Paragraphs (f) and (g) of this section do not apply to operations conducted by a holder of a certificate issued under FAR 121 or FAR 135.

91.211. Supplemental oxygen

(a) *General.* No person may operate a civil aircraft of US registry:

(1) At cabin pressure altitudes above 12,500 ft (MSL) up to and including 14,000 ft (MSL) unless the required minimum flight crew is provided with and uses supplemental oxygen for that part of the flight at those altitudes that is of more than 30 minutes duration. (2) At cabin pressure altitudes ....

(b) *Pressurized cabin aircraft.* (1) No person may operate a civil aircraft of US registry with a pressurized cabin:

(i) At flight altitudes above flight level 250 unless at least a 10-minute supply of supplemental oxygen, in addition to any oxygen required to satisfy Paragraph (a) of this section, is available for each occupant of

---

[30] *Category II operations*, with respect to the operation of aircraft, means a straight-in ILS approach to the runway of an airport under a Category II ILS instrument approach procedure issued by the Administrator or other appropriate authority.

[31] *Category III operations*, with respect to the operation of aircraft, means an ILS approach to, and landing on, the runway of an airport using a Category III ILS instrument approach procedure issued by the Administrator or other appropriate authority.

the aircraft for use in the event that a descent is necessitated by loss of cabin pressurization; and (ii) at flight altitudes above ....

91.213. Inoperative instruments and equipment

(a) Except as provided in Paragraph (d) of this section, no person may takeoff an aircraft with inoperative instruments or equipment installed unless the following conditions are met:

(**1**) An approved MEL exists for that aircraft. (2) The aircraft has within it a letter of authorization, issued by the FAA Flight Standards district office.... (3) The approved MEL must (i) be prepared in accordance with the limitations specified in Paragraph (b) of this section ....

(b) The following instruments and equipment may not be included in an MEL:

(**1**) Instruments and equipment that are either specifically or otherwise required by the airworthiness requirements under which the aircraft is type certificated and which are essential for safe operations under all operating conditions. (2) Instruments and equipment required by an Airworthiness Directive to be in operable condition unless the Airworthiness Directive provides otherwise. (3) Instruments and equipment required for specific operations by this FAR.

(c) A person authorized to use an approved MEL issued for a specific aircraft under Subpart K of this FAR 91, FAR 121, 125, or 135 must use that MEL to comply with the requirements in this section.

(d) Except for operations conducted in accordance with Paragraph (a) or (c) of this section, a person may takeoff an aircraft in operations conducted under this part with inoperative instruments and equipment without an approved MEL provided:

(**1**) The flight operation is conducted in a (i) rotorcraft, nonturbine-powered airplane, glider, lighter-than-air aircraft, PPC, or WSC aircraft, for which a master MEL has not been developed ....

(e) Notwithstanding any other provision of this section, an aircraft with inoperable instruments or equipment may be operated under a special flight permit issued in accordance with FAR 21.197 and 21.199 of FAR 91.

The titles of the other paragraphs of Subpart C are:

91.215 ATC transponder and altitude reporting equipment and use.

91.219 Altitude alerting system or device: Turbojet-powered civil airplanes.

91.221 Traffic alert and collision avoidance system equipment and use.

91.223 Terrain awareness and warning system.

SUBPART D: SPECIAL FLIGHT OPERATIONS

91.309. Towing: gliders and unpowered ultralight vehicles

(a) No person may operate a civil aircraft towing a glider or unpowered ultralight vehicle unless:

...

(2) The towing aircraft is equipped with a tow-hitch of a kind, and installed in a manner that is approved by the Administrator. (3) The towline used has breaking strength not less than 80 percent of the maximum certificated operating weight of the glider....

  (i) A safety link is installed at the point of attachment of the towline to the glider....

  (ii) A safety link is installed at the point of attachment of the towline to the towing aircraft....

SUBPART G: ADDITIONAL EQUIPMENT AND OPERATING REQUIREMENTS FOR LARGE AND TRANSPORT CATEGORY AIRCRAFT

91.601. Applicability

This subpart applies to operation of Large and Transport category US-registered civil aircraft.

91.603. Aural speed warning device

No person may operate a Transport category airplane in air commerce unless that airplane is equipped with an aural speed warning device that complies with FAR 25.1303(c)(1).

91.609. Flight recorders and cockpit voice recorders

(a) No holder of an air carrier operating certificate or an operating certificate may conduct any operation under this part with an aircraft listed in the holder's operations specifications or current list of aircraft used in air transportation unless that aircraft complies with any applicable flight recorder and cockpit voice recorder requirements....

(f) In complying with this section, an approved cockpit voice recorder having an erasure feature may be used, so that at any time during the operation of the recorder, information recorded more than 15 minutes earlier may be erased or otherwise obliterated.

SUBPART H: FOREIGN AIRCRAFT OPERATIONS AND OPERATIONS OF US-REGISTERED CIVIL AIRCRAFT OUTSIDE OF THE UNITED STATES; AND RULES GOVERNING PERSONS ON BOARD SUCH AIRCRAFT

91.701. Applicability

(a) This subpart applies to the operations of civil aircraft of US registry outside the United States and the operations of foreign civil aircraft within the United States.

(b) Section 91.702 of this subpart also applies to each person on board an aircraft operated as follows:

  (1) A US registered civil aircraft operated outside the United States. (2) Any aircraft operated outside the United States....

91.711. Special rules for foreign civil aircraft

(a) *General.* In addition to the other applicable regulations of this part, each person operating a foreign civil aircraft within the United States shall comply with this section....

(c) *IFR.* No person may operate a foreign civil aircraft under IFR unless (1) that aircraft is equipped with:

(i) Radio equipment allowing two-way radio communication with ATC when it is operated in controlled airspace; and (ii) radio navigational equipment appropriate to the navigational facilities to be used;....

(e) *Flight at and above FL 240.* If VOR navigational equipment is required under Paragraph (c)(1)(ii) of this section, no person may operate a foreign civil aircraft within the 50 States and the District of Columbia at or above FL 240, unless the aircraft is equipped with distance measuring equipment (DME) or a suitable RNAV....

SUBPART I: OPERATING NOISE LIMITS

91.801. Applicability: in relation to FAR 36

(a) This subpart prescribes operating noise limits and related requirements that apply, as follows, to the operation of civil aircraft in the United States:

(1) Sections 91.803, 91.805, 91.807, 91.809, and 91.811 apply to civil subsonic jet (turbojet) airplanes with maximum weights of more than 75,000 lb....

91.815. Agricultural and fire-fighting airplanes: noise operating limitations

(a) This section applies to propeller-driven, small airplanes having Standard airworthiness certificates that are designed for "agricultural aircraft operations" (as defined in FAR 137.3 of this FAR 91, as effective on 1 January 1966) or for dispensing fire-fighting materials....

91.817. Civil aircraft sonic boom

(a) No person may operate a civil aircraft in the United States at a true flight Mach number greater than 1 except in compliance with conditions and limitations in an authorization to exceed Mach 1 issued to the operator under Appendix B of this FAR.

(b) In addition, no person may operate a civil aircraft for which the maximum operating limit speed M_{M0} exceeds a Mach number of 1, to or from an airport in the United States, unless:

(1) Information available to the flight crew includes flight limitations that ensure that flights entering or leaving the United States will not cause a sonic boom to reach the surface within the United States.

(2) The operator complies with the flight limitations prescribed in Paragraph (b)(1) of this section or complies with conditions and limitations in an authorization to exceed Mach 1 issued under Appendix B of this FAR 91.

8.7.2. FAR 121. Operating Requirements: Domestic, flag, and supplemental operations

SUBPART A: GENERAL

121.1. Applicability

This part prescribes rules governing:

(a) The domestic, flag, and supplemental operations of each person who holds or is required to hold an Air Carrier Certificate or Operating Certificate under FAR 119.

(b) Each person employed or used by a certificate holder conducting operations under this part, including maintenance, preventive maintenance, and alteration of aircraft.

(c) Each person who applies for provisional approval....

(d) Nonstop Commercial Air Tours conducted for compensation or hire in accordance with §119.1(e)(2) of this chapter must comply with drug and alcohol requirements..... An operator who does not hold an air carrier certificate or an operating certificate is permitted to use a person who is otherwise authorized to perform aircraft maintenance or preventive maintenance duties and who is not subject to antidrug and alcohol misuse prevention programs.......

SUBPART G: MANUAL REQUIREMENTS

121.131. Applicability

This subpart prescribes requirements to prepare and maintain manuals by all certificate holders.

121.141. Airplane flight manual

(a) Each certificate holder shall keep a current approved airplane flight manual for each type of airplane that it operates....

121.159. Single-engine airplanes prohibited

No certificate holder may operate a single-engine airplane under this part.

SUBPART I: AIRPLANE PERFORMANCE OPERATING LIMITATIONS

121.171. Applicability

(a) This subpart prescribes airplane performance operating limitations for all certificate holders....

121.173. General

(a) Except as provided in Paragraph (c) of this section, each certificate holder operating a reciprocating engine-powered airplane shall comply with Paragraphs 121.175 to 121.187.

(b) Except as provided in Paragraph (c) of this section, each certificate holder operating a turbine engine-powered airplane shall comply with the applicable provisions of Paragraphs 121.189 to 121.197, except when it operates:

(1) A turbopropeller-powered airplane...

The titles of other paragraphs of this subpart are:

SUBPART J: SPECIAL AIRWORTHINESS REQUIREMENTS

121.211. Applicability

(a) This subpart prescribes special airworthiness requirements applicable to certificate holders as stated in Paragraphs (b) to (e) of this section....

The titles of the paragraphs of this subpart are:

SUBPART K: INSTRUMENT AND EQUIPMENT REQUIREMENTS
121.301. Applicability
This subpart prescribes instrument and equipment requirements for all certificate holders.

The titles of paragraphs of this subpart are:

8.7.3. FAR 125. Certification and Operations: Airplanes having a seating capacity of 20 or more passengers or a maximum payload capacity of 6000 lb or more; and rules governing persons on board such aircraft

SUBPART A: GENERAL

125.1. Applicability

(a) Except as provided in Paragraphs (b), (c), and (d) of this section, this part prescribes rules governing the operations of US-registered civil airplanes that have a seating configuration of 20 or more passengers or a maximum payload capacity of 6000 lb or more when common carriage is not involved.

(b) The rules of this part do not apply to the operations of airplanes specified in Paragraph (a) of this section, when:

   (1) They are required to be operated under Part 121, 129, 135, or 137 of this chapter. (2) They have been issued Restricted, Limited, or Provisional airworthiness certificates, special flight permits, or Experimental certificates. (3) They are being operated by a Part 125 certificate holder without carrying passengers or cargo under Part 91 for training, ferrying, positioning, or maintenance purposes ....

(c) The rules of this FAR, except Paragraph 125.247, do not apply to the operation of airplanes specified in Paragraph (a) when they are operated outside the United States by a person who is not a citizen of the United States ....

SUBPART E: SPECIAL AIRWORTHINESS REQUIREMENTS

125.111. General

(a) Except as provided in Paragraph (b) of this section, no certificate holder may use an airplane powered by airplane engines rated at more than 600 HP each for maximum continuous operation unless that airplane meets the requirements of Paragraphs 125.113 to 125.181.

(b) If the Administrator determines that, for a particular model of airplane used in cargo service, literal compliance with any requirement under Paragraph (a) of this section would be extremely difficult and that compliance would not contribute materially to the objective sought, the Administrator may require compliance with only those requirements that are necessary to accomplish the basic objectives of this part.

(c) This section does not apply to any airplane certificated under:

   (1) CAR 4b in effect after 31 October 1946; (2) FAR 25; or (3) Special Civil Air Regulation 422, 422A, or 422B.

125.113. Cabin interiors

(a) On the first major overhaul of an airplane cabin or refurbishing of the cabin interior, all materials in each compartment used by the crew or passengers which do not meet the following requirements must be replaced with materials that meet these requirements ....

(b) Except as provided in Paragraph (a) of this section, each compartment used by the crew or passengers must meet the following requirements:

(1) Materials must be at least flash resistant. (2) The wall and ceiling linings and the covering of upholstering, floors, and furnishings must be flame resistant. (3) Each compartment where smoking is to be allowed must be equipped ....

(c) Thermal/acoustic insulation materials ....

125.117. Ventilation

Each passenger or crew compartment must be suitably ventilated. Carbon monoxide concentration may not be more than one part in 20,000 parts of air, and fuel fumes may not be present ....

125.119. Fire precautions

(a) Each compartment must be designed so that, when used for storing cargo or baggage, it meets the following requirements ....

(b) *Class A.* Cargo and baggage compartments are classified in the "A" category if a fire therein would be readily discernible to a member of the crew while at that crew member's station and all parts of the compartment are easily accessible in flight. There must be a hand fire extinguisher available for each Class A compartment.

(c) *Class B.* Cargo and baggage compartments are classified in the "B" category if enough access is provided while in flight to enable a member of the crew to effectively reach all the compartments and its contents with a hand fire extinguisher and the compartment is so designed that, when the access provisions are being used, no hazardous amount of smoke, flames, or extinguishing agent enters any compartment occupied by the crew or passengers ....

(d) *Class C.* Cargo and baggage compartments are classified in the "C" category if they do not conform to the requirements for the "A," "B," "D," or "E" categories ....

(e) *Class D.* Cargo and baggage compartments are classified in the D category if they are so designed and constructed that a fire occurring therein will be completely confined without endangering the safety of the airplane or the occupants ....

(f) *Class E.* On airplanes used for the carriage of cargo only, the cabin area may be classified as a Class E compartment ....

(1) It must be completely lined with fire-resistant material. (2) It must have a separate system of an approved type smoke or fire detector to give warning at the pilot or flight engineer station. (3) It must have a means to shut off the ventilating air flow ....

125.121. Proof of compliance with Paragraph 125.119

Compliance with those provisions of Paragraph 125.119 that refer to compartment accessibility, the entry of hazardous quantities of smoke or extinguishing agent into compartment occupied by the crew or passengers, and the dissipation of the extinguishing agent in Class C compartments must be shown by tests in flight ....

125.121. Propeller deicing fluid

If combustible fluid is used for propeller deicing, the certificate holder must comply with Paragraph 125.153.

The titles of the remaining paragraphs of this subpart are:

125.125 Pressure crossfeed arrangements
125.127 Location of fuel tanks
125.129 Fuel system lines and fittings
125.131 Fuel lines and fittings in designated fire zones
125.133 Fuel valves
125.135 Oil lines and fittings in designated fire zones
125.137 Oil valves
125.139 Oil system drains
125.141 Engine breather lines
125.143 Firewalls
125.145 Firewall construction
125.147 Cowling
125.149 Engine accessory section diaphragm
125.151 Power plant fire protection
125.153 Flammable fluids
125.155 Shutoff means
125.157 Lines and fittings
125.159 Vent and drain lines
125.161 Fire-extinguishing systems
125.163 Fire-extinguishing agents
125.165 Extinguishing agent container pressure relief
125.167 Extinguishing agent container compartment temperature
125.169 Fire-extinguishing system materials
125.171 Fire-detector systems
125.173 Fire detectors
125.175 Protection of other airplane components against fire
125.177 Control of engine rotation
125.179 Fuel system independence
125.181 Induction system ice prevention
125.183 Carriage of cargo in passenger compartments
125.185 Carriage of cargo in cargo compartments
125.187 Landing gear: Aural warning device
125.189 Demonstration of emergency evacuation procedures

SUBPART F: INSTRUMENT AND EQUIPMENT REQUIREMENTS
125.201. Inoperable instruments and equipment

(a) No person may takeoff an airplane with inoperable instruments or equipment installed unless the following conditions are met:

(1) An approved MEL exists for that airplane ....

The titles of the remaining paragraphs of this subpart are:

125.201 Inoperable instruments and equipment

125.203 Radio and navigational equipment
125.204 Portable electronic devices
125.205 Equipment requirements: Airplanes under IFR
125.206 Pitot heat indication systems
125.207 Emergency equipment requirements
125.209 Emergency equipment: Extended over-water operations
125.211 Seat and safety belts
125.213 Miscellaneous equipment
125.215 Operating information required
125.217 Passenger information
125.219 Oxygen for medical use by passengers
125.221 Icing conditions: Operating limitations
125.223 Airborne weather radar equipment requirements
125.224 Collision avoidance system
125.225 Flight recorders
125.226 Digital flight data recorders
125.227 Cockpit voice recorders
125.228 Flight data recorder: filtered data
  Appendix A: Additional Emergency Equipment
(a) Means for emergency evacuation........
(b) Interior emergency exit marking........
(c) Lighting for interior emergency exit markings....
(d) Emergency light operation........
(e) Emergency exit operating handles.....
(f) Emergency exit access......
(g) Exterior exit markings......
(h) Exterior emergency lighting and escape route......
(i) Floor level exits......
(j) Additional emergency exits......
(k) Turbojet-powered airplane, ventral exit and tailcone exit.....
  Appendix C Ice Protection
  Appendix D Airplane Flight Recorder Specification
  Appendix E Airplane Flight Recorder Specifications

8.7.4. FAR 129. Operations: Foreign air carriers and foreign operators of US-registered aircraft engaged in common carriage

SUBPART A: GENERAL
129.1. Applicability and definitions
(a) *Foreign air carrier operations in the United States.* This part prescribes rules governing the operation within the United States of each foreign air carrier holding the following:
  (1) A permit issued by the Civil Aeronautics Board or the US Department of Transportation under 49 USC 41301 to 41306 (formerly section 402

of the Federal Aviation Act of 1958, as amended), or (2) other appropriate economic or exemption authority issued by the Civil Aeronautics Board or the US Department of Transportation.

(b) *Operations of US-registered aircraft solely outside the United States.* In addition to the operations specified under Paragraph (a) of this section, Paragraphs 129.14, 129.16, 129.20, 129.32, and 129.33 also apply to US-registered aircraft operated solely outside the United States in common carriage by a foreign person or foreign air carrier.

(c) *Definitions.* For the purpose of this part:

(1) *Foreign person* means any person who is not a citizen of the United States and who operates a US-registered aircraft in common carriage solely outside the United States. (2) *Years in service* means the calendar time elapsed since an aircraft was issued its first US or foreign airworthiness certificate.

129.13. Airworthiness and registration certificates

(a) Except as provided in Paragraph 129.28(b) of this part, no foreign air carrier may operate any aircraft within the United States unless that aircraft carries current registration and airworthiness certificates issued or validated by the country of registry and displays the nationality and registration markings of that country.

(b) No foreign air carrier may operate a foreign aircraft within the United States except in accordance with the limitations on maximum certificated weights prescribed for that aircraft and that operation by the country of manufacture of the aircraft.

129.17. Aircraft communication and navigation equipment for operations under IFR or over-the-top

(a) *Aircraft navigation equipment requirements—General.* No foreign air carrier may conduct operations under IFR or over-the-top unless—(1) The en route navigation aids necessary for navigating the aircraft along the route (e.g., ATS routes, arrival and departure routes, and instrument approach procedures, including missed approach procedures if a missed approach routing is specified in the procedure) are available and suitable for use by the aircraft navigation equipment required by this section; (2) The aircraft used in those operations is equipped with at least the following......

129.18. Collision avoidance system

Effective 1 January 2005, any airplane you, as a foreign air carrier, operate under Part 129 must be equipped and operated according to the following table: ...

129.20. Digital flight data recorders

No person may operate an aircraft under this part that is registered in the United States unless it is equipped with one or more approved flight recorders that use a digital method of recording and storing data ....

129.24. Cockpit voice recorders

No person may operate an aircraft under this part that is registered in the United States unless it is equipped with an approved cockpit voice recorder that meets the standards of TSO—C123a or later revision. The cockpit voice recorder must record the information that would be required to be recorded if the aircraft were operated under FAR 121, 125, or 135 of this chapter, and must be installed by the compliance times required by that FAR, as applicable to the aircraft.

8.7.5. FAR 135. Operating Requirements: Commuter and on-demand operations and rules governing persons on board such aircraft

SUBPART A: GENERAL

135.1. Applicability

(a) This part prescribes rules governing:
  (1) The commuter or on-demand operations of each person who holds or is required to hold an Air Carrier Certificate or Operating Certificate under FAR 119.
  (2) Each person employed or used by a certificate holder conducting operations under this part, including the maintenance, preventative maintenance, and alteration of an aircraft.
  (3) The transportation of mail by aircraft conducted under a postal service contract ....
  (4) Each person who applies for provisional approval of an Advanced Qualification Program curriculum, curriculum segment ....
  (5) Nonstop Commercial Air Tour flights conducted for compensation or hire in accordance with §119.1(e)(2) of this chapter that begin and end at the same airport and are conducted within a 25 statute-mile radius of that airport; provided further that these operations must comply only with the drug and alcohol testing requirements......
  (6) Each person who is on board an aircraft being operated under this FAR.
  (7) Each person who is an applicant for an Air Carrier Certificate or an Operating Certificate under FAR 119, when conducting proving tests.
  (8) Commercial Air tours conducted by holders of operations specifications issued under this part must comply with the provisions of part 136, Subpart A of this chapter by 11 September 2007.

135.25. Aircraft requirements

...

(d) A certificate holder may operate in common carriage, and for the carriage of mail, a civil aircraft that is leased or chartered to it without crew and is registered in a country that is a party to the Convention on International Civil Aviation if:
  (1) The aircraft carries an appropriate airworthiness certificate issued by the country of registration and meets the registration and identification

requirements of that country. (2) The aircraft is of a type design that is approved under a US type certificate and complies with all the requirements of this chapter (14 CFR Chapter I) that would be applicable to that aircraft registered in the United States, including the requirements that must be met for issuance of a US standard airworthiness certificate (including type design conformity, condition for safe operation, and the noise, fuel venting, and engine emission requirements of this chapter), except that a US registration certificate and a US standard airworthiness certificate will not be issued for the aircraft....

SUBPART C-AIRCRAFT AND EQUIPMENT
135.141. Applicability
This subpart prescribes aircraft and equipment requirements for operations under this part. The requirements of this subpart are in addition to the aircraft and equipment requirements of FAR 91. However, this part does not require the duplication of any equipment required by this chapter.

The titles of the other paragraphs of this subpart are:

135.143 General requirements.

135.144 Portable electronic devices.

135.145 Aircraft proving and validation tests.

135.147 Dual controls required.

135.149 Equipment requirements: General.

135.150 Public address and crewmember interphone systems.

135.151 Cockpit voice recorders.

135.152 Flight data recorders.

135.153 Ground proximity warning system.

135.154 Terrain awareness and warning system.

135.155 Fire extinguishers: Passenger-carrying aircraft.

135.157 Oxygen equipment requirements.

135.158 Pitot heat indication systems.

135.159 Equipment requirements: Carrying passengers under VFR at night or under VFR over-the-top conditions.

135.161 Communication and navigation equipment for aircraft operations under VFR over routes navigated by pilotage.

135.163 Equipment requirements: Aircraft carrying passengers under IFR.

135.165 Communication and navigation equipment: Extended over-water or IFR operations.

135.167 Emergency equipment: Extended over-water operations.

135.168 [Reserved].

135.169 Additional airworthiness requirements.

135.170 Materials for compartment interiors.

135.171 Shoulder harness installation at flight crewmember stations.

135.173 Airborne thunderstorm detection equipment requirements.

135.175 Airborne weather radar equipment requirements.

135.177 Emergency equipment requirements for aircraft having a passenger seating configuration.

135.178 Additional emergency equipment of more than 19 passengers.

135.179 Inoperable instruments and equipment.

135.180 Traffic Alert and Collision Avoidance System

135.181 Performance requirements: Aircraft operated over-the-top or in IFR conditions.

135.183 Performance requirements: Land aircraft operated over water.

135.185 Empty weight and center of gravity: Currency requirement.

SUBPART I—AIRPLANE PERFORMANCE OPERATING LIMITATIONS
135.361. Applicability

(a) This subpart prescribes airplane performance operating limitations applicable to the operation of the categories of airplanes listed in §135.363 when operated under this part.

135.365 Large transport category airplanes: Reciprocating engine powered: Weight limitations.

135.367 Large transport category airplanes: Reciprocating engine powered: Takeoff limitations.

135.369 Large transport category airplanes: Reciprocating engine powered: En route limitations: All engines operating.

135.371 Large transport category airplanes: Reciprocating engine powered: En route limitations: One-engine inoperative.

135.373 Part 25 transport category airplanes with four or more engines: Reciprocating engine powered: En route limitations: Two engines inoperative.

135.375 Large transport category airplanes: Reciprocating engine powered: Landing limitations: Destination airports.

135.377 Large transport category airplanes: Reciprocating engine powered: Landing limitations: Alternate airports.

135.379 Large transport category airplanes: Turbine engine powered: Takeoff limitations.

135.381 Large transport category airplanes: Turbine engine powered: En route limitations: One-engine inoperative.

135.383 Large transport category airplanes: Turbine engine powered: En route limitations: Two engines inoperative.

135.385 Large transport category airplanes: Turbine engine powered: Landing limitations: Destination airports.

135.387 Large transport category airplanes: Turbine engine powered: Landing limitations: Alternate airports.

135.389 Large nontransport category airplanes: Takeoff limitations.

135.391 Large nontransport category airplanes: En route limitations: One-engine inoperative.

135.393 Large nontransport category airplanes: Landing limitations: Destination airports.

135.395 Large nontransport category airplanes: Landing limitations: Alternate airports.

135.397 Small transport category airplane performance operating limitations.

135.398 Commuter category airplanes performance operating limitations.

135.399 Small nontransport category airplane performance operating limitations.

8.7.6. FAR 137. Agricultural Aircraft Operations

SUBPART A: GENERAL

137.1. Applicability

(a) This part prescribes rules governing:

  (1) Agricultural aircraft operations within the United States and

  (2) The issue of commercial and private agricultural aircraft operator certificates for those operations.

(b) In a public emergency, a person conducting agricultural aircraft operations under this part may, to the extent necessary, deviate from the operating rules of this part for relief and welfare activities approved by an agency of the United States or of a State or local government.

(c) Each person who, under the authority of this section, deviates from a rule of this part. …

SUBPART B: CERTIFICATION RULES

137.11. Certificate required

(a) Except as provided in Paragraphs (c) and (d) of this section, no person may conduct agricultural aircraft operations without, or in violation of, an agricultural aircraft operator certificate issued under this part.

(b) Notwithstanding FAR, an operator may, if he complies with this part, conduct agricultural aircraft operations with a rotorcraft with external dispensing equipment in place without a rotorcraft external-load operator certificate.

(c) A federal, state, or local government conducting agricultural aircraft operations with public aircraft need not comply with this subpart.

(d) The holder of a rotorcraft external-load operator certificate under FAR 133 conducting an agricultural aircraft operation, involving only the dispensing of water on forest fires by rotorcraft external-load means, need not comply with this subpart.

SUBPART C: OPERATING RULES

137.31. Aircraft requirements

No person may operate an aircraft unless that aircraft:

(a) Meets the requirements of Paragraph 137.19(d)[32] and

(b) Is equipped with a suitable and properly installed shoulder harness for use by each pilot.

---

[32] 137(d) *Aircraft*. The applicant must have at least one certificated and airworthy aircraft, equipped for agricultural operation.

8.8. JAA OPERATIONAL STANDARDS (ADDITIONAL AIRWORTHINESS REQUIREMENTS)

NOTE: See Appendix 8.8/8.9

8.8.1. JAR-OPS 1. Commercial Air Transportation (Aeroplanes)

SUBPART A: APPLICABILITY

JAR-OPS 1.001 Applicability

(a) JAR-OPS Part 1 prescribes requirements applicable to the operation of any civil aeroplane for the purpose of commercial air transportation by any operator whose principal place of business and [if any, its registered office] is in a JAA Member State. JAR-OPS 1 does not apply

(1) to aeroplanes when used in military, customs, and police services; (2) to parachute dropping and fire-fighting flights, and to associated positioning and return flights in which the persons carried are those who would normally be carried on parachute dropping or fire-fighting; or (3) to flights immediately before, during, or immediately after an aerial work activity provided these flights are connected with that aerial work activity and in which, excluding crew members, no more than six persons indispensable to the aerial work activity are carried. ...

(b) The requirements in JAR-OPS Part 1 are applicable:

(1) For operators of aeroplanes over 10 ton Maximum Takeoff Mass or with maximum-approved passenger seating configuration of 20 or more, or with mixed fleets of aeroplanes above and below this discriminant, ...

(2) For operators of all other aeroplanes, ...

SUBPART B: GENERAL

JAR-OPS 1.030 MELs—Operator's responsibilities

(a) An operator shall establish, for each aeroplane, an MEL approved by the authority. ...

(b) An operator shall not operate an aeroplane other than in accordance with the MEL unless permitted by the authority. Any such permission will in no circumstances permit operation outside the constraints of the MMEL.

JAR-OPS 1.060 Ditching

An operator shall not operate an aeroplane with an approved passenger seating configuration of more than 30 passengers on over-water flights at a distance from land suitable for making an emergency landing, greater than 120 minutes at cruising speed, or 400 nautical miles, whichever is the lesser, unless the aeroplane complies with the ditching requirements prescribed in the applicable airworthiness code.

SUBPART F: PERFORMANCE GENERAL[33]
JAR-OPS 1.470 Applicability

(a) An operator shall ensure that multiengine aeroplanes powered by turbopro-peller engines with a maximum-approved passenger seating configuration of more than nine or a maximum takeoff mass exceeding 5700 kg, and all multiengine turbojet-powered aeroplanes are operated in accordance with Subpart G (Performance Class A).

(b) An operator shall ensure that propeller-driven aeroplanes with a maximum approved passenger seating configuration of nine or less, and a maximum takeoff mass of 5700 kg or less are operated in accordance with Subpart H (Performance Class B).

(c) An operator shall ensure that aeroplanes powered by reciprocating engines with a maximum-approved passenger seating configuration of more than nine or a maximum takeoff mass exceeding 5700 kg are operated in accor-dance with Subpart I (Performance Class C).

(d) Where full compliance with the requirements of the appropriate subpart cannot be shown due to specific design characteristics (e.g., supersonic aero-planes or seaplanes), the operator shall apply approved performance stan-dards that ensure a level of safety equivalent to that of the appropriate subpart.

SUBPART G: PERFORMANCE CLASS A
JAR-OPS 1.485 General

(a) An operator shall ensure that, for determining compliance with the require-ments of this subpart, the approved performance data in the aeroplane flight manual is supplemented as necessary with other data acceptable to the authority if the approved performance data in the aeroplane flight manual is insufficient with respect to items such as:

(1) Accounting for reasonably expected adverse operating conditions such as takeoff and landing on contaminated runways and (2) consideration of engine failure in all flight phases.

(b) An operator shall ensure that, for the wet and contaminated runway case, performance data determined in accordance with JAR 25X1591 or equiva-lent acceptable to the authority is used.

The titles of the other Subpart G paragraphs are:

JAR-OPS 1.490	Takeoff
JAR-OPS 1.495	Takeoff obstacle clearance
JAR-OPS 1.500	En route—One engine inoperative
JAR-OPS 1.505	Appendix 8 En route—Aeroplanes with three or more engines, two engines inoperative
JAR-OPS 1.510	Landing—Destination and alternate aerodromes
JAR-OPS 1.515	Landing—Dry runways
JAR-OPS 1.520	Landing—Wet and contaminated runways

---

[33] Performance Class A is defined in JAR-OPS 1 Subpart G; Performance Class B is defined in JAR-OPS 1 Subpart H; Performance Class C is defined in JAR-OPS 1 Subpart I.

SUBPART H: PERFORMANCE CLASS B
JAR-OPS 1.525 General

(a) An operator shall not operate a single-engine aeroplane:

    (1) At night or (2) in Instrument Meteorological Conditions except under Special VFRs.

(b) An operator shall treat two-engine aeroplanes that do not meet the climb requirements of Appendix 1 to JAR-OPS 1.525(b) as single-engine aeroplanes. The titles of the other paragraphs of Subpart H are

JAR-OPS 1.530	Takeoff
JAR-OPS 1.535	Takeoff obstacle clearance—Multiengine aeroplanes
JAR-OPS 1.540	En route—Multiengine aeroplanes
JAR-OPS 1.542	En route—Single-engine aeroplanes
JAR-OPS 1.545	Landing—Destination and alternate aerodromes
JAR-OPS 1.550	Landing—Dry runway
JAR-OPS 1.555	Landing—Wet and contaminated runways

SUBPART I: PERFORMANCE CLASS C
JAR-OPS 1.560 General

An operator shall ensure that, for determining compliance with the requirements of this subpart, the approved performance data in the aeroplane flight manual is supplemented, as necessary, with other data acceptable to the authority if the approved performance data in the aeroplane flight manual is insufficient.

    The titles of the other Subpart I paragraphs are

JAR-OPS 1.565	Takeoff
JAR-OPS 1.570	Takeoff obstacle clearance
JAR-OPS 1.575	En route—All engines operating
JAR-OPS 1.580	En route—One engine inoperative
JAR-OPS 1.585	En route—Aeroplanes with three or more engines, two engines inoperative

SUBPART K: INSTRUMENTS AND EQUIPMENT
JAR-OPS 1.630 General introduction

(a) An operator shall ensure that a flight does not commence unless the instruments and equipment required under this subpart are

    (1) Approved, except as specified in Subparagraph (c), and installed in accordance with the requirements applicable to them, including the minimum performance standard and the operational and airworthiness requirements and (2) in operable condition for the kind of operation being conducted except as provided in the MEL (JAR-OPS 1.030 refers).

(b) Instruments and equipment minimum performance standards are those prescribed in the applicable JTSO as listed in JAR-TSO, unless different performance standards are prescribed in the operational or airworthiness codes....

(c) The following items shall not be required to have an equipment approval: (1) Fuses referred to in JAR-OPS 1.635; (2) electric torches referred to in JAR-OPS 1.640(a)(4); (3) an accurate timepiece referred to in JAR-OPS 1.650(b) and 1.652(b); (4) chart holder referred to in JAR-OPS 1.652(n); (5) first-aid kits referred to in JAR-OPS 1.745....

(d) If equipment is to be used by one flight crew member at his station during flight, it must be readily operable from his station. When a single item of equipment is required to be operated by more than one flight crew member it must be installed so that the equipment is readily operable from any station at which the equipment is required to be operated.

(e) Those instruments that are used by any one flight crew member shall be so arranged as to permit the flight crew member to see the indications readily from his station, with the minimum practicable deviation from the position and line of vision that he normally assumes when looking forward along the flight path. Whenever a single instrument is required in an aeroplane operated by more than one flight crew member it must be installed so that the instrument is visible from each applicable flight crew station.

JAR-OPS 1.635 Circuit protection devices

An operator shall not operate an aeroplane in which fuses are used unless ...

JAR-OPS 1.640 Aeroplane operating lights

An operator shall not operate an aeroplane unless it is equipped with:

(a) For flight by day:

(1) Anticollision light system; (2) lighting supplied from the aeroplane's electrical system to provide adequate illumination for all instruments and equipment essential to the safe operation of the aeroplane; (3) lighting supplied from the aeroplane's electrical system to provide illumination in all passenger compartments; and (4) an electric torch for each required crew member readily accessible to crew members when seated at their designated station.

(b) For flight by night, in addition to equipment specified in Paragraph (a) above:

(1) Navigation/position lights; (2) two landing lights or a single light having two separately energized filaments; and (3) lights to conform to the international regulations for preventing collisions at sea if the aeroplane is a seaplane or an amphibian.

JAR-OPS 1.645 Windshield wipers

An operator shall not operate an aeroplane with a maximum certificated takeoff mass of more than 5700 kg unless it is equipped at each pilot station with a windshield wiper or equivalent means to maintain a clear portion of the windshield during precipitation.

JAR-OPS 1.650 Day VFR operations: Flight and navigational instruments and associated equipment

An operator shall not operate an aeroplane by day in accordance with VFRs unless it is equipped with the flight and navigational instruments and associated equipment and, where applicable, under the conditions stated in the following subparagraphs:

(a) A magnetic compass.

(b) An accurate timepiece showing the time in hours, minutes, and seconds.

(c) A sensitive pressure altimeter calibrated in feet with a subscale setting, calibrated in hectopascals/millibars, adjustable for any barometric pressure likely to be set during flight.

(d) An airspeed indicator calibrated in knots.

(e) A vertical speed indicator.

(f) A turn and slip indicator, or a turn coordinator incorporating a slip indicator.

(g) An attitude indicator ....

JAR-OPS 1.652 IFR or night operations: Flight and navigational instruments and associated equipment

An operator shall not operate an aeroplane in accordance with IFRs or by night in accordance with VFRs unless it is equipped with the flight and navigational instruments and associated equipment and, where applicable, under the conditions stated in the following subparagraphs:

(a) A magnetic compass.

(b) An accurate timepiece showing the time in hours, minutes, and seconds.

(c) Two sensitive pressure altimeters calibrated in feet with subscale settings, calibrated in hectopascals/millibars, adjustable for any barometric pressure likely to be set during flight....

(d) An airspeed indicating system with heated pitot tube or equivalent means for preventing malfunctioning due to either condensation or icing, including a warning indication of pitot heater failure. ...

(e) A vertical speed indicator.

(f) A turn and slip indicator.

(g) An attitude indicator. ...

JAR-OPS 1.655 Additional equipment for single pilot operation under IFR

An operator shall not conduct single pilot IFR operations unless the aeroplane is equipped with an autopilot with at least altitude hold and heading mode.

JAR-OPS 1.660 Altitude alerting system

(a) An operator shall not operate a turbine propeller-powered aeroplane with a maximum certificated takeoff mass in excess of 5700 kg or having a maximum approved passenger seating configuration of more than nine seats or a turbojet-powered aeroplane unless it is equipped with an altitude alerting system capable of:

(1) Alerting the flight crew on approaching a preselected altitude and (2) alerting the flight crew by at least an aural signal, when deviating from a preselected altitude, except for aeroplanes with a maximum certificated takeoff mass of 5700 kg or less having a maximum approved passenger seating configuration of more than nine and first issued with an individual certificate of airworthiness in a JAA Member State before 1 April 1972 and already registered in a JAA Member State on 1 April 1995.

The titles of the other Subpart K paragraphs are

JAR-OPS 1.665	Ground proximity warning system and terrain awareness warning system
JAR-OPS 1.668	Airborne collision avoidance system
JAR-OPS 1.670	Airborne weather radar equipment
JAR-OPS 1.675	Equipment for operations in icing conditions
JAR-OPS 1.680	Cosmic radiation detection equipment
JAR-OPS 1.685	Flight crew interphone system
JAR-OPS 1.690	Crew member interphone system
JAR-OPS 1.695	Public address system
JAR-OPS 1.700	Cockpit voice recorders
JAR-OPS 1.705	Cockpit voice recorders
JAR-OPS 1.710	Cockpit voice recorders
JAR-OPS 1.715	Flight data recorders
JAR-OPS 1.720	Flight data recorders
JAR-OPS 1.725	Flight data recorders
JAR-OPS 1.727	Combination recorder
JAR-OPS 1.730	Seats, seat safety belts, harnesses, and child restraint devices
JAR-OPS 1.731	Fasten seat belt and no smoking signs
JAR-OPS 1.735	Internal doors and curtains
JAR-OPS 1.745	First-aid kits
JAR-OPS 1.750	Intentionally blank
JAR-OPS 1.755	Emergency medical kit
JAR-OPS 1.760	First-aid oxygen
JAR-OPS 1.765	Intentionally blank
JAR-OPS 1.770	Supplemental oxygen—Pressurized aeroplanes
JAR-OPS 1.775	Supplemental oxygen—Nonpressurized aeroplanes
JAR-OPS 1.780	Crew protective breathing equipment
JAR-OPS 1.785	Intentionally blank
JAR-OPS 1.790	Hand fire extinguishers
JAR-OPS 1.795	Crash axes and crowbars
JAR-OPS 1.800	Marking of break-in points
JAR-OPS 1.805	Means for emergency evacuation
JAR-OPS 1.810	Megaphones

JAR-OPS 1.815	Emergency lighting
JAR-OPS 1.820	Emergency locator transmitter
JAR-OPS 1.825	Life jackets
JAR-OPS 1.830	Life-rafts and survival ELTs for extended over-water flights
JAR-OPS 1.835	Survival equipment
JAR-OPS 1.840	Seaplanes and amphibians—Miscellaneous equipment

SUBPART L: COMMUNICATION AND NAVIGATION EQUIPMENT

The titles of the Subpart L paragraphs are

JAR-OPS 1.845	General introduction
JAR-OPS 1.850	Radio equipment
JAR-OPS 1.855	Audio selector panel
JAR-OPS 1.860	Radio equipment for operations under VFR over routes navigated by reference to visual landmarks
JAR-OPS 1.865	Communication and navigation equipment for operations under IFR, or under VFR over routes not navigated by reference to visual landmarks
JAR-OPS 1.866	Transponder equipment
JAR-OPS 1.870	Additional navigation equipment for operations in MNPS airspace
JAR-OPS 1.872	Equipment for operation in defined airspace with reduced vertical separation minima

8.8.2. JAR-OPS 3. Commercial Air Transportation (Helicopters)

SUBPART A: APPLICABILITY

JAR-OPS 3.001 Applicability

(a) JAR-OPS Part 3 prescribes requirements applicable to the operation of any civil helicopter for the purpose of commercial air transportation by any operator whose principal place of business is in a JAA Member State. JAR-OPS Part 3 does not apply:

(1) To helicopters when used in military, customs, police services, and SAR or

(2) To parachute dropping and fire-fighting flights, and to associated positioning and return flights in which the only persons carried are those who would normally be carried on parachute dropping or fire-fighting flights or

(3) To flights immediately before, during, or immediately after an aerial work activity provided these flights are connected with that aerial work activity and in which, excluding crew members, no more than six persons indispensable to the aerial work activity are carried....

The scheme of JAR-OPS 3 is similar to JAR-OPS 1.

The **additional airworthiness requirements** can be found in the following paragraphs:

SUBPART B: GENERAL
JAR-OPS 3.030   MELs—Operator's responsibilities

SUBPART F: PERFORMANCE GENERAL
JAR-OPS 3.470   Applicability
JAR-OPS 3.475   General

SUBPART G: PERFORMANCE CLASS 1
JAR-OPS 3.485   General
JAR-OPS 3.490   Takeoff
JAR-OPS 3.495   Takeoff Flight Path
JAR-OPS 3.500   En-route—Critical power unit inoperative
JAR-OPS 3.510   Landing

SUBPART H: PERFORMANCE CLASS 2
JAR-OPS 3.515   General
JAR-OPS 3.517   Applicability
JAR-OPS 3.520   Takeoff
JAR-OPS 3.525   Takeoff Flight Path
JAR-OPS 3.530   En-route—Critical power unit inoperative
JAR-OPS 3.535   Landing

SUBPART I: PERFORMANCE CLASS 3
JAR-OPS 3.540   General
JAR-OPS 3.545   Takeoff
JAR-OPS 3.550   En-route
JAR-OPS 3.555   Landing

SUBPART K: INSTRUMENTS AND EQUIPMENT
JAR-OPS 3.630   General introduction
JAR-OPS 3.640   Helicopter operating lights
JAR-OPS 3.647   Equipment for operations requiring a radio communication and/or radio navigation system
JAR-OPS 3.650   Day VFR operations—Flight and navigational instruments and associated equipment
JAR-OPS 3.652   IFR or night operations—Flight and navigational instruments and associated equipment
JAR-OPS 3.655   Additional equipment for single pilot operation under IFR
JAR-OPS 3.660   Radio Altimeters
JAR-OPS 3.670   Airborne Weather Radar Equipment

JAR-OPS 3.675	Equipment for operations in icing conditions
JAR-OPS 3.685	Flight crew interphone system
JAR-OPS 3.690	Crew member interphone system
JAR-OPS 3.695	Public address system
JAR-OPS 3.700	Cockpit voice recorders-1
JAR-OPS 3.705	Cockpit voice recorders-2
JAR-OPS 3.715	Flight data recorders-1
JAR-OPS 3.720	Flight data recorders-2
JAR-OPS 3.730	Seats, seat safety belts, harnesses, and child restraint devices
JAR-OPS 3.731	Fasten seat belt and no smoking signs
JAR-OPS 3.745	First-Aid Kits
JAR-OPS 3.775	Supplemental oxygen—nonpressurized helicopters
JAR-OPS 3.790	Hand fire extinguishers
JAR-OPS 3.800	Marking of break-in points
JAR-OPS 3.810	Megaphones
JAR-OPS 3.815	Emergency lighting
JAR-OPS 3.820	Automatic Emergency Locator Transmitter
JAR-OPS 3.825	Life Jackets
JAR-OPS 3.827	Crew Survival Suits
JAR-OPS 3.830	Life-rafts and survival ELTs or extended over-water flights
JAR-OPS 3.835	Survival equipment
JAR-OPS 3.837	Additional requirements for helicopters operating to or from helidecks located in a hostile sea area
JAR-OPS 3.840	Helicopters certificated for operating on water—Miscellaneous equipment
JAR-OPS 3.843	All helicopters on flights over water—Ditching

SUBPART L: COMMUNICATION AND NAVIGATION EQUIPMENT

JAR-OPS 3.845	General introduction
JAR-OPS 3.850	Radio equipment
JAR-OPS 3.855	Audio-selector panel
JAR-OPS 3.860	Radio equipment for operations under VFR over routes navigated by reference to visual landmarks
JAR-OPS 3.865	Communication and Navigation equipment for operations under IFR or under VFR over routes not navigated by reference to visual landmarks

8.8.3. JAR-AWO All Weather Operations

JAR-AWO consists of Annex to the EASA Executive Director Decision 2003/06/RM dated 17 October 2003 (also called **CS-AWO**).

8.8.4. CS-AWO Certification Specification for All Weather Operations

The general layout of the document is reported here.

BOOK 1: AIRWORTHINESS CODE

Subpart 1: Automatic Landing Systems

Subpart 2: Airworthiness Certification of Aeroplanes for Operations with Decision Heights Below 60 m (200 ft) Down to 30 m (100 ft) —Category 2 Operations

Subpart 3: Airworthiness Certification of Aeroplanes for Operations with Decision Heights Below 30 m (100 ft) or No Decision Height—Category 3 Operations

Subpart 4: Directional Guidance for Takeoff in Low Visibility

BOOK 2: ACCEPTABLE MEANS OF COMPLIANCE

(for the above mentioned subparts)

The following paragraphs report the applicability of the four subparts.

Subpart 1. Automatic Landing Systems

General: CS-AWO 100 applicability and terminology

(a) Subpart 1 of this airworthiness code is applicable to aeroplanes that are capable of automatic landing carried out in association with an ILS, a Microwave Landing System (MLS), or both. In addition, the automatic landing system must meet the requirements of CS-25.1329.

(b) The term "automatic landing system" in this CS-AWO refers to the airborne equipment, which provides automatic control of the aeroplane during the approach and landing. It includes all the sensors, computers, actuators, and power supplies necessary to control the aeroplane to touchdown. It also includes provisions to control the aeroplane along the runway during the landing rollout. In addition, it includes the indications and control necessary for its management and supervision by the pilot.

Subpart 2. Airworthiness Certification of Aeroplanes for Operations with Decision Heights Below 60 m (200 ft) and Down to 30 m (100 ft)—Category 2 Operations

General: CS-AWO 200 applicability and terminology

(a) Subpart 2 of this airworthiness code is applicable to aeroplanes for which certification is sought to allow the performance of approaches with decision heights below 60 m (200 ft) down to 30 m (100 ft)—Category 2 operations, using a precision approach system as defined in Annex 10 of the Chicago Convention, that is, an ILS, or an MLS which has outputs indicating the magnitude and sense of deviation from a preset azimuth and elevation angle giving equivalent operational characteristics to that of a conventional ILS.

(b) Terminology:

(1) The term "approach system" used here refers only to the airborne system. It includes the equipment listed in CS-AWO 221 and all related sensors, instruments, and power supplies.

(2) "Decision height" is the wheel height above the runway elevation by which a go-around must be initiated unless adequate visual reference has been established and the aircraft position and approach path have been visually assessed as satisfactory to continue the approach and landing in safety. Where it is used in this Subpart 2 it means the minimum decision height at which compliance with the requirements of this Subpart 2 have been demonstrated.

(3) A go-around is the transition from an approach to a stabilized climb.

(4) "Failure condition" and terms describing the probabilities and effects of failure...

Subpart 3. Airworthiness Certification of Aeroplanes for Operations with Decision Height Below 30 m (100 ft)—Category 3 Operations
General: CS-AWO 300 applicability and terminology

(a) Subpart 3 of this airworthiness code is applicable to aeroplanes for which certification is sought to allow the performance of approaches with decision heights below 30 m (100 ft) or with no decision height—Category 3 operations, using a precision approach system as defined in Annex 10 of the Chicago Convention, that is, an ILS, or an MLS that has outputs indicating the magnitude and sense of deviation from a preset azimuth and elevation angle giving equivalent operational characteristics to that of a conventional ILS. The criteria are divided, where necessary, into those applicable to the following types of operation:

(1) Decision heights below 30 m (100 ft) but not less than 15 m (50 ft).

(2) Decision heights below 15 m (50 ft).

(3) No decision height.

(b) Terminology:

(1) The term "landing system" used here refers only to the airborne system. It includes the equipment listed in JAR-AWO 321 and also all related sensors, instruments, and power supplies.

(2) Automatic landing system: the airborne equipment that provides automatic control of the aeroplane during the approach and landing.

(3) Fail-passive automatic landing system: an automatic landing system is fail passive if, in the event of a failure, there is no significant out-of-trim condition or deviation of flight path or attitude but the landing is not completed automatically. For a fail-passive automatic landing system, the pilot assumes control of the aircraft after a failure....

Subpart 4. Directional Guidance for Takeoff in Low Visibility
CS-AWO 400 applicability and terminology

(a) Subpart 4 of this airworthiness code is applicable to aeroplanes for which certification is sought to allow the performance of takeoff in lower visibilities than those that are sufficient to ensure that the pilot will at all times have sufficient visibility to complete or abandon the takeoff safely. It is only concerned with directional guidance during the ground-borne portion of the takeoff (i.e., from start to main wheel lift-off, or standstill in the event of abandoned takeoff).

(b) Takeoff guidance system: a takeoff guidance system provides directional guidance information to the pilot during the takeoff or abandoned takeoff. It includes all the airborne sensors, computers, controllers, and indicators necessary for the display of such guidance.

Guidance normally takes the form of command information, but it may alternatively be situation (or deviation) information.

8.9. EASA OPERATIONAL STANDARDS (ADDITIONAL AIRWORTHINESS REQUIREMENTS)

(See Appendix 8.8/8.9)

8.9.1. General

The Annex III to Council Regulation (EEC) No. 3922/91, containing common technical requirements and administrative procedures applicable to commercial transportation by aeroplane, is now replaced by the Annex III to the Commission Regulation (EC) No. 8/2008, based on a set of harmonized rules adopted by the JAA, the already mentioned JAR-OPS 1 "Requirements for Commercial Air Transportation (Aeroplanes)" (Amendment 8).

However, JAR-OPS 1 has since progressed to Amendment 13. The European Commission has therefore put in place a process to update the Annex before the date of its applicability (16 July 2008).

The new Annex III, *"Common technical requirements and administrative procedures applicable to commercial transportation by aircraft"* is now **OPS 1: Commercial Air Transportation (Aeroplanes)**.

OPS 1 prescribes requirements applicable to the operation of any civil aeroplane for the purpose of commercial air transport by any operator whose principal place of business and, if any, registered office is in a Member State.

An operator shall not operate an aeroplane for the purpose of commercial air transport other than in accordance with EU-OPS.

8.9.2. Additional airworthiness requirements for commercial air transportation (aeroplanes)

With reference to the considerations in Sections 8.9.1 and 8.8.3, we can conclude that it is still valid for EASA what is written in Sections 8.8.1 and 8.8.3.

8.9.3. Additional airworthiness requirements for commercial air transportation (helicopters)

EU-OPS 3 does not yet exists, but **JAR-OPS 3** is applicable to Commercial Air Transportation involving helicopters that are used by an operator based in a JAA Member State.

Then, for additional airworthiness requirements, see Section 8.8.2.

APPENDIX 8.4 EASA PART 21 CERTIFICATES OF AIRWORTHINESS

Classification	Notes	Certificates of Airworthiness
21A.173(a) Certificates of airworthiness issued to aircraft, which conform to a type certificate issued in accordance with Part 21 (TC, see 21A.21)		21A.183, 21B.325
21A.173(b) Restricted (TC, see 21A.23)		21A.184, 21B.325
Permit to fly	Special-purpose flight of aircraft not currently meeting applicable airworthiness regulations, but capable of safe flight 1. development; 2. showing compliance with regulations or certification specifications; 3. design organizations or production organizations crew training; 4. production flight testing of new production aircraft; 5. flying aircraft under production between production facilities; 6. flying the aircraft for customer acceptance; 7. delivering or exporting the aircraft; 8. flying the aircraft for Authority acceptance; 9. market survey, including customer's crew training; 10. exhibition and air show; 11. flying the aircraft to a location where maintenance or airworthiness review are to be performed, or to a place of storage; 12. flying an aircraft at a weight in excess of its maximum-certificated takeoff weight for flight beyond the normal-range over-water, or over-land areas where adequate landing facilities or appropriate fuel are not available; 13. record breaking, air racing, or similar competition; 14. flying aircraft meeting the applicable airworthiness requirements before conformity to the environmental requirements has been found; and 15. for noncommercial flying activity on individual noncomplex aircraft or types for which a certificate of airworthiness or restricted certificate of airworthiness are not appropriate.	21A.185 Subpart P (Sections A and B)

APPENDIX 8.5 FAR 21 AIRWORTHINESS CERTIFICATES

Classification	Category	Notes	Airworthiness Certificates
Standard 21.175(a)	— Normal — Utility — Acrobatic — Commuter — Transport — Manned free balloons — Special classes (TC, see 21.21)		21.183
Special 21.175(b)	Primary (TC, see 21.24)	Aircraft flown for pleasure and personal use	21.184
	Restricted (TC, see 21.25)	Aircraft for the following special purposes operations: — Agricultural — Forest and wildlife conservation — Aerial surveying — Patrolling (pipelines, power lines) — Weather control — Aerial advertising — Other operations specified by the Administrator	21.185
	Limited (TC, see Order 8130-2F Paragraph 117)	Operate military aircraft converted to civilian use	21.189
	Light-Sport	Operate a light-sport aircraft, other than a gyroplane.	21.190
	Provisional (TC, see Subpart C)	Special purpose operations of aircraft with provisional TC	Subp. I
	Special Flight Permit	Special-purpose flight of aircraft not currently meeting applicable airworthiness regulations, but capable of safe flight	21.197, 21.199
	Experimental	Aircraft for the following special purposes operations: (a) Research and development (b) Showing compliance with regulations (c) Crew training (d) Exhibition (e) Air racing (f) Market surveys (g) Operating amateur-built aircraft (h) Operating primary kit-built aircraft (i) Operating a Light-Sport aircraft	21.191 21.193 21.195 21.195

APPENDIX 8.7 FAA OPERATIONAL STANDARDS (ADDITIONAL AIRWORTHINESS REQUIREMENTS)

FARs	Applicability	NOTES
91	(a) Except as provided in Paragraphs (b) and (c) of this section and FAR 91.701 and 91.703, FAR 91 prescribes rules governing the operation of aircraft within the United States, including the waters within three nautical miles of the US coast. (b) Each person operating an aircraft in the airspace overlying the waters between 3 and 12 nautical miles from the coast of the United States must comply with FAR 91.1 to 91.21.......	FAR 91 does not apply to moored balloons, kites, unmanned rockets, and unmanned free balloons, which are governed by FAR 101, and ultralight vehicles operated in accordance with FAR 103. **Additional airworthiness requirements in: Subparts C, D, G, H, and I Su**
121	FAR 121 prescribe rules governing: (a) The **domestic, flag,** and **supplemental operations** of each person who holds or is required to hold an **Air Carrier Certificate or Operating Certificate under FAR 119.** (b) Each person employed	**Additional airworthiness requirements in Subparts G, I, J, and K**
125	(a) Except as provided in Paragraphs (b), (c), and (d) of this section, FAR125 prescribe rules governing the operations of US-registered civil **airplanes** that have a seating configuration of **20 or more passengers** or a **maximum payload capacity of 6000 lb or more** when **common carriage is not involved.**	(b) The rules of this part do not apply to the operations of airplanes specified in Paragraph (a) of this section, when: (1) They are required to be operated under FAR 121, 129, 135, or 137 of this chapter. (2) They have been issued Restricted, Limited, or Provisional airworthiness certificates, special flight permits, or Experimental certificates. (3) They are being operated by an FAR 125 certificate holder without carrying passengers or cargo under FAR 91 for training, ferrying, positioning, or maintenance purposes .... (c) The rules of this FAR, except Paragraph 125.247, do not apply to the operation of airplanes specified in Paragraph (a) when they are operated outside the United States by a person who is not a citizen of the United States ...

FARs	Applicability	NOTES
		Additional airworthiness requirements in Subparts E, F and Appendices A, C, D, E
129	(a) *Foreign air carrier operations in the United States.* FAR 129 prescribes rules governing the operation within the United States of each foreign air carrier holding the following: (1) A permit issued by the Civil Aeronautics Board or the US Department of Transportation .... (b) *Operations of US-registered aircraft solely outside the United States.* In addition to the operations specified under Paragraph (a) of this section, Paragraphs 129.14, 129.16, 129.20, 129.32, and 129.33 also apply to US-registered aircraft operated solely outside the United States in common carriage by a foreign person or foreign air carrier.	**Additional airworthiness requirements in Subpart A**
135	(a) This part prescribes rules governing: (1) The **commuter or on-demand operations** of each person who holds or is required to hold an **Air Carrier Certificate or Operating Certificate under FAR 119.** (2) Each person employed or used by a certificate holder conducting operations under this part, including the maintenance, preventative maintenance, and alteration of an aircraft. (3) The transportation of mail by aircraft conducted under a **postal service contract** .... (4) Each person who applies for provisional approval of an Advanced Qualification Program curriculum, curriculum segment .... (5) **Nonstop Commercial Air Tour flights** conducted for compensation or hire in accordance with §119.1(e)(2) of this chapter that begin and end at the same airport and are conducted within a 25 statute-mile radius of that airport; provided further that these operations must comply only with the drug and alcohol testing requirements......	**Additional airworthiness requirements in Subparts A, C, and I**
137	(a) This part prescribes rules governing: (1) **Agricultural aircraft** operations within the United States; and (2) The issue of commercial and private agricultural aircraft operator certificates for those operations.	(b) In a public emergency, a person conducting agricultural aircraft operations under this part may, to the extent necessary, deviate from the operating rules of this part for relief and welfare

FARs	Applicability	NOTES
		activities approved by an agency of the United States or of a State or local government. **(c)** Each person who, under the authority of this section, deviates from a rule of this part....
		Additional airworthiness requirements in Subparts B and C

APPENDIX 8.8/8.9 JAA/EASA OPERATIONAL STANDARDS (ADDITIONAL AIRWORTHINESS REQUIREMENTS)

	Applicability	NOTES
JAR/ EASA OPS 1	**JAR/EASA-OPS 1** prescribes requirements applicable to the operation of **any civil aeroplane** for the purpose of **commercial air transportation** by any **operator** whose principal place of business and, if any, its registered office is in a Member State, hereafter called **operator**.	**OPS 1 does not apply:** (1) to aeroplanes when used in military, customs, and police services nor (2) to parachute dropping and fire-fighting flights, and to associated positioning and return flights in which the persons carried are those who would normally be carried on parachute dropping or fire-fighting nor (3) to flights immediately before, during, or immediately after an aerial work activity provided these flights are connected with that aerial work activity and in which, excluding crew members, no more than 6 persons indispensable to the aerial work activity are carried.
		Additional airworthiness requirements in: Subparts B, F, G, H, I, K, and L Su
JAR- OPS 3	**JAR-OPS 3** prescribes requirements applicable to the operation of **any civil helicopter** for the purpose of **commercial air transportation** by any **operator** whose principal place of business is in a JAA Member State.	JAR-OPS 3 does not apply: **(1)** to helicopters when used in military, customs, police services, and SAR nor **(2)** to parachute dropping and fire-fighting flights, and to associated positioning and return flights in which the only persons carried are

	Applicability	NOTES
		those who would normally be carried on parachute dropping or fire-fighting flights; nor (3) to flights immediately before, during, or immediately after an aerial work activity, provided these flights are connected with that aerial work activity and in which, excluding crew members, no more than six persons indispensable to the aerial work activity are carried.
		Additional airworthiness requirements in Subparts B, F, G, H, I, K, and L
JAR/ CS AWO	• **SUBPART 1 AUTOMATIC LANDING SYSTEMS.** • **SUBPART 2 AIRWORTHINESS CERTIFICATION OF AEROPLANES FOR OPERATIONS WITH DECISION HEIGHTS BELOW 60 M (200 FT) AND DOWN TO 30 M (100 FT). CATEGORY 2 OPERATIONS.** • **SUBPART 3 AIRWORTHINESS CERTIFICATION OF AEROPLANES FOR OPERATIONS WITH DECISION HEIGHTS BELOW 30 M (100 FT) OR NO DECISION HEIGHT. CATEGORY 3 OPERATIONS.** • **SUBPART 4 DIRECTIONAL GUIDANCE FOR TAKEOFF IN LOW VISIBILITY**	**Additional airworthiness requirements in Subparts 1, 2, 3, and 4**

Chapter | nine

Continued Airworthiness and Operation

9.1. CONTINUED AIRWORTHINESS

Safety must be ensured for all flight operations, and aircraft must constantly be maintained in an airworthy state. This means that all maintenance operations listed in the relevant manuals and Airworthiness Directives[1] (ADs) must be performed.

Continued airworthiness also depends on the particular organizations of operators and maintenance.

Therefore, in extreme synthesis, continued airworthiness is made by

(1) Maintenance

(2) In a more general sense, certification of operators.[2]

9.1.1. Maintenance

From an airworthiness point of view, there is no such concept as an "old"[3] aircraft: the term applied is "used" aircraft. This means that the aircraft's age could influence its commercial value, but not its airworthiness conditions.

Typically, and this also applies to maintenance, we need to know:

(1) What to do

(2) How to do it

(3) Where to do it

(4) Who does it.

These points are detailed below.

(1) The term "maintenance" refers to preventive maintenance, alterations and repairs, and introduction of ADs. Airworthiness should depend on the maintenance programs, which also establishes the replacement of time

---

[1] See the "ADs" section in this chapter (Section 9.2).

[2] See the "EASA certification of operators" and "FAA certification of operators" sections in this chapter.

[3] We will see an exception in the case of "older airplanes," related to maintenance procedures only, without any decrease in safety (Section 9.3).

243

change items, the overhaul of engines, propellers, and various parts and appliances.

In Chapter 5, we illustrated that the JAR/FAR 21/EASA Part 21 require Instructions for Continued Airworthiness (ICA) as a part of the product type certification, and also for the issue of Supplemental type certificates, for approval of changes to type design and major repairs.

Therefore, those instructions become the basic tool for the maintenance of the aircraft because they establish the basic maintenance program.

However, as for the aircraft configuration and the additional airworthiness requirements for operation, maintenance programs must also conform to the requirements of what we have called the "operational standards" and other maintenance standards (such as FAR 43 and EASA Part M). This will now be explained in more detail.

(2) The above-mentioned ICA, alongside the preventive maintenance programs, also contain the necessary instructions for this activity.[4] For extraordinary maintenance, such as repairs, we have also seen that repair manuals are normally provided and, if not available or not covering the particular repair case, a repair design has to be approved.

The operational standards and other maintenance standards (such as FAR 43 and EASA Part M) also establish the rules to be complied with for maintenance in relation to the type of aircraft and the kind of operation involved.

(3) Apart from those particular cases, discussed in Chapter 8, of special certifications allowing the aircraft owner to perform the maintenance himself or herself, the authorities issue requirements for aircraft operators with particular attention to the maintenance organizations. This will now be discussed further.

(4) Associated with the certification of maintenance organization is the certification and training of personnel authorized to perform maintenance operation and issue "release to service certificates" on completion of maintenance.

9.1.2. EASA continued airworthiness/maintenance

NOTE: See Appendix 9.1.2

The EU Commission has approved EC Regulation No. 2042/2003 "On the continuing airworthiness of aircraft and aeronautical products, parts and appliances, and on the approval of organizations and personnel involved in these tasks."

An excerpt of this Regulation is given below.

OBJECTIVE AND SCOPE

(1) This Regulation establishes common technical requirements and administrative procedures for ensuring the continuing airworthiness of aircraft, including any component for installation thereto, which are as follows:

---

[4] In Chapter 5, the "Instructions for Continued Airworthiness" section contains an example of what kind of instruction must be produced.

(a) Registered in a Member State or

(b) Registered in a third country and used by an operator for which a Member State ensures oversight of operations.

(2) Paragraph 1 shall not apply to aircraft, the regulatory safety oversight of which has been transferred to a third country and which are not used by a Community operator, or to aircraft referred to in Annex II to the basic Regulation.

(3) The provisions of this Regulation related to commercial air transport are applicable to licensed air carriers as defined by Community law.

CONTINUING AIRWORTHINESS REQUIREMENTS

(1) The continuing airworthiness of aircraft and components shall be ensured in accordance with the provisions of **Annex I (Part M)**.

(2) Organizations and personnel involved in the continuing airworthiness of aircraft and components, including maintenance, shall comply with the provisions of Annex I and where appropriate with those specified in Articles 4 and 5.

(3) By derogation from Paragraph 1, the continuing airworthiness of aircraft holding a permit to fly shall be granted on the basis of the specific continuing airworthiness arrangements as defined in the permit to fly issued in accordance with the Part 21.

MAINTENANCE ORGANIZATION APPROVALS (ARTICLE 4)

Organizations involved in the maintenance of large aircraft or of aircraft used for commercial air transport, and components intended for fitment thereto, shall be approved in accordance with the provisions of **Annex II (Part 145)** ...

CERTIFYING STAFF (ARTICLE 5)[5]

Certifying staff shall be qualified in accordance with the provisions of **Annex III (Part 66)** ...

TRAINING ORGANIZATION REQUIREMENTS (ARTICLE 6)

Organizations involved in the training of personnel referred to in Article 5 shall be approved in accordance with **Annex IV (Part 147)** ...

The above-mentioned Annexes are all provided with Acceptable Means of Compliance (AMC) and Guidance Material (GM).

To give an idea of their content, we will now consider some significant extracts of the four Annexes.

9.1.2.1. ANNEX I, PART M

1. Section A—Technical Requirements

Subpart A: General

M.A.101 Scope. This section establishes the measures to be taken to ensure that airworthiness is maintained, including maintenance. It also specifies the

---

[5] "Certifying staff" means personnel responsible for the release of an aircraft or a component after maintenance.

conditions to be met by the individuals or organizations involved in such continuing airworthiness management.

Subpart B: Accountability
M.A.201 Responsibilities

(a) The owner is responsible for the continuing airworthiness of an aircraft and shall ensure that no flight takes place unless:

(1) The aircraft is maintained in an airworthy condition,

(2) Any operational and emergency equipment fitted is correctly installed and serviceable or clearly identified as unserviceable,

(3) The airworthiness certificate remains valid, and

(4) The maintenance of the aircraft is performed in accordance with the approved maintenance program as specified in M.A.302.

⋮

(e) To satisfy the responsibilities of Paragraph (a), the owner of an aircraft may contract the tasks associated with continuing airworthiness to an approved continuing airworthiness management organization (CAMO) as specified in M.A. Subpart G (CAMO hereinafter) in accordance with Appendix I. In this case, the CAMO assumes responsibility for the proper accomplishment of these tasks.

(f) In the case of large aircraft, to satisfy the responsibilities of Paragraph (a), the owner of an aircraft shall ensure that the tasks associated with continuing airworthiness are performed by an approved CAMO. A written contract shall be made in accordance with Appendix I. In this case, the CAMO assumes responsibility for the proper accomplishment of these tasks.

(g) Maintenance of large aircraft, aircraft used for commercial air transport and components thereof shall be carried out by a Part 145 approved maintenance organization.

(h) In the case of commercial air transport, the operator is responsible for the continuing airworthiness of the aircraft it operates and shall

(1) Be approved, as a part of the air operator certificate (AOC) issued by the competent authority, pursuant to M.A. Subpart G for the aircraft it operates;

(2) Be approved in accordance with Part 145 or contract such an organization; and

(3) Be ensured that Paragraph (a) is satisfied.

(i) When an operator is requested by a Member State to hold a certificate for its operational activities, other than for commercial air transport, it shall

(1) Be appropriately approved, pursuant to M.A. Subpart G, for the management of the continuing airworthiness of the aircraft it operates or contract such an organization;

(2) Be appropriately approved in accordance with M.A. Subpart F or Part 145, or contract such organizations; and

(3) Be ensured that Paragraph (a) is satisfied.

Subpart C: Continuing Airworthiness
M.A.302 Maintenance program

(a) Every aircraft shall be maintained in accordance with a maintenance program approved by the competent authority, which shall be periodically reviewed and amended accordingly.

(b) The maintenance program and any subsequent amendments shall be approved by the competent authority.

(c) The maintenance program must establish compliance with:
- **(1)** Instructions for continuing airworthiness issued by type certificate and Supplementary type certificate holders and any other organization that publishes such data in accordance with Part 21;
- **(2)** Instructions issued by the competent authority, if they differ from Subparagraph 1 or in the absence of specific recommendations; or
- **(3)** Instructions defined by the owner or the operator and approved by the competent authority if they differ from Subparagraphs 1 and 2.

(d) The maintenance program shall contain details, including frequency, of all maintenance to be carried out, including any specific tasks linked to specific operations. ...

M.A.303 Airworthiness Directives.
Any applicable AD must be carried out within the requirements of that AD, unless otherwise specified by the Agency.

Subpart F: Maintenance Organization
M.A.601 Scope.
This subpart establishes the requirements to be met by an organization to qualify for the issue or continuation of an approval for the maintenance of aircraft and components not listed in M.A.201(f) and (g) (large aircraft).

M.A.615 Privileges of the organization.
The organization may
- **(1)** maintain any aircraft and/or component for which it is approved at the locations specified in the approval certificate and in the manual;
- **(2)** maintain any aircraft and/or component for which it is approved at any other location subject to such maintenance being only necessary to rectify arising defects; and
- **(3)** issue certificates of release to service on completion of maintenance, in accordance with M.A.612 or M.A.613.

Subpart G: Continuing Airworthiness Management Organization
M.A.701 Scope.
This subpart establishes the requirements to be met by an organization to qualify for the issue or continuation of an approval for the management of continuing aircraft airworthiness, coordinating the compliance of aircraft with maintenance program, ADs, and service bulletins.

M.A.711 Privileges of the organization

(a) An approved CAMO may
- **(1)** manage the continuing airworthiness of noncommercial air transport aircraft as listed on the approval certificate;
- **(2)** manage the continuing airworthiness of commercial air transport aircraft when listed on its AOC (Air Operator Certificate); and

(**3**) arrange to carry out any task of continuing airworthiness within the limitation of its approval with another organization that is working under its quality system.

(**b**) An approved CAMO may additionally be approved to

(**1**) issue an airworthiness review certificate (ARC) or

(**2**) make a recommendation for the airworthiness review to a Member State of Registry.

(**c**) An organization shall be registered in one of the Member States to be granted the privilege pursuant to Paragraph (b).

M.A.712 Quality system

(**a**) To ensure that the approved CAMO continues to meet the requirements of this subpart, it shall establish a **Quality System** and designate a **Quality Manager** to monitor compliance with, and the adequacy of, procedures required to ensure airworthy aircraft. Compliance monitoring shall include a feedback system to the accountable manager to ensure corrective action as necessary.

Subpart H: Certificate of Release to Service

M.A.801 Aircraft certificate of release to service

(**a**) Except for aircraft released to service by a Part 145 organization, the certificate of release to service (CRS) shall be issued according to this Subpart.

(**b**) A CRS shall be issued before flight at the completion of any maintenance. When satisfied that all maintenance required has been properly carried out, a CRS shall be issued:

(**1**) By appropriate certifying staff on behalf of the M.A. Subpart F-approved maintenance organization or

(**2**) Except for complex maintenance tasks listed in Appendix VII, by certifying staff in compliance with the requirements of Part 66 or

(**3**) By the M.A.803 pilot-owner. ...

M.A.803 Pilot-owner authorization

(**a**) The pilot-owner is the person who owns or jointly owns the aircraft being maintained and holds a valid pilot license with the appropriate type or class rating.

(**b**) For any privately operated aircraft of simple design with a maximum takeoff mass of less than 2730 kg, glider, and balloon, the pilot-owner may issue the CRS after limited pilot-owner maintenance listed in Appendix VIII.

(**c**) Limited pilot-owner maintenance shall be defined in the M.A.302 aircraft maintenance program. ...

2. Section B—Procedure for Competent Authorities

Subpart A: General

M.B.101 Scope. This section establishes the administrative requirements to be followed by the competent authorities in charge of the application and the enforcement of Section A of this Part.

M.B.102 Competent authority

(a) *General*. A Member State shall designate a competent authority with allocated responsibilities for the issuance, continuation, change, suspension, or revocation of certificates and for the oversight of continuing airworthiness. This competent authority shall establish documented procedures and an organizational structure....

Subpart F: Maintenance Organization

M.B.603 Issue of approval

(a) The competent authority shall issue to the applicant an **EASA Form 3 Approval Certificate** (Appendix V—*Approval Certificate, Part M, Section A, Subpart F: Maintenance Organization*), which includes the extent of approval, when the maintenance organization is in compliance with the applicable paragraphs of this Part. ...

Subpart G: Continuing Airworthiness Management Organization

M.B.703 Issue of approval

(a) The competent authority shall issue to the applicant an **EASA Form 14 Approval Certificate** (Appendix VI—*Approval Certificate, Part M, Section A, Subpart G: Continuing Airworthiness Management Organization*), which includes the extent of approval, when the CAMO is in compliance with M.A. Subpart G. ...

3. General Remarks About Part M

To better understand this complex document, we can summarize some of its main features.

First of all, some useful definitions:

Large aircraft means an aircraft, classified as an aeroplane with a maximum takeoff mass of more than 5700 kg, or a multiengined helicopter (Definition from Article 2 of EC 2042/2003).

Commercial air transport means aircraft operations carrying passengers or freight, but is not intended to cover Aerial Work or Corporate Aviation.

Aerial Work means an aircraft operation in which an aircraft is used for specialized services such as agriculture, construction, photography, surveying, observation and patrol, search and rescue, aerial advertisement, and so on.

Organization means a natural person, a legal person or part of a legal person. Such an organization may be established at more than one location whether or not within the territory of the Member States;

Part M addresses the issue of the continuing airworthiness of all aircraft (large and nonlarge, used in commercial or noncommercial air operations) by

- Defining responsibilities
- Describing what is necessary to manage the continuing airworthiness of aircraft

- Regulating aircraft maintenance
- Mandating a release to service after maintenance
- Setting forth a control process through an airworthiness review resulting in the issue of a certificate validating the airworthiness certificate
  For all aircraft used in **commercial air operations**, it is specified:
- Responsibilities: the operator of an aircraft is responsible for the airworthiness of the aircraft.
- Continuing Airworthiness Management: the operator must also be approved for the management of the continuing airworthiness of the aircraft according to Part M Subpart G.
- Maintenance: the aircraft must be maintained by a Part 145-approved maintenance organization.
- Release to service: after maintenance, the operator must ensure that a CRS is issued for the maintenance requested by a person authorized by a Part 145 maintenance organization approved for the work accomplished.

Subpart F: Maintenance organization. This subpart describes the approval procedure for maintenance organizations for nonlarge/small aircraft (5700 kg and below and single-engine helicopter used in noncommercial air operations). It is a simplified Part 145 approval.

Subpart G: Continuing airworthiness management organization. This subpart describes the approval procedure for CAMOs.

This subpart requires facilities, data, and competent staff; it also describes the tasks for which these organizations are approved for; it gives the general rules for record keeping.

Any organization approved to this subpart may also have the privilege to carry out **airworthiness reviews**. These periodic reviews are carried out to ensure that the aircraft's continuing airworthiness has been properly carried out and that the aircraft can be considered as airworthy at the time of the inspection. The content of these reviews is incorporated in this subpart.

A very important feature of these organizations is the establishment of a Quality System (M.A.712) to ensure, through an independent audit process, that the approved CAMO continues to meet the requirements of this subpart.

Subpart H: Certificate of Release to Service—CRSM.A. 801
M.A.801 Aircraft CRS

(a) Except for aircraft released to service by a Part-145 organization, the CRS shall be issued according to this subpart.
(b) A CRS shall be issued before flight at the completion of any maintenance. When satisfied that all maintenance required has been properly carried out, a CRS shall be issued
   (1) By appropriate certifying staff on behalf of the M.A. Subpart F-approved maintenance organization or

(2) Except for complex maintenance tasks listed in Appendix 7, by certifying staff in compliance with the requirements of Part-66 or

(3) By the M.A.803 pilot owner.[6]

NOTE: A pilot owner is the person who owns or jointly owns the aircraft being maintained and holds a valid pilot license with the appropriate type or class rating.

Airworthiness Review Certificate. With effect from 28th September 2008, all aircraft subject to EASA regulation must, under EU law, be issued with a nonexpiring Certificate of Airworthiness supported by an ARC.

A nonexpiring Certificate of Airworthiness conforms to EASA Part 21A.181 "Duration and continued validity"

(a) An airworthiness certificate shall be issued for *an unlimited duration.* It shall remain valid subject to:

(1) compliance with the applicable type design and continuing airworthiness requirements; and ......

Then, unlike the past C of A issued by the EU Airworthiness Authorities, the new EASA C of A is *nonexpiring* and its validity is dependent on the validity of the associated ARC.

According to M.A.901 "Aircraft airworthiness review": "To ensure the validity of the aircraft airworthiness certificate, an *airworthiness review* of the aircraft and its continuing airworthiness records must be carried out periodically."

The expiring date of the ARC is the date by which a new airworthiness review of the aircraft must be carried out, within the rules contained in Part M.

For instance, when an aircraft is continuously managed and maintained by an organization approved in accordance with Part M Subpart G, this activity is carried out by a Subpart G organization without the intervention of the competent authority.

NOTE: The Agency's Executive Director decision of November 2003 provides the **Annex I Acceptable Means of Compliance to Part M** (Last amendment 5 May 2010).

9.1.2.2. ANNEX II, PART 145

1. Section A

145.A.10 Scope. This section establishes the requirements to be met by an organization to qualify for the issue or continuation of an approval for the maintenance of aircraft and components.

---

[6] The following constitutes the limited pilot maintenance referred to in M.A.803, provided it does not involve complex maintenance tasks and is carried out in accordance with M.A.402: (some examples)

(1) Removal, installation of wheels.

(2) Replacing elastic shock absorber cords on landing gear.

(3) Servicing landing gear shock struts by adding oil, air, or both.

(4) ....

145.A.20 Terms of approval. The organization shall specify the scope of work deemed to constitute approval in its exposition (Appendix II to this part contains a table of all classes and ratings).

145.A.25 Facility requirements. The organization shall ensure that
(a) Facilities are provided appropriate for all planned work, ensuring in particular, protection from the weather elements. Specialized workshops and bays are segregated as appropriate. ...

145.A.30 Personnel requirements
(a) The organization shall appoint an accountable manager who has corporate authority for ensuring that all maintenance required by the customer can be financed and carried out to the standard required by this part. ...
(b) The organization shall nominate a person or group of persons, whose responsibilities include ensuring that the organization complies with this part. Such person(s) shall ultimately be responsible to the accountable manager. ...

145.A.40 Equipment, tools, and material
(a) The organization shall have available and use the necessary equipment, tools, and material to perform the approved scope of work. ...

145.A.45 Maintenance data
(a) The organization shall hold and use applicable current maintenance data in the performance of maintenance, including modifications and repairs. "Applicable" means relevant to any aircraft, component, or process specified in the organization's approval class rating schedule and in any associated capability list...

145.A.50 Certification of maintenance
(a) A CRS shall be issued by appropriately authorized certifying staff on behalf of the organization when it has been verified that all maintenance ordered has been properly carried out by the organization in accordance with the procedures specified in 145.A.70, taking into account the availability and use of the maintenance data specified in 145.A.45 and that there are no noncompliances that are known that hazard seriously the flight safety. ...

145.A.70 Maintenance organization exposition
(a) "**Maintenance organization exposition**" means the document or documents that contain the material specifying the scope of work deemed to constitute approval and showing how the organization intends to comply with this Part. The organization shall provide the competent authority with a maintenance organization exposition (MOE), containing the following information: ...

145.A.75 Privileges of the organization. In accordance with the exposition, the organization shall be entitled to carry out the following tasks:

(a) Maintain any aircraft and/or component for which it is approved at the locations identified in the approval certificate and in the exposition.

(b) Arrange for maintenance of any aircraft or component for which it is approved at another organization that is working under the quality system of the organization. This refers to work being carried out by an organization not itself appropriately approved to carry out such maintenance under this part and is limited to the work scope permitted under 145.A.65(b) procedures. This work scope shall not include a base maintenance check of an aircraft or a complete workshop maintenance check or overhaul of an engine or engine module.

(c) Maintain any aircraft or any component for which it is approved at any location subject to the need for such maintenance arising either from the unserviceability of the aircraft or from the necessity of supporting occasional line maintenance, subject to the conditions specified in the exposition.

(d) Maintain any aircraft and/or component for which it is approved at a location identified as a line-maintenance location capable of supporting minor maintenance and only if the organization exposition both permits such activity and lists such locations.

(e) Issue certificates of release to service in respect of completion of maintenance in accordance with 145.A.50.

145.A.80 Limitations on the organization. The organization shall only maintain an aircraft or component for which it is approved when all the necessary facilities, equipment, tooling, material, maintenance data, and certifying staff are available.

2. Section B: Procedure for Competent Authorities

145.B.01 Scope. This section establishes the administrative procedures that the competent authority shall follow when exercising its tasks and responsibilities regarding issuance, continuation, change, suspension, or revocation of Part 145 maintenance organization approvals. …

145.B.10 Competent authority

1. *General.* The Member State shall designate a competent authority with allocated responsibilities for the issuance, continuation, change, suspension, or revocation of a maintenance approval. This competent authority shall establish documented procedures and an organizational structure.

NOTE: the paragraph continues with reference to the authority's resources involved in Part 145 organizations, their number, qualification, and training.

145.B.25 Issue of approval

(1) The competent authority shall formally approve the exposition and issue to the applicant a Form 3 **Approval Certificate**, which includes the approval ratings. The competent authority shall only issue a certificate when the organization is in compliance with Part 145.

(2) The competent authority shall indicate the conditions of the approval on the Form 3 Approval Certificate.

(3) The reference number shall be included on the Form 3 Approval Certificate in a manner specified by the Agency.

3. General Remarks About Part 145

EASA Part 145 is the Implementing Regulation issued by EASA for the aircraft maintenance sector (**Maintenance Organization Approval**) establishing the requirements to be met by an organization to qualify for the issuing or continuation of an approval for the maintenance of aircraft and components.

To obtain approval to be an aeronautical repair station, an organization must write, submit, and keep updated an **MOE.** To support their MOE, they must have a documented set of procedures. The organization must also have a compliance matrix to show how they meet the requirements of Part 145.

When maintenance facilities are located in more than one Member State, the investigation and continued oversight of the approval must be carried out in conjunction with the competent authorities from the Member States in whose territory the other maintenance facilities are located.

According to Part M, maintenance of large aircraft, aircraft used for commercial air transport and components thereof shall be carried out by a Part 145-approved maintenance organization. But such an organization may maintain **any aircraft and/or component** for which it is approved at the locations identified in the approval certificate and in the MOE.

An important feature of Part 145 is the guidance on how the smallest organizations could satisfy the intent of this part.

According to the GM, the smallest maintenance organization would only be involved with a limited number of light aircraft, or aircraft components, used for commercial air transport. It is therefore a matter of scale: light aircraft do not demand the same level of resources, facilities, or complex maintenance procedures as the large organization.

For example, when only one person is employed (in fact having the certifying function and others), this organization approved under Part 145 may use the alternatives provided in the GM limited to … (*the GM provides the list*). The minimum requirement for the organization is one full-time person who meets the requirements of Part 66 for certifying staff and holds the position of "accountable manager, maintenance engineer and is also certifying staff." No other person may issue a CRS and therefore if absent, no maintenance may be released during such absence.

NOTE: The Agency's Executive Director decision of November 2003 provides the **Annex II Acceptable Means of Compliance to Part 145** and **Annex III Guidance Material to Part 145** (Last amendment 5 May 2010).

9.1.2.3. ANNEX III, PART 66

1. Section A

Subpart A: Aircraft Maintenance License Aeroplanes and Helicopters

66.A.1 Scope

(a) This section establishes the requirements for the issue of an aircraft maintenance license and conditions of its validity and use, for aeroplanes and helicopters of the following categories: A, B1, B2, and C.

(b) Categories A and B1 are subdivided into subcategories relative to combinations of aeroplanes, helicopters, turbine, and piston engines. The subcategories are A1 and B1.1 Aeroplanes Turbine; A2 and B1.2 Aeroplanes Piston; A3 and B1.3 Helicopters Turbine; and A4 and B1.4 Helicopters Piston.

66.A.20 Privileges

(a) Subject to compliance with Paragraph (b), the following privileges shall apply to

(1) A Category A aircraft maintenance license permits the holder to issue certificates of release to service following minor scheduled line maintenance and simple defect rectification within the limits of tasks specifically endorsed on the authorization. The certification privileges shall be restricted to work that the license holder has personally performed in a Part 145 organization.

(2) A Category B1 aircraft maintenance license shall permit the holder to issue certificates of release to service following maintenance, including aircraft structure, power plant, and mechanical and electrical systems. Replacement of avionic line replaceable units, requiring simple tests to prove their serviceability, shall also be included in the privileges. Category B1 shall automatically include the appropriate A subcategory.

(3) A Category B2 aircraft maintenance license shall permit the holder to issue certificates of release to service following maintenance on avionic and electrical systems.

(4) A Category C aircraft maintenance license shall permit the holder to issue certificates of release to service following base maintenance on aircraft. The privileges apply to the aircraft in its entirety in a Part 145 organization.

(b) The holder of an aircraft maintenance license may not exercise certification privileges unless:

(1) In compliance with the applicable requirements of Part M and/or Part 145.

(2) In the preceding 2-year period, he/she has either had 6 months of maintenance experience in accordance with the privileges granted by the aircraft maintenance license, or met the provision for the issue of the appropriate privileges.

(3) He/she is able to read, write, and communicate to an understandable level in the language(s) in which the technical documentation and procedures necessary to support the issue of the CRS are written.

66.A.30 Experience requirements

(a) An applicant for an aircraft maintenance license shall have acquired:

(1) For Category A and Subcategories B1.2 and B1.4: (i) 3 years of practical maintenance experience on operating aircraft, if the applicant has no previous relevant technical training or (ii) 2 years of practical maintenance experience on operating aircraft and completion of training considered relevant by the competent authority as a skilled worker, in a technical trade or (iii) 1 year of practical maintenance experience on operating aircraft and completion of a Part 147-approved basic training course....

2. Section B—Procedure for Competent Authorities

Subpart A: General

66.B.05 Scope. This section establishes the administrative requirements to be followed by the competent authorities in charge of the application and the enforcement of Section A of this part.

Subpart B: Issue of an Aircraft Maintenance License. This subpart provides the procedures to be followed by the competent authority to issue or vary or to permit continuity of the aircraft maintenance license.

66.B.100 Procedure for the issue of an aircraft maintenance license by the competent authority

(a) On receipt of EASA Form 19 and any supporting documentation, the competent authority shall verify EASA Form 19 for completeness and ensure that the experience claimed meets the requirement of this part.

(b) The competent authority shall verify an applicant's examination status and/ or confirm the validity of any credits to ensure that all required modules of Appendix 1 have been met as required by this Part B. ...

NOTE: The Agency's Executive Director decision of November 2003 provides the **Annex IV Acceptable Means of Compliance to Part-66** and **Annex V Guidance Material to Part-66**. (Last amendment 5 May 2010).

9.1.2.4. ANNEX IV, PART 147

1. Section A

Subpart A: General

147.A.05 Scope. This section establishes the requirements to be met by organizations seeking approval to conduct training and examination as specified in Part 66.

147.A.10 General. A training organization shall be an organization or part of an organization registered as a legal entity.

Subpart B: Organization Requirements

147.A.100 Facility requirements

(a) The size and structure of facilities shall ensure protection from the prevailing weather elements and proper operation of all planned training and examination on any particular day.

(b) Fully enclosed appropriate accommodation separate from other facilities shall be provided for the instruction of theory and the conduct of knowledge examinations. ...

147.A.105 Personnel requirements

(a) The organization shall appoint an accountable manager who has corporate authority for ensuring that all training commitments can be financed and carried out to the standard required by this part.

(b) A person or group of persons, whose responsibilities include ensuring that the maintenance training organization is in compliance with the requirements of this part, shall be nominated. Such person(s) must be responsible to the accountable manager. The senior person or one person from the group of persons may also be the accountable manager who is subject to meet the requirements for the accountable manager as defined in Paragraph (a). ...

147.A.115 Instructional equipment

(a) Each classroom shall have appropriate presentation equipment of a standard that ensures that students can easily read presentation text/drawings/diagrams and figures from any position in the classroom.

Presentation equipment shall include representative synthetic training devices to assist students in their understanding of the particular subject matter where such devices are considered beneficial for such purposes. ...

147.A.140 Maintenance training organization exposition

(a) The organization shall provide an exposition for use by the organization describing the organization and its procedures and containing the following information:

    (1) A statement signed by the accountable manager confirming that the maintenance training organization exposition and any associated manuals define the maintenance training organization's compliance with this part and shall be complied with at all times. ...

147.A.145 Privileges of the maintenance training organization

(a) The maintenance training organization may carry out the following as permitted by and in accordance with the maintenance training organization exposition:

    (1) Basic training courses to the Part 66 syllabus or part thereof.

    (2) Aircraft type/task training courses in accordance with Part 66.

    (3) The examinations on behalf of the competent authority, including the examination of students who did not attend the basic or aircraft type training course at the maintenance training organization.

    (4) The issue of certificates in accordance with Appendix III following successful completion of the approved basic or aircraft type training courses and examinations specified in Subparagraphs (a)(1), (a)(2), and (a)(3), as applicable. ...

2. Section B—Procedure for Competent Authorities

Subpart A: General

147.B.05 Scope. This section establishes the administrative requirements to be followed by the competent authorities in charge of the application and the enforcement of Section A of this part.

Subpart B: Issue of an Approval.

This subpart provides the requirements to issue or vary the maintenance training organization approval.

147.B.100 General

(a) An application for maintenance training organization initial approval or variation of a maintenance training organization approval shall be made on a form and in a manner established by the competent authority.

(b) The maintenance training organization approval shall be granted to the organization by the competent authority. ...

NOTE: The Agency's Executive Director decision of November 2003 provides the **Annex VI Acceptable Means of Compliance to Part 147** and **Annex VII Guidance Material to Part 147** (Last amendment 5 May 2010).

9.1.2.5. GENERAL REMARKS ABOUT EASA CONTINUED AIRWORTHINESS/MAINTENANCE

In Section 9.1.2., we have a summary of the EASA requirements for the continued airworthiness/maintenance as an implementation of the EC Regulation No. 2042/2003.

These requirements are included in the following parts:

- Part M—Continuing Airworthiness.
- Part 145—Maintenance Organization Approval.
- Part 66—Certifying staff.
- Part 147—Training organization requirements.

We have seen how these requirements are intertwined: maintenance organizations must be based on Part M requirements, the operating personnel have to be licensed according to Part 66 and through a training organization in compliance with Part 147.

This matter is really complex and this book can only provide basic generic information in line with what has been discussed in other cases.

Because the common interpretation and uniform implementation of these requirements is very important for civil aviation, workshops and other channels of information are provided by EASA, JAA, and national authorities for either privates or organizations involved in the operation of aircraft.

9.1.3. JAR-OPS 1 and JAR-OPS 3 requirements for maintenance

For JAR-OPS 1, the Subpart M is now the following:

SUBPART M: AEROPLANE MAINTENANCE
JAR-OPS 1.875 General

(a) An operator shall not operate an aeroplane unless it is maintained and released to service by an organization appropriately approved/accepted in accordance with Commission Regulation (EC) No. 2042/2003 Part 145, except that preflight inspections need not necessarily be carried out by the Part 145 organization.

(b) Aeroplane continuing airworthiness requirements needed to comply with the operator certification requirements in JAR-OPS 1.180 are those set up in Commission Regulation (EC) No. 2042/2003, Part M (hereinafter abbreviated to Part M for convenience).

The rest of this subpart has been withdrawn due to the implementation of Commission Regulation (EC) No. 2042/2003 Part M.

NOTE: JAR-OPS 3 presents the same arrangement for the Subpart M.

9.1.4. EASA certification of air operators[7]

In the "applicability" of JAR-OPS 1 and 3 in Chapter 8, we explained that these standards are applicable to any civil aeroplane and helicopter aeroplane for the purpose of commercial air transportation by any operator whose principal location of business is in a JAA Member State.

We similarly discussed the content of these standards from the point of view of additional requirements for airworthiness and maintenance.

All the requirements of these standards lead to the issue of an **AOC**, according to Subpart C of JAR OPS.

As already explained in Chapter 8 (Paragraph 8.9.1), the Commission Regulation (EC) 8/2008 replaced the Annex III to Council Regulation (EEC) No. 3922/91 by a new Annex III based on JAR-OPS 1.

The new Annex III, "Common technical requirements and administrative procedures applicable to commercial transportation by aircraft" is now **OPS 1: Commercial Air Transportation (Aeroplanes)**.

The **OPS 1** contains the prescription for the certification of operators, and in particular their organization, operational procedures, manuals, crew employment and training, equipments, aircraft adequacy and maintenance, transport of dangerous goods, and protection against acts of unlawful interference. The operator is required to establish a Quality System to monitor compliance with, and the adequacy of, procedures to ensure safe operational practices and airworthy aircraft.

OPS 3 do not exist yet, but **JAR-OPS 3** is applicable to Commercial Air Transportation involving helicopters used by operators based in a Member State.

OPS 2, when developed, will be applicable to General Aviation (GA) operations (including Aerial Work)[8] involving aeroplanes used by operators based in a Member State.

---

[7] *Operator* means any legal or natural person, operating or proposing to operate one or more aircraft.

[8] *Aerial Work* means an aircraft operation in which an aircraft is used for specialized services such as agriculture, construction, photography, surveying, observation and patrol, search and rescue, aerial advertisement, and so on.

OPS 4, when developed, will be applicable to GA operations (including Aerial Work) involving helicopters used by operators based in a Member State.

That means that, for the time being, the national authorities approve operator's organizations for which EU-OPS (or future OPS 2 and 4) are still not available, with the prescriptions applicable in each single state.

In the specific case of maintenance, the operators certificated according to the OPS 1 must rely on a maintenance organization approved according to **EASA Part 145 "Approved Maintenance Organization"**.

The operator is not obliged to perform all the maintenance operations inside his own organization; the company can collaborate with other (Part 145) approved organizations. Of course, this has to be clearly established in the operator's procedures.

We quote, with some remarks, a few paragraphs of OPS 1, significant for the scope of this book, without comments on the purely operational requirements that are nevertheless fundamental for the achievement of an AOC.

SUBPART B: GENERAL
OPS 1.035 Quality system

(a) An operator shall establish one quality system and designate one Quality Manager to monitor compliance with, and adequacy of, procedures required to ensure safe operational practices and airworthy aeroplanes. Compliance monitoring must include a feedback system to the Accountable Manager to ensure corrective action as necessary.

(b) The quality system must include a Quality Assurance (QA) Program that contains procedures designed to verify that all operations are being conducted in accordance with all applicable requirements, standards, and procedures.

(c) The quality system and the Quality Manager must be acceptable to the Authority.

(d) The quality system must be described in relevant documentation.

(e) Notwithstanding Subparagraph (a) above, the Authority may accept the nomination of two Quality Managers, one for operations and the other for maintenance, provided that the operator has designated one Quality Management Unit to ensure that the quality system is applied uniformly throughout the entire operation.

NOTE: The terms used in the context of the requirement for an operator's Quality System have the following meanings:

(i) Accountable Manager. The person acceptable to the Authority who has corporate authority for ensuring that all operations and maintenance activities can be financed and carried out to the standard required by the Authority, and any additional requirements defined by the operator.

(ii) Quality Assurance. All those planned and systematic actions necessary to provide adequate confidence that operational and maintenance practices satisfy given requirements.

(iii) Quality Manager. The manager, acceptable to the Authority, responsible for the management of the Quality System, monitoring function, and requesting corrective actions.

In the case of small/very small operators, the posts of the Accountable Manager and the Quality Manager may be combined. However, in this event, quality audits should be conducted by independent personnel.

The "small" operator may decide to use internal or external auditors or a combination of the two. In these circumstances, it would be acceptable for external specialists and/or qualified organizations to perform the quality audits on behalf of the Quality Manager.

SUBPART C: OPERATOR CERTIFICATE AND SUPERVISION
OPS 1.180 Issue, variation, and continued validity of an AOC

(a) An operator will not be granted an AOC, or a variation to an AOC, and that AOC will not remain valid unless:

(1) Aeroplanes operated have a standard Certificate of Airworthiness issued in accordance with Commission Regulation (EC) No. 1702/2003 of 24 September 2003 laying down implementing rules for the airworthiness and environmental certification of aircraft and related products, parts, and appliances, as well as for the certification of design and production organizations by a Member State. Standard Certificates of Airworthiness issued by a Member State other than the State responsible for issuing the AOC will be accepted without further showing when issued in accordance with Part 21;

(2) The maintenance system has been approved by the Authority in accordance with Part M, Subpart G; and

(3) He has satisfied the Authority that he has the ability to: (i) establish and maintain an adequate organization; (ii) establish and maintain a quality system in accordance with OPS 1.035; (iii) comply with required training programs; (iv) comply with maintenance requirements, consistent with the nature and extent of the operations specified including the relevant items prescribed in OPS 1.175 (g) to (o); and (v) comply with OPS 1.175.

SUBPART M: AEROPLANE MAINTENANCE
OPS 1.875 General

(a) An operator shall not operate an aeroplane unless it is maintained and released to service by an organization appropriately approved/accepted in accordance with Part 145 except that preflight inspections need not necessarily be carried out by the Part 145 organization.

(b) Aeroplane continuing airworthiness requirements needed to comply with the operator certification requirements in OPS 1.180 are those set up in Part M.

NOTE: An approval for the CAMO is propaedeutical to the issue of an AOC.

When an operator is not appropriately approved in accordance with Part 145, the operator shall establish a written maintenance contract between the operator and a Part 145-approved organization or another operator, detailing the functions specified under Part M.

9.1.5. FAA continued airworthiness/maintenance

NOTE: See Appendix 9.1.5

The requirements for FAA continued airworthiness are much more articulated than the corresponding EASA documents. On comparing them with the standards listed for EASA continued airworthiness, we find the following correspondences:

(1) General rules for maintenance, including organizations and personnel, involved in continuing airworthiness can be found in FAR 43.

(2) Approval of organizations involved in maintenance can be found in FAR 145.

(3) The certification of personnel involved in maintenance operation is regulated by FAR 65.

(4) The certification of an organization seeking approval to conduct training of personnel is regulated by FAR 147.

Furthermore, some of the "operational standards" we considered in the section "FAA operational standards (additional airworthiness requirements)" in Chapter 8 (Section 8.7) prescribe maintenance requirements for the operators subject to these standards. We can quote the following: FAR 91, FAR 121, FAR 125, FAR 129, and FAR 135.

It is worth remembering that there are plenty of Advisory Circulars and FAA Orders to provide guidance on these standards.

Extracts of the above-mentioned standards are given below.

NOTE: To give an idea of the content of these operational standards prescribing requirements for continued airworthiness and maintenance, we quote the most noteworthy articles of these standards, often only partially or referring to the titles. This is done for practical reasons and for reference; however, we are not suggesting that this could replace the good practice of reading the original texts in full.

9.1.5.1. FAR 43. MAINTENANCE, PREVENTIVE MAINTENANCE, REBUILDING, AND ALTERATION

43.1. Applicability

(a) Except as provided in Paragraphs (b) and (d) of this section, this part prescribes rules governing the maintenance, preventive maintenance, rebuilding, and alteration of any:

(1) Aircraft having a US airworthiness certificate,

(2) Foreign-registered civil aircraft used in common carriage or carriage of mail under the provisions of Part 121 or 135 of this chapter, and

(3) Airframe, aircraft engines, propellers, appliances, and component parts of such aircraft.

(b) This part does not apply to any aircraft for which the FAA has issued an Experimental certificate, unless the FAA has previously issued a different kind of airworthiness certificate for that aircraft.
(c) This part applies to all life-limited parts that are removed from a type-certificated product, segregated, or controlled as provided in Paragraph 43.10.
(d) This part applies to any aircraft issued a Special airworthiness certificate in the Light-Sport category except: ...

The content of this standard is as follows:

Appendix A is of particular interest and an excerpt is given below.[9]

---

[9] Subparagraph (b) dealing with repairs has been considered in Chapter 5, "FAA repairs."

Appendix A to Part 43: Major alterations, major repairs, and preventive maintenance

(a) *Major alterations*:

(1) *Airframe major alterations.* Alterations of the following parts and alterations of the following types, when not listed in the aircraft specifications issued by the FAA, are airframe major alterations: (i) Wings, (ii) Tail surfaces, (iii) Fuselage, (iv) Engine mounts, (v) Control system, (vi) Landing gear, ... (xiii) Changes to the wing or to fixed or movable control surfaces that affect flutter and vibration characteristics.

(2) *Power plant major alterations.* The following alterations of a power plant when not listed in the engine specifications issued by the FAA are power plant major alterations. (i) Conversion of an aircraft engine from one approved model to another, involving any changes in compression ratio, propeller reduction gear, impeller gear ratios, or the substitution of major engine parts that requires extensive rework and testing of the engine. (ii) Changes to the engine by replacing aircraft engine structural parts with parts not supplied by the original manufacturer or parts not specifically approved by the Administrator. ... (vi) Conversions of any sort for the purpose of using fuel of a rating or grade other than that listed in the engine specifications.

(3) *Propeller major alterations.* The following major alterations of a propeller when not authorized in the propeller specifications issued by the FAA are (i) changes in blade design; (ii) changes in hub design; (iii) changes in the governor or control design; (iv) installation of a propeller governor or feathering system; (v) installation of propeller deicing system; and (vi) installation of parts not approved for the propeller.

(4) *Appliance major alterations.* Alterations of the basic design not made in accordance with recommendations of the appliance manufacturer or in accordance with an FAA ADs are appliance major alterations. In addition, changes in the basic design of radio communication and navigation equipment approved under type certification or a Technical Standard Order that have an effect on frequency stability, noise level, sensitivity, selectivity, distortion, spurious radiation, AVC characteristics, or ability to meet environmental test conditions and other changes that have an effect on the performance of the equipment are also major alterations.

9.1.5.2. FAR 145. REPAIR STATIONS

Subpart A: General

145.1 Applicability. This part describes how to obtain a repair station certificate. This part also contains the rules a certificated repair station must follow related to its performance of maintenance, preventive maintenance, or alterations of an aircraft, airframe, aircraft engine, propeller, appliance, or component part to which Part 43 applies. It also applies to any person who holds, or is required to hold, a repair station certificate issued under this part.

Subpart B: Certification
145.53 Issue of certificate

(a) Except as provided in Paragraph (b) of this section, a person who meets the requirements of this part is entitled to a repair station certificate with appropriate ratings, prescribing such operations specifications (OpSpecs) and limitations as necessary in the interest of safety.

(b) If the person is located in a country with which the United States has a bilateral aviation safety agreement, the FAA may find that the person meets the requirements of this part based on a certification from the civil aviation authority of that country. This certification must be made in accordance with implementation procedures signed by the Administrator or the Administrator's designee.

Subpart E: Operating Rules
145.201 Privileges and limitations of certificate

(a) A certificated repair station may

(1) Perform maintenance, preventive maintenance, or alterations in accordance with FAR 43 on any article for which it is rated and within the limitations in its OpSpecs.

(2) Arrange for another person to perform the maintenance, preventive maintenance, or alterations of any article for which the certificated repair station is rated. If that person is not certificated under FAR 145, the certificated repair station must ensure that the noncertificated person follows a quality control (QC) system equivalent to the system followed by the certificated repair station.

(3) Approve for return to service any article for which it is rated after it has performed maintenance, preventive maintenance, or an alteration in accordance with FAR 43.

(b) A certificated repair station may not maintain or alter any article for which it is not rated, and may not maintain or alter any article for which it is rated if it requires special technical data, equipment, or facilities that are not available to it.

(c) A certificated repair station may not approve for return to service:

(1) Any article unless the maintenance, preventive maintenance, or alteration was performed in accordance with the applicable approved technical data or data acceptable to the FAA,

(2) Any article after a major repair or major alteration unless the major repair or major alteration was performed in accordance with applicable approved technical data, and

(3) Any experimental aircraft after a major repair or major alteration performed under Paragraph 43.1(b) unless the major repair or major alteration was performed in accordance with methods and applicable technical data acceptable to the FAA.

9.1.5.3. FAR 65. CERTIFICATION: AIRMEN OTHER THAN CREW MEMBERS

Subpart A: General

65.1 Applicability. This part prescribes the requirements for issuing the following certificates and associated ratings and the general operating rules for the holders of those certificates and ratings:

(a) Air-traffic control-tower operators.

(b) Aircraft dispatchers.

(c) Mechanics.

(d) Repairmen.

(e) Parachute riggers.

Subpart D: Mechanics

65.95 Inspection authorization: Privileges and limitations

(a) The holder of an inspection authorization may

(1) Inspect and approve for return to service any aircraft or related part or appliance (except any aircraft maintained in accordance with a continuous airworthiness program under FAR 121r) after a major repair or major alteration to it in accordance with FAR 43 [New], if the work was done in accordance with technical data approved by the Administrator; and

(2) Perform an annual, or perform or supervise a progressive, inspection according to FAR 43.13 and 43.15. …

9.1.5.4. FAR 147. AVIATION MAINTENANCE TECHNICIAN SCHOOLS

Subpart A: General

147.1 Applicability. This part prescribes the requirements for issuing **aviation maintenance technician school certificates** and associated ratings and the general operating rules for the holders of those certificates and ratings.

Subpart B: Certification Requirements

147.11 Ratings. The following ratings are issued under this part:

(a) Airframe.

(b) Power plant.

(c) Airframe and power plant.

9.1.6. FAA operational standards (requirements for maintenance)

An excerpt of these requirements is reported here.

9.1.6.1. FAR 91

Subpart E: Maintenance, Preventive Maintenance, and Alterations

91.401 Applicability

(a) This subpart prescribes rules governing the maintenance, preventive maintenance, and alterations of US-registered civil aircraft operating within or outside of the United States.

(b) Sections 91.405, 91.409, 91.411, 91.417, and 91.419 of this subpart do not apply to an aircraft maintained in accordance with a continuous airworthiness maintenance program (CAMP) as provided in FAR 121, 129, or paragraphs 91.1411 or 135.411(a)(2).

(c) Sections 91.405 and 91.409 of this part do not apply to an airplane inspected in accordance with FAR 125.

91.403 General

(a) The owner or operator of an aircraft is primarily responsible for maintaining that aircraft in an airworthy condition, including compliance with FAR 39.[10]

(b) No person may perform maintenance, preventive maintenance, or alterations on an aircraft other than that prescribed in this subpart and other applicable regulations, including FAR 43.[11]

(c) No person may operate an aircraft for which a manufacturer's maintenance manual or ICA has been issued that contains an airworthiness limitations section unless the mandatory replacement times, inspection intervals, and related procedures specified in that section or alternative inspection intervals and related procedures set forth in an operations specification approved by the Administrator under FAR 121 or 135 or in accordance with an inspection program approved under Paragraph 91.409(e) have been complied with.

91.405 Maintenance required. Each owner or operator of an aircraft

(a) Shall have that aircraft inspected as prescribed in Subpart E of this part and shall between required inspections, except as provided in Paragraph (c) of this section, have discrepancies repaired as prescribed in FAR 43 of this chapter.

(b) Shall ensure that maintenance personnel make appropriate entries in the aircraft maintenance records indicating that the aircraft has been approved for return to service.

(c) Shall have any inoperative instrument or item of equipment, permitted to be inoperative by FAR 91.213(d)(2), repaired, replaced, removed, or inspected at the next required inspection.

(d) When listed discrepancies include inoperative instruments or equipment, shall ensure that a placard has been installed as required by FAR 43.11.

91.407 Operation after maintenance, preventive maintenance, rebuilding, or alteration

(a) No person may operate any aircraft that has undergone maintenance, preventive maintenance, rebuilding, or alteration unless:

  (1) It has been approved for return to service by a person authorized under FAR 43.7. ...,

---

[10] ADs.

[11] Maintenance, Preventive Maintenance, Rebuilding, and Alteration.

91.409 Inspections

(a) Except as provided in Paragraph (c) of this section, no person may operate an aircraft unless, within the preceding 12 calendar months, it has had:

(1) An annual inspection in accordance with FAR 43 and has been approved for return to service by a person authorized by FAR 43.7; or

(2) An inspection for the issuance of an airworthiness certificate in accordance with FAR 21. ...

(b) Except as provided in Paragraph (c) of this section, no person may operate an aircraft carrying any person (other than a crew member) for hire, and no person may give flight instruction for hire in an aircraft which that person provides, unless within the preceding 100 hours of time in service the aircraft has received an annual or 100-hour inspection and been approved for return to service.

(c) Paragraphs (a) and (b) of this section do not apply to:

(1) An aircraft that carries a special flight permit, a current Experimental certificate, or a Light-Sport or Provisional airworthiness certificate. ...

(d) *Progressive inspection.* Each registered owner or operator of an aircraft desiring to use a progressive inspection program must submit a written request to the FAA Flight Standards district office having jurisdiction over the area in which the applicant is located, and shall provide. ...

(e) Large airplanes (to which FAR 125 is not applicable), turbojet multi-engine airplanes, turbopropeller-powered multiengine airplanes, and turbine-powered rotorcraft. No person may operate a large airplane, turbojet multiengine airplane, turbopropeller-powered multiengine airplane, or turbine-powered rotorcraft unless the replacement times for life-limited parts specified in the aircraft specifications, type data sheets ...

(g) *Inspection program approved under Paragraph (e) of this section.* Each operator of an airplane or turbine-powered rotorcraft desiring to establish or change an approved inspection program under Paragraph (f)(4) of this section must submit the program for approval. ...

91.410 Special maintenance program requirements

(a) No person may operate an Airbus Model A300 (excluding the 600 series), British Aerospace Model BAC 1-11, Boeing Model, 707, 720, 727, 737 or 747, McDonnell Douglas Model DC-8, DC-9/MD-80 or DC-10, Fokker Model F28, or Lockheed Model L-1011 airplane beyond applicable flight cycle implementation time specified below. ...

91.411 Altimeter system and altitude reporting equipment tests and inspections. No person may operate an airplane, or helicopter, in controlled airspace under IFR unless:

(1) Within the preceding 24 calendar months, each static pressure system, each altimeter instrument, and each automatic pressure altitude reporting system have been tested and inspected. ...

91.413 ATC transponder tests and inspections

(a) No persons may use an ATC transponder that is specified in FAR 91.215(a), FAR 121.345(c), or FAR 135.143(c) unless, within the preceding 24 calendar months, the ATC transponder has been tested and inspected ...

9.1.6.2. FAR 121
Subpart L: Maintenance, Preventive Maintenance, and Alterations
121.361 Applicability

(a) Except as provided by Paragraph (b) of this section, this subpart prescribes requirements for maintenance, preventive maintenance, and alterations for all certificate holders. ...

121.367 Maintenance, preventive maintenance, and alterations programs. Each certificate holder shall have an inspection program and a program covering other maintenance, preventive maintenance, and alterations that ensures that:

(a) Maintenance, preventive maintenance, and alterations performed by it, or by other persons, are performed in accordance with the certificate holder's manual. ...

121.368 Aging airplane inspections and records reviews

...

(b) *Operation after inspection and records review.* After the dates specified in this paragraph, a certificate holder may not operate an airplane under this part unless the Administrator has notified the certificate holder that the Administrator has completed the aging airplane inspection and records review required by this section. During the inspection and records review, the certificate holder must demonstrate to the Administrator that the maintenance of age-sensitive parts and components of the airplane have been adequate and timely enough to ensure the highest degree of safety. ...

121.370 Special maintenance program requirements

(a) No certificate holder may operate an Airbus Model A300 (excluding the 600 series), British Aerospace Model BAC 1-11, Boeing Model 707, 720, 727, 737, or 747, McDonnell Douglas Model DC-8, DC-9/MD-80 or DC-10, Fokker Model F28, or Lockheed Model L-1011 airplane beyond the applicable flight cycle implementation time specified below. ...

9.1.6.3. FAR 125
Subpart G: Maintenance
125.241 Applicability. This subpart prescribes rules, in addition to those prescribed in other parts of this chapter, for the maintenance of airplanes, airframes, aircraft engines, propellers, appliances, each item of survival and emergency equipment, and their component parts operated under this part.

125.247 Inspection programs and maintenance

(a) No person may operate an airplane subject to this FAR unless:

    (1) The replacement times for life-limited parts specified in the aircraft type certificate data sheets, or other documents approved by the Administrator, are complied with; (2) defects disclosed between inspections, or as a result of inspection, have been corrected in accordance with FAR 43; and (3) the airplane, including airframe, aircraft engines, propellers, appliances, and survival and emergency equipment, and their component parts, is inspected in accordance with an inspection program approved by the Administrator.

(b) The inspection program specified in Paragraph (a)(3) of this section must include at least the following ...

125.248 Special maintenance program requirements

(a) No person may operate an Airbus Model A300 (excluding the 600 series), British Aerospace Model BAC 1-11, Boeing Model 707, 720, 727, 737 or 747, McDonnell Douglas Model DC-8, DC-9/MD-80 or DC-10, Fokker Model F28, or Lockheed Model L-1011 beyond the applicable flight cycle implementation time specified below ...

9.1.6.4. FAR 129

129.14 Maintenance program and minimum equipment list requirements for US-registered aircraft

(a) Each foreign air carrier and each foreign person operating a US-registered aircraft within or outside the United States in common carriage shall ensure that each aircraft is maintained in accordance with a program approved by the Administrator. ...

129.32 Special maintenance program requirements

(a) No foreign air carrier or foreign persons operating a US-registered airplane may operate an Airbus Model A300 (excluding 600 series), British Aerospace Model BAC 1-11, Boeing Model 707, 720, 727, 737 or 747, McDonnell Douglas Model DC-8, DC-9/MD-80 or DC-10, Fokker Model F28, or Lockheed Model L-1011 beyond the applicable flight cycle implementation time specified below ...

9.1.6.5. FAR 135

Subpart J: Maintenance, Preventive Maintenance, and Alterations

135.411 Applicability

(a) This subpart prescribes rules in addition to those in other parts of this chapter for the maintenance, preventive maintenance, and alterations for each certificate holder as follows:

    (1) Aircraft that are type certificated for a passenger seating configuration, excluding any pilot seat, of nine seats or less, shall be maintained under FAR 91 and 43, and FAR 135.415, 135.416, 135.417, 135.421, and 135.422. An approved aircraft inspection program may be used under

FAR 135.419. (2) Aircraft that are type certificated for a passenger seating configuration, excluding any pilot seat, of 10 seats or more, shall be maintained under a maintenance program in FAR 135.415, 135.416, 135.417, and 135.423 to 135.443.

(b) A certificate holder who is not otherwise required may elect to maintain its aircraft under Paragraph (a)(2) of this section.

(c) Single-engine aircraft used in passenger-carrying IFR operations shall also be maintained in accordance with Paragraphs 135.421(c), (d), and (e).

135.419 Approved aircraft inspection program

(a) Whenever the Administrator finds that the aircraft inspections required or allowed under FAR 91 of this chapter are not adequate to meet this part, or on application by a certificate holder, the Administrator may amend the certificate holder's OpSpecs under Paragraph 135.17 ...

135.421 Additional maintenance requirements

(a) Each certificate holder who operates an aircraft type certificated for a passenger seating configuration, excluding any pilot seat, of nine seats or less, must comply with the manufacturer's recommended maintenance programs, or a program approved by the Administrator, for each aircraft engine, propeller, rotor, and each item of emergency equipment required by this chapter.

(c) For each single-engine aircraft to be used in passenger-carrying IFR operations, ...

(e) No certificate holder may operate a single-engine aircraft under IFR, carrying passengers, unless the certificate holder records and maintains in the engine maintenance records the results of each test, observation, and inspection required by the applicable engine trend monitoring program specified in (c)(1) and (2) of this section.

135.422 Aging airplane inspections and records reviews for multiengine airplanes certificated with nine or fewer passenger seats

(a) *Applicability.* This section applies to multiengine airplanes certificated with nine or fewer passenger seats, operated by a certificate holder in a scheduled operation under this part, ...

(b) *Operation after inspections and records review.* After the dates specified in this paragraph, a certificate holder may not operate a multiengine airplane in a scheduled operation under this part unless the Administrator has notified the certificate holder that the Administrator has completed the aging airplane inspection and records review required by this section. ...

135.425 Maintenance, preventive maintenance, and alteration programs.
Each certificate holder shall have an inspection program and a program covering other maintenance, preventive maintenance, and alterations that ensures that:

(a) Maintenance, preventive maintenance, and alterations performed by it, or by other persons, are performed under the certificate holder's manual ...

(b) Competent personnel and adequate facilities and equipment are provided for the proper performance of maintenance, preventive maintenance, and alterations; and

(c) Each aircraft released to service is airworthy and has been properly maintained for operation under this part.

9.1.7. FAA air operators certification and fractional ownership

NOTE: See Appendix 9.1.7

    NOTE: See definitions in Chapter 8 Paragraph 8.6.2.1 and in this section.

    FAR 119 *"Certification: Air Carriers and Commercial Operators"* includes the certification and OpSpecs requirements for persons who operate in common carriage under FAR 121 *"Operating Requirements: Domestic, Flag, and Supplemental Operations"*; and FAR 135, *"Operating Requirements: Commuter and On Demand Operations and Rules Governing Persons on Board Such Aircraft."* FAR 119 also contains definitions pertinent to operations that do not involve common carriage.

    Common Carriage. An applicant is engaged in common carriage if the applicant "holds out" to the public (by advertising or other means) to transport persons or property for compensation or hire.

    Noncommon Carriage. Operations not involving common carriage include the following definitions or exceptions. These definitions or exceptions are contained in FAR 119 and in sections of FAR 91, *"General Operating and Flight Rules."*

(1) *Noncommon carriage* involves the carriage of persons or property for compensation. or hire but there is *no* holding out. Noncommon carriage operations require the issuance of an **operating certificate**. Operations would be conducted under FAR 125, *"Certification and Operations: Airplanes Having a Seating Capacity of 20 or More Passengers or a Maximum Payload Capacity of 6000 Pounds or More; and Rules Governing Persons on Board Such Aircraft"*; or FAR 135, depending on the type of aircraft, seating configuration, and payload capacity.

(2) *Private carriage* involves the carriage of persons or property for compensation or hire with limitations on the number of contracts. (In this situation, the customer seeks an operator to perform the desired service and enters into an exclusive, mutual agreement as opposed to the operator seeking customers.) Private carriage operations require the issuance of an **operating certificate**. Operations would be conducted under FAR 125 or FAR 135, depending on the type of aircraft, seating configuration, and payload capacity.

(3) *Direct air carrier* is defined by FAR 119 as a person who provides or offers to provide air transportation and who has control over the operational functions performed in providing that transportation. The FAA issues certificates to these direct air carriers.

(4) Operations in which persons or cargo are transported without compensation or hire. These operations are conducted under FAR 91 and do not require a certificate.

(5) There are exceptions from the certification requirements of FAR 119 and the operating rules of FAR 121 and (refer to the cited regulations for the complete regulatory content). For example, FAR 91 section 91.501 lists certain operations not involving common carriage that may be conducted under FAR 91 or FAR 135. These operations involve the transportation of persons or property and may involve compensation. Section 91.501 sets conditions on the amount and types of compensation for certain of these operations; for example, aerial work, sales demonstration flights, fractional ownership, and so on.

NOTE: FAR 119 certification requirements do not apply to **fractional ownership** (see Section 9.1.7.2) or operations conducted under FAR 129, 133, 137, or 139.

Section 119.1(e) lists operations not requiring air carrier or commercial operator certification; for example, student instruction, ferry or training flights, aerial work, sightseeing in hot air balloons, FAR 133 rotorcraft external load, and so on.

Types of certificates under FAR 119. There are two basic types of AOCs for operations in **common carriage**:

(A) An *air carrier certificate* is issued to applicants who plan to conduct interstate, foreign, or overseas transportation, or to carry mail.

(B) An *operating* certificate is issued to applicants who plan to conduct intrastate transportation.

Determining appropriate operating rule and kind of operation for FAR 119. Once the type of certificate is determined, the following step is to determine the appropriate operating rule and kinds of operation.

There are two operating rules appropriate to air carriers and commercial operators. An applicant will operate under FAR 135 or FAR 121, or both, depending on whether the operation is scheduled and the size and type of aircraft used.

There are 5 **kinds of operation**.

Domestic, flag, and *supplemental operations* apply to operations conducted under FAR 121; and *commuter* and *on-demand operations* describe operations under FAR 135.

To determine the appropriate operating rule and kind of operation, it is necessary to determine whether the applicant will conduct *scheduled* or *nonscheduled operations*.

Scheduled operations include passenger operations in which the departure location and time and the arrival location are offered in advance by the operator. Scheduled operations can also carry cargo. However, an all-cargo operation is defined as nonscheduled.

Nonscheduled operations include:

(a) Passenger carrying operations in which the departure time and the departure and arrival locations are specifically negotiated with the customer or the customer's representative

(b) All-cargo operations
(c) Scheduled passenger operations in aircraft (other than turbojet-powered airplanes) that have nine or fewer passenger seats and 7500-pound payload or less that operate with a frequency of less than 5 round trips a week on at least one route between two or more points according to a published flight schedule
(d) Passenger operations conducted as a Public charter under FAR 380.

Once it has been determined whether the operation is scheduled or nonscheduled, the next step is to determine the appropriate operating rule and kinds of operation. An operator can conduct operations under FAR 121 or FAR 135, or both. However, the applicant will only be issued with one certificate. The OpSpecs will detail the operating rules and kinds of operation. The definitions for kinds of operation are contained in FAR (see also Chapter 8, Paragraph 8.6.2.1).

Appendix 9.3 provides a summary of the appropriate operating rule and kinds of operation with some example of aircraft type, size, seating configuration, and payload capacity.

For air carriers and commercial operators, FAR 119 defines which operating rule will apply to the operation of their aircraft. FAR 119 references passenger seat configuration and payload capacity to determine the applicable operating rules. In general, on-demand operation of airplanes having a passenger seat configuration of 30 seats or fewer, excluding each crew member seat, *and* a payload capacity of 7500 pounds or less are conducted under FAR 135. On-demand operations of multiengine airplanes with a passenger seat configuration of more than 30 seats *or* a payload capacity of more than 7500 pounds are conducted under FAR 121.

FAR 125 prescribes rules governing the operations of US-registered airplanes that have a seating configuration of 20 or more passenger seats, or a maximum payload capacity of 6000 pounds or more when common carriage is not involved.

We will now report the general content and an excerpt of some relevant paragraphs of FAR 119.

9.1.7.1. FAR 119. CERTIFICATION: AIR CARRIERS AND COMMERCIAL OPERATORS

Subpart A: General
119.1   Applicability
119.3   Definitions
119.5   Certifications, authorizations, and prohibitions
119.7   Operations specifications
119.9   Use of business names

119.1 Applicability
(a) This part applies to each person operating or intending to operate civil aircraft:
(1) As an air carrier or commercial operator, or both, in air commerce; or

(2) When common carriage is not involved, in operations of US-registered civil airplanes with a seat configuration of 20 or more passengers, or a maximum payload capacity of 6000 lb or more.

(b) This part prescribes

(1) The types of **AOCs** issued by the FAA, including **air carrier certificates** and **operating certificates**.

(2) The certification requirements an operator must meet to obtain and hold a certificate authorizing operations under FAR 121, 125, or/and OpSpecs for each kind of operation to be conducted and each class and size of aircraft to be operated under FAR 121 or 135.

(3) The requirements an operator must meet to conduct operations under FAR 121, 125, or 135 and in operating each class and size of aircraft authorized in its OpSpecs.

(4) Requirements affecting wet leasing of aircraft and other arrangements for transportation by air.

(5) Requirements for obtaining deviation authority to perform operations under a military contract and obtaining deviation authority to perform an emergency operation.

(6) Requirements for management personnel for operations conducted under FAR 121 or FAR 135.

(c) Persons subject to this part must comply with the other requirements of this chapter, except where those requirements are modified by or where additional requirements are imposed by FAR 119, 121, 125, or 135.

(d) This FAR does not govern operations conducted under FAR 91, Subpart K (when common carriage is not involved) nor does it govern operations conducted under FAR 129, 133, 137, or 139.

(e) Except for operations when common carriage is not involved conducted with airplanes having a passenger-seat configuration of 20 seats or more, excluding any required crew member seat, or a payload capacity of 6000 lb or more, this part does not apply to

(1) Student instruction.

(2) Nonstop sightseeing flights conducted with aircraft having a passenger seat configuration of 30 or fewer, excluding each crew member seat, and a payload capacity of 7500 lb or less, that begin and end at the same airport, and are conducted within a 25 statute mile radius of that airport. ...

(3) Ferry or training flights.

(4) Aerial work operations. ...

119.5 Certifications, authorizations, and prohibitions

(a) A person authorized by the Administrator to conduct operations as a direct air carrier will be issued an **air carrier certificate**.

(b) A person who is not authorized to conduct direct air carrier operations, but who is authorized by the Administrator to conduct operations as a US commercial operator, will be issued an **operating certificate**.

(c) A person who is not authorized to conduct direct air carrier operations, but who is authorized by the Administrator to conduct operations when common carriage is not involved as an operator of US-registered civil airplanes with a seat configuration of 20 or more passengers, or a maximum payload capacity of 6000 lb or more, will be issued an **operating certificate**. ...

Subpart B: Applicability of Operating Requirements to Different Kinds of Operations Under FAR 121, 125, and 135

Subpart C: Certification, Operations Specifications, and Certain Other Requirements for Operations Conducted Under FAR 121 or FAR 135

9.1.7.2. FRACTIONAL OWNERSHIP

Fractional ownership is a practice of dividing the value of an aircraft into percentage shares to be sold to individual owners with the privilege to use the aircraft for a certain period of time (hours, days, or weeks). The owner could also benefit from the share of income derived from the general use of the aircraft.

Fractional ownership has an important role in the activities of GA.

Fractional ownership programs are subject to an FAA oversight program similar to the one provided to air carriers, with the exception of line checks and en-route inspections.

FAA aviation safety inspectors conduct scheduled and unscheduled inspections and surveillance of personnel, aircraft, records, and other documents to ensure compliance with the regulations.

Part 91, Subpart K, establishes regulatory requirements for fractional ownership programs and their program managers and owners. This regulation defines the program and program elements, allocates operational control responsibilities and authority to the owners and program manager, and provides increased operational and maintenance safety requirements for fractional ownership programs.

Fractional ownership programs are not issued with a certificate but with Management Specifications under FAR 91 K.

9.1.7.3. OPERATOR CERTIFICATE

Operator certificates are issued for particular aerial operations such as *Agricultural Aircraft Operation* under FAR 137 and *Rotorcraft External-Load Operation* under FAR 133.

FAR 91.147 provides another example of *Operator* conducting nonstop passenger-carrying flights in an airplane or helicopter for compensation or hire that begin and end at the same airport and are conducted within a 25-statute mile radius of that airport. Flights should be made in accordance with FAR119.1(e)(2), FAR 135.1(a)(5), or FAR 121.1(d).[12]

9.2. AIRWORTHINESS DIRECTIVES

Besides the ordinary actions aimed at maintaining the continued airworthiness of a product, sometimes it is also essential to intervene with extraordinary measures.

If the authority reveals an unsafe condition in an aircraft, such as a deficiency of an engine, propeller, part, or appliance installed on this aircraft,

---

[12] *FAR119.1(e)(2)*. Nonstop Commercial Air Tours conducted in an airplane or helicopter having a standard airworthiness certificate and passenger-seat configuration of 30 seats or fewer and a maximum payload capacity of 7500 pounds or less that begin and end at the same airport, and are conducted within a 25-statute mile radius of that airport, ... *FAR 135.1(a)(5)*. Nonstop Commercial Air Tour flights conducted for compensation or hire in accordance with FAR 119.1(e)(2) .... *FAR 121.1(d)*. Nonstop Commercial Air Tours conducted for compensation or hire in accordance with FAR119.1(e)(2).........

which exists or has the potential to develop on similar types of aircraft then the authority issues an **AD**.

This is a document that mandates actions to be performed on an aircraft to restore an acceptable level of safety.

9.2.1. EASA ADs

According to its statute, the Agency is responsible for the design of products, parts, and appliances designed, manufactured, or used under the regulatory oversight of the EU Member States. In that context, it will issue ADs to ensure the continuing airworthiness of such products, parts, and appliances. In doing so, the Agency only exercises the responsibilities of a state of design or those related to the design of such products, parts, and appliances of a state of registry. ADs are therefore addressed to the holders of the design approvals affected by such ADs.

The dissemination of ADs to aircraft owners is a responsibility of the state of registry and does not belong to the Agency.

In the case of products, parts, and appliances for which the Agency only exercises the design responsibilities of the state of registry, its policy is to endorse automatically the ADs issued by the state of design. This does not apply if the Agency itself issues a different AD before the date at which it comes into effect of the state of design AD.

It is a common practice for imported products, parts, and appliances to rely on the state of design to first detect whether unsafe conditions require the issuing of an AD.

Only those ADs issued by the Agency itself are published.

The processes for issuing ADs are included in the *Continuing Airworthiness of Type Design Procedure (CAP) C.P006-01* of March 2008.

9.2.2. FAA ADs

FAR 39 provides a legal framework for FAA's system of ADs.

The FAA issues three types of ADs:

(1) Notice of Proposed Rulemaking (NPRM): A standard AD process is to issue an NPRM followed by a Final Rule. After an unsafe condition is discovered, a proposed solution is published as an NPRM, which solicits public comment on the proposed action. After the comment period closes, the final rule is prepared, taking into account all the comments received, with the rule perhaps being changed as warranted by the comments. The preamble to the Final Rule AD will state if no changes were made or if there were no comments received.

(2) Final Rule with request for comments: In certain cases, the critical nature of an unsafe condition may warrant the immediate adoption of a rule without prior notice and solicitation of comments. This is an exception to the standard process. If the time by which the terminating action must be accomplished is too short to allow a public comment (i.e., less than 60 days) then finding of impracticability is justified for the terminating of

the action, and this can be issued as an immediately adopted rule. The immediately adopted rule is then published in the Federal Register with a request for comments. The Final Rule AD may be changed later if substantive comments are received.

(3) Emergency ADs: An Emergency AD is issued when an unsafe condition exists that requires immediate action by an owner/operator. The intent of an Emergency AD is to rapidly correct an urgent safety of flight situation. An Emergency AD may be distributed by fax, letter, or other methods.

An AD is considered to be no longer in effect when it is superseded by a new AD. The superseding AD identifies the AD that is no longer in effect. There are no compliance requirements for an AD that has been superseded.

9.3. OLDER AIRCRAFT

Older aircraft are also known as "aging aircraft."

Aircraft are designed and built to provide for many years of service. If an aircraft is to remain airworthy and safe operating throughout a long in-service life, it must be operated in accordance with the recommendations of the manufacturer and cared for with sound inspection and maintenance practices.

We will ultimately consider transport aeroplanes, on the whole the most long-lived aircraft.[13]

Service experience has revealed that aging aeroplanes need more care and special attention during the maintenance processes and, at times, more frequent inspection of structural components is required for damage due to environmental deterioration, accidental damage, and fatigue. Hence, manufacturers have to provide operators with programs of continued airworthiness in which virtually every component of an aeroplane is involved in some form of preservation, inspection, maintenance, preventive maintenance, overhaul, repair, and/or replacement activity.

Safety of operation through continued airworthiness demands increasing vigilance as an aeroplane ages.

Maintenance information needs to be continually updated. Open communication should exist between the **owner/operator**, who should notify the **manufacturer** as soon as a new situation arises, and the **authority**. Such communication and cooperation will facilitate the maintenance of an entire fleet in a constant airworthy condition.

Thus, the manufacturer should prepare and distribute recommendations on the need for increased inspection vigilance, updating programs of continued airworthiness, while the authority will review and approve such programs eventually issuing ADs to enforce them.

It is also important to consider the possibility that the aeroplane could be used in a manner significantly different from the original intended mission

---

[13] It is not rare to see aeroplanes having totaled 80,000–100,000 flight cycles.

profile. Low-altitude operation, such as pipeline patrol and training operations, will subject the airplanes to more fatigue damage than high-altitude cruise. Furthermore, airplane operations on distances shorter than those immediately foreseen by the manufacturer lead to an increase in the cycle/flight hours rate, with consequent alteration of the structure fatigue life.

9.3.1. FAA GM

Background. To address aging aircraft concerns, in October 1991, the US Congress enacted Title IV of Public Law 102-143, known as the "Aging Aircraft Safety Act of 1991." The law instructed the Administrator to prescribe regulations that would ensure the continuing airworthiness of aging aircraft. The law also instructed the Administrator to conduct inspections and review the maintenance and other records of each aircraft, which an air carrier uses to provide air transportation. These inspections and record reviews were intended to enable the Administrator to decide whether aging aircraft are in a safe condition and properly maintained for air transportation operation. The law also required the Administrator to establish procedures to be followed to perform such inspections.

In addition to imposing obligations on the Administrator, the law stated that air carriers must demonstrate that the maintenance of their aircraft's age-sensitive parts and components has been adequate and well timed, and operators must make their aircraft and books available for inspection.

As a result of these statutory requirements, the FAA published a final rule titled "Aging Airplane Safety" that specifies mandatory aging aircraft inspections for certain airplanes according to their time-in-service, as well as requirements for damage-tolerance-based inspections and procedures to be included in the maintenance or inspection programs of certain airplanes. The rule also prohibits operation of those airplanes after specified deadlines unless damage-tolerance-based inspections and procedures are included in the maintenance or inspection programs under which the airplanes are maintained. This requirement was implemented to ensure the continuing airworthiness of aging airplanes operated in air transportation by assessing the damage tolerance of older airplane structures.

The Aging Airplane Safety rule requires all airplanes operated under FAR 121 of all US-registered multiengine airplanes operated under FAR 129, and all multiengine airplanes used in scheduled operations under FAR 135, to undergo record reviews and inspections by the Administrator after their 14th year in service and to ensure that the maintenance of their age-sensitive parts and components has been adequate and well timed.

Subsequently, the FAA issued AC 91-56A, "Continuing Structural Integrity Program for Large Transport Category Airplanes" applicable to aeroplanes that have a gross weight more than 75,000 lb and certificated under fail-safe and fatigue requirements *prior to Amendment 25-45 of FAR 25.*

This AC provides GM to manufacturers and operators of transport category airplanes for use in developing a continuing structural integrity program to ensure safe operation of older airplanes throughout their operational life.

The procedures set forth by this AC are applicable to the Large Transport category airplanes operated under Subpart D of FAR 91 and FAR 121 and 125.

Actually, an interesting aspect of these procedures is that aeroplanes certificated before the 1970s had to comply with less-stringent fatigue requirements[14] than those contained nowadays in JAR/FAR 25/EASA CS-25 and relevant advisory material of ACJs and ACs.

This AC, besides the usual recommendation for an exchange of field service information among operators, manufacturers, and the FAA, deals with the development of a **Supplemental Structural Inspection Program** (SSIP) to be implemented before analysis, tests, and/or service experience, indicating that a significant increase in inspection and/or modification is necessary to maintain the structural integrity of the aeroplane. In the absence of other data as a guideline, the program should be initiated not later than when the high-time or high-cycle aeroplane in the fleet reaches one-half its design service goal.

Then, a **Supplemental Inspection Document (SID)** should be developed for FAA review and approval. The manufacturer should revise the SID whenever additional information shows a specific need for it.

The program of the SID for the structures to be evaluated, the type of damage considered (fatigue, corrosion, service, and production damage), and the inspection and/or modification criteria should, to the extent practicable, be in accordance with the damage-tolerance principles of the current FAR 25 standards.

The above-mentioned AC provides guidelines for development of the SID.

The AC 91-56A is now superseded by the AC 91-56 B that applies to design approval holders and operators of transport category airplanes. This AC may also be used by design approval holders and operators of normal, acrobatic, utility, and commuter category airplanes. This guidance may be useful for design approval holders that choose to certificate a small airplane according to the damage tolerance requirements of FAR 23 and also for small airplane design approval holders and operators who choose to develop a structural integrity program as a nonmandatory operational safeguard against the effects of structural aging.

As already mentioned, the previous version of this AC (AC 91-56A) provided guidance to support the development of a damage-tolerance-based SSIP for large-transport airplanes certified under the fail-safe and fatigue requirements of Civil Air Regulations 4b or Part 25, prior to Amendment 25−45. This guidance was traditionally applied to large-transport airplane models such as Airbus Model A300; British Aerospace Model BAC 1-11; Boeing Models B-707/720, B-727, B-737, B-747 and so on. For these models, ADs were issued to mandate the implementation of damage-tolerance-based SSIPs. These airplanes have a maximum takeoff gross weight of more than 75,000 pounds.

In addition to these airplane models, the guidance in the previous version of this AC (which is included in this revision) has been successfully used to

---

[14] See Chapter 4, "Fatigue strength."

develop an SSIP for airplanes with a maximum takeoff gross weight of less than 75,000 pounds. As this guidance was determined to be applicable to smaller airplane models, the term "Large Transport Category" was removed from the title of this revision to the AC.

Besides SSIPs, this AC discusses the following additional elements of a continuing structural integrity program:

- Repairs, Alterations, and Modifications
- Mandatory Modification Program
- Corrosion Prevention and Control Program
- Repair Assessment Program

Additional background information can be found in Appendix 4 of AC 120-93, *Damage Tolerance Inspections for Repairs and Alterations.*

Also, this AC provides guidelines for the development of the SID.

9.3.1.1. PARTS AND MATERIALS SUBSTITUTION FOR VINTAGE AIRPLANES

Today, vintage airplanes need safety enhancing upgrades and modifications to maintain the continued airworthiness of the aircraft. These same vintage airplanes often have little of the required data needed to get FAA approval of such modifications.

Many vintage airplanes no longer have factory support for replacement parts. An approved duplicate replacement part or the data describing the original part (form, fit, and function) are difficult to find or no longer exist. Additionally, some of the materials used today for hoses and fabrics are better than those used when the vintage airplanes were originally built.

Confusion exists about what makes a "correct" replacement part. Usually, there is little or no documentation regarding the replacement part's suitability for installation on a specific airplane model. This lack of information makes it difficult to approve many substitutions.

The AC No. 23—27 provides guidance for substantiating parts or materials substitutions to maintain the safety of old or out of production GA airplanes or other GA aircraft in which the parts or materials are either difficult or impossible to obtain. The AC does not include specific approval for installation, but provides guidelines to follow when collecting information needed for an FAA approval.

9.3.2. JAA GM

A JAA Administrative GM on "Continued Airworthiness of Aging Aircraft Structures" was issued on 6 December 2002. This document provides guidelines for JAA Member States recommending a common approach for continued airworthiness of aging aircraft structures in advance of rule publication to ensure a safe operation of older aeroplanes throughout their operational life. The document has been drafted by a JAA Study Group (European Aging Aircraft Working Group—EAAWG), and reviews the existing published material cooperating with the FAA, with the aim to ensure a consistent transatlantic approach for all transport category aircraft.

Subsequently, the JAA issued NPA 20-10. This NPA is based on the technical agreement reached by EAAWG. The related issues of the FAA (draft) NPRMs were last revised at various dates between 1999 and 2002.

The proposals contained in this NPA are intended to achieve a common approach to the continued airworthiness of (aging) aircraft structure requirements of JAR and FAR to maintain the safety provided by the regulations, without reducing it below a level that is acceptable to both authorities and industry.

The harmonization of JAR 25 and FAR 25 and the adoption of a common approach to operational maintenance would generate cost savings by minimizing any duplication of certification and maintenance activities.

9.3.3. EASA GM

AMC 20-20 CONTINUING STRUCTURAL INTEGRITY PROGRAM
(EFFECTIVE: 26 DECEMBER 2007)

This AMC provides guidance to type-certificate holders, STC holders, repair approval holders, maintenance organizations, operators, and competent authorities in developing a continuing structural integrity program to ensure safe operation of aging aircraft throughout their operational life, including provision to preclude Widespread Fatigue Damage.

The AMC is primarily aimed at large aeroplanes that are operated in Commercial Air Transport or are maintained under Part M. However, this material is also applicable to other aircraft types.

9.4. EXTENDED OPERATIONS

We have previously described additional airworthiness requirements for operation, that is, requirements to be complied with to obtain a certificate of airworthiness allowing certain kinds of operation. Very often these requirements are likely to alter a type design after the type certification.

We now consider the case in which this is likely to be taken into account from the beginning of the design, because the aeroplane is designed for that particular type of operation.

It is quite normal to notice different types of airplane with more than two engines, but also twin-engine aeroplanes in fleets used for long-range operations such as crossing the Atlantic or Pacific Oceans.

At present, one of the most tenacious competitions between Airbus and Boeing is about the new generation of long-range, twin-engine aeroplanes, the A350 and the B787.

Two or more engines are statistically much better than one, but what about emergencies during long-range operations?

This question is at the foundation of the requirements for "extended operations" (ETOPS).

As mentioned in the previous chapters, for the main purpose of this book we will limit our considerations to the basic concepts.

9.4.1. FAR 121 Extended Operations (ETOPS and Polar Operations)

9.4.1.1. GENERAL

We will first consider Subparagraph (a) of FAR 121.161.

121.161 AIRPLANE LIMITATIONS: TYPE OF ROUTE

(a) *Except as provided in Paragraph (e)*[15] *of this section, **unless approved** by the Administrator in accordance with Appendix P of this part and authorized in the certificate holder's OpSpecs, no certificate holder may operate a turbine-engine-powered airplane over a route that contains a point—*

    (1) *Farther than a flying time from an Adequate Airport*[16] *(at a one-engine-inoperative cruise speed under standard conditions in still air) of **60 minutes for a two-engine airplane** or **180 minutes for a** passenger-carrying **airplane with more than two engines;***

    (2) *Within the North Polar Area; or*

    (3) *Within the South Polar Area.*

Extended operations

Since 1985, the acronym, ETOPS, has been defined as "extended twin-engine operations" and has been limited to FAR 121 airplanes with only two engines. Current regulations have extended these applications to all passenger-carrying airplanes operating in both FAR 121 and 135, and the acronym has now been redefined to mean "**extended operations.**" This is to acknowledge the similarity of certain long-range passenger-carrying operations of all airplanes operating today, and the common issues that impact such operations.

The advisory circular **AC 120-42 B** provides certificate holders with guidance for obtaining operational approval to conduct ETOPS **under FAR 121.161**. The FAA may authorize ETOPS with **two-engine** airplanes over a route that contains a point **farther than 60** minutes flying time from an adequate airport at **an approved one-engine inoperative** cruise speed under standard conditions in still air.

The FAA may also authorize ETOPS with passenger-carrying airplanes **with more than two engines** over a route that contains a point **farther than 180** minutes flying time from an adequate airport at **an approved one-engine inoperative** cruise speed under standard conditions in still air.

The AC provides guidance for obtaining authorization to conduct operations under part 121 in **Polar Areas** as well.

This AC 120-42 B is a very complex document of which we will report some significant concepts.

---

[15] Temporary provision expired on February 2008.

[16] "Adequate Airport": see definitions in FAR 121.7 and Appendix 1 of AC 120-42 B.

9.4.1.2. APPLICABLE REGULATIONS

All two-engine airplanes and three- and four-engine passenger-carrying airplanes **operated under FAR 121** are required to comply with **121.161**.

To conduct ETOPS, the specified airplane-engine combination must be certificated to the airworthiness standards of transport-category airplanes and be approved for ETOPS.

Airplane certification guidance for ETOPS can be found in the following.

FAR 21: 21.4.

FAR 25: 25.3, 25.1535, and Appendix K.

FAR 121: 121.7, 121.97, 121.99, 121.106, 121.135, **121.161**, 121.162, 121.191, 121.197, 121.374, 121.410, 121.415, 121.565, 121.624, 121.625, 121.631, 121.633, 121.646, 121.687, 121.689, 121.703, 121.704, 121.705, and Appendix P.

FAR 33: 71, 201, and Appendix A.

9.4.1.3. BACKGROUND ON ETOPS

AC 120-42 in 1985, and AC 120-42A in 1988, recognized the increasing reliability of turbojet engines and helped to establish type design and operational practices for safe and reliable long-range operations with two-engine airplanes. As the technology and reliability of two-engine airplanes continued to improve, due in large measure to the requirements of these documents, such operations became compatible with those long-range operations typically associated with three- and four-engine airplanes. At the same time, this technology brought two-engine airplanes to the arena of long-range operations, the infrastructure to support such operations was changing. Political and funding priorities forced the closure or reduction in basic services of a number of airports, military, and civilian in remote areas that historically had been used as diversion airports for routes over oceanic and/or desolate land areas. The increasing use of polar flights, while creating economic benefits, has also brought new challenges to the operation. The risks associated with these areas' remoteness, harsh climate and terrain, and their unique operational issues, needed to be addressed to maintain an equivalent level of safety in the operation.

These issues began to significantly impact the viability of all long-range two-engine airplane operations under current regulations, and likewise began to erode the basic safety net that long-range operations in three- and four-engine airplanes had relied on. Because of these pressures and the increasing commonality of all long-range operations, the data began to show that ETOPS requirements and processes are generally **applicable to all long-range passenger-carrying operations**, including those by three- and four-engine airplanes, and would improve the safety and viability of such operations. All long-range passenger-carrying airplanes, regardless of the number of engines, needed a viable diversion airport in the case of onboard fire, medical emergency, or catastrophic decompression. Ensuring adequate fire-fighting coverage at these airports, and fuel planning to account for depressurization, is a sound operational practice for all airplanes, including three- and four-engine airplanes. Likewise, planning for the maximum allowable diversion and worst-case scenarios should account for all airplane time-critical systems.

Unlike the ETOPS guidance provided for two-engine airplanes, there has been no regulatory framework governing the long-range operations of three- and four-engine airplanes.

Consequently, the FAA has found that there is a need for all passenger-carrying operations beyond 180 minutes from an adequate airport to adopt many of the ETOPS requirements that have been based on sound safety principles and successfully proven over many years of operations. Accordingly, the FAA revised the Paragraph 121.161 to include passenger-carrying airplanes with more than two engines in these long-range operations.

9.4.1.3.1. Preclude and protect

The whole premise of ETOPS has been to preclude a diversion and, if it were to occur, to have programs in place to protect the diversion. Under this concept, propulsion systems are designed and tested to ensure an acceptable level of in-flight shutdowns (IFSD), and other airplane systems are designed and tested to ensure their reliability. Two-engine airplane maintenance practices are enhanced to better maintain and monitor the condition of the engines and systems significant to ETOPS. The design of these enhanced practices has been a major factor in the joint development of the FAA's and industry's aggressive steps to develop a foundation to resolve problems with airplane systems and engines to minimize the potential for procedural and human errors, thereby precluding a diversion.

However, despite the best design, testing, and maintenance practices, situations occur that may require an airplane to divert. Regardless of whether the diversion is for technical (airplane system- or engine-related) or nontechnical reasons, the certificate holder must have a flight operations plan to protect that diversion. For example, such a plan must include ensuring that pilots are knowledgeable about diversion airport alternates and weather conditions (FAR 121.631), have the ability to communicate with the certificate holder's dispatch office and air traffic control (FAR 121.99 and 121.122), and have sufficient fuel to divert to the alternate (FAR 121.646). Under the "**preclude and protect**" concept, various failure scenarios need to be considered. For example, during the design of the airplane, time-limited systems such as cargo compartment fire suppression/containment capability are considered. Fuel planning must account for the possibility of decompression or the failure of an engine with considerations for in-flight icing conditions. Best options under these scenarios should be provided to the pilot before and during the flight.

9.4.1.3.2. ETOPS Areas of Operation

ETOPS areas of operation are defined in FAR 121.7 to be areas beyond a certain distance from adequate airports measured by an airplane's one-engine inoperative cruise speed under standard conditions in still air. Because of the impact such distances might have on the diversion time of an airplane, regulatory guidance has been established for the planning, operational, and equipage requirements for such operations. A certificate holder must apply to the FAA for approval to operate in an ETOPS area using the methodologies in this AC or other means approved by the FAA. When approval is granted, the ETOPS authority for a specific ETOPS area of operations will be noted in the certificate holder's OpSpecs.

NOTE: The AC provides plenty of instructions on this subject.

9.4.1.3.3. ETOPS in-service experience requirements

When AC 120-42 was first released in 1985, two-engine ETOPS was a new concept and ETOPS approvals were sought on airframe–engine combinations (AEC) that were already in service. Hence, it was logical to establish criteria for approvals based on **in-service experience**. At that same time, the FAA recognized the possibility that other approval methods could be developed without in-service experience, and accordingly, provided statements that recognized those options.

The basic two-engine in-service requirements have been retained and are discussed in **Appendix 3** of the present AC. Achieving these levels of experience, combined with the required levels of engine reliability, **is an acceptable means** of attaining ETOPS approval for operators of two-engine airplanes.

At the time AC 120-42A was drafted, the FAA recognized that a reduction of two-engine in-service experience requirements or substitution of in-service experience on another airplane would be possible. Any reduction was to be based on an evaluation of the certificate holder's ability and competence to achieve the necessary reliability for the particular AEC in ETOPS. For example, a reduction in in-service experience would be considered for a certificate holder who could show extensive in-service experience with a related engine on another airplane that had achieved acceptable reliability. Eventually, specific GM (AC 120-42A, Appendix 7, Accelerated ETOPS Operational Approval)[17] was developed by the FAA permitting ETOPS without accumulating in-service experience in the airplane-engine combination. Most subsequent ETOPS approvals have been granted under these guidelines and this method is retained in Appendix 3.

9.4.1.3.4. Operational reliability and systems suitability requirements

The safety of long-range operations such as ETOPS depends on the reliability of **all** airplane systems including the propulsion systems. Time-limited systems such as cargo compartment fire suppression/containment capability must be considered (FAR 121.633). The certificate holder must also have an established program that monitors the reliability of systems significant to ETOPS (FAR 121.374).

---

[17] Accelerated ETOPS: see Note 19.

To achieve and maintain the required engine reliability standards, the certificate holder operating a two-engine airplane in ETOPS should assess the proposed maintenance and reliability program's ability to maintain a satisfactory level of airplane system's reliability for the particular airplane—engine combination.

Required ETOPS maintenance practices must also minimize the potential for procedural and human errors that could be detrimental to the safety of the operation. Fuel planning must account for the possibility of a depressurization and/or failure of an engine with considerations for in-flight icing conditions (FAR 121.646).

System failures or malfunctions occurring during extended range operations could affect flight crew member workload and procedures. Although the demands on the flight crew member may increase, a manufacturer applying for ETOPS type-design approval must consider crew workload, operational implications, and the crew's and passengers' physiological needs during continued operation with failure effects for the longest diversion time for which it seeks approval.

The manufacturer must also conduct flight tests to validate the adequacy of the airplane's flying qualities and performance, and the flight crew's ability to safely conduct an ETOPS diversion with expected system failures and malfunctions. An ETOPS operator should carefully consider the possible adverse effects that changes in airplane equipment or operating procedures may have on the original evaluations conducted when the airplane was approved for ETOPS before implementing such changes.

9.4.1.4. REQUIREMENTS FOR ETOPS AUTHORIZATION
9.4.1.4.1. ETOPS requirements
The FAA may approve ETOPS for various areas of operation in accordance with the requirements and limitations specified in FAR 121, Appendix P ETOPS must be authorized in the certificate holder's OpSpecs and conducted in compliance with those sections of FAR 121 applicable to ETOPS.

9.4.1.4.2. Maintenance requirements for two-engine ETOPS authorization
The certificate holder conducting ETOPS with two-engine airplanes must comply with the ETOPS maintenance requirements as specified in FAR 121.374.

The basic maintenance program for the airplane being considered for ETOPS is a **CAMP** (Continuous Airworthiness Maintenance Program) that may currently be approved for a non-ETOPS certificate holder for a particular make and model airplane—engine combination. The basic CAMP must be a maintenance and an inspection program that contains the ICA (Instruction for Continuous Airworthiness) based on the manufacturer's maintenance program, or those contained in a certificate holder's maintenance manual approved in its OpSpecs. The certificate holder must review the CAMP to ensure that it provides an adequate basis for development of an ETOPS maintenance program. The certificate holder's ETOPS CAMP must include specific ETOPS requirements that will be incorporated as supplemental requirements to the basic CAMP. These supplemental requirements include the enhanced maintenance and training processes that will ensure that ETOPS airplanes achieve and maintain the level of performance and reliability necessary for ETOPS operations.

The certificate holder must develop an **ETOPS Maintenance Document** with clear instructions for the personnel involved in ETOPS.

The certificate holder must also develop an ETOPS **predeparture service check** to verify that the airplane and certain significant items are airworthy and ETOPS capable.

9.4.1.4.3. ETOPS Maintenance Training Requirements

The certificate holder is responsible for ensuring that all maintenance personnel who perform maintenance on its ETOPS airplanes, including repair stations, vendors, and contract maintenance, have received adequate technical training for the specific airplane—engine combination it intends to operate in ETOPS.

9.4.1.4.4. ETOPS Flight Operations Requirements

Airplane Performance Data. The certificate holder may not dispatch an airplane on an ETOPS flight unless it makes performance data available to its flight crew members and dispatchers who support all phases of ETOPS operations, including divert scenarios.

En-Route Airport Information. In accordance with FAR 121.97, the certificate holder must maintain current status information on the operational capabilities of the airports designated for use as ETOPS alternates.

Other Instructions. The AC provides instructions about how to dispatch an aeroplane in ETOPS, Flight Planning Limitation, the characteristics and the minima required for the alternate airports, the necessary fuel supply, the communications, the dispatch/flight release, and so on.

9.4.1.4.5. Flight Operations Training Requirements

The certificate holder-approved training program for ETOPS should prepare flight crew members to evaluate probable propulsion and airframe systems malfunctions and failures for diversion decision making. The goal of this training should be to establish flight crew members competency in dealing with the most probable operating contingencies.

NOTE: The AC provides a list of the specific ETOPS requirements for the training program.

The FAA reviews the Training and the Operating Manuals to verify the adequacy of the information provided by these manuals.

9.4.1.5. APPLICATIONS TO CONDUCT ETOPS
9.4.1.5.1. ETOPS Qualifications

To receive approval to conduct ETOPS, the certificate holder must satisfy the following conditions:

(a) **Airplane.** The specified airplane—engine combination listed in the certificate holder's application must have been certificated to the airworthiness standards of transport category airplanes and must be approved for ETOPS.

(1) **Two Engine.** Airplane—engine combinations already approved for ETOPS under previous FAA guidance can continue to be used in ETOPS operations under FAR 121. No recertification under FAR 25.1535 is required. Two-engine airplanes with existing type certificates on 15 February 2007, may be approved for up to 180 minutes ETOPS without meeting requirements for fuel system pressure and flow, low-fuel alerting, and engine oil-tank design contained in FAR 25.1535.

(2) **More than Two Engines.** Airplanes with more than two engines that are to be used in ETOPS and are manufactured prior to 17 February 2015 may operate in ETOPS without type-design approval under the revised FAR 25.1535. Airplanes with more than two engines manufactured on or after 17 February 2015 must meet the requirements of ETOPS type design.

(b) **Flight Operations and Maintenance Requirements.** The certificate holder must show compliance with the flight operations requirements and the maintenance requirements discussed in this AC.

(c) **Training Requirements.** The certificate holder should show that it has trained its personnel to achieve competency in ETOPS and must show compliance with the flight operations and maintenance training requirements discussed in this AC.

(d) **Requirements for ETOPS Approval.** Before the FAA grants ETOPS operational approval to an applicant for two-engine ETOPS, the certificate holder must be able to demonstrate the ability to achieve and maintain the level of propulsion system reliability that is required for the ETOPS-approved airplane—engine combination to be used (Appendix P to part 121).

The certificate holder must also demonstrate that it can operate the particular airframe and other airplane systems at levels of reliability appropriate for the intended operation. This can be achieved directly by a successful in-service operational history or by successfully validating all the required ETOPS processes according to the Accelerated ETOPS Application Method in Appendix 3 of this AC.

(e) **Accelerated ETOPS Application.** An applicant for an initial operating certificate who is applying for ETOPS authority at entry into service under the Accelerated ETOPS Application method must comply with the same requirements for certificate holders outlined in this AC. It should be understood that validation of an applicant with no previous operational experience should be more robust than would be necessary for a certificate holder with operational experience.

9.4.1.5.2. Application for ETOPS Authorization
9.4.1.5.2.1. Two-Engine Airplanes
(1) **Up to 180-Minute ETOPS.** An applicant requesting ETOPS up to 180 minutes for two-engine operations may select one of the following

two application methods best suited to their proposed operation (see Appendix 3):

(**a**) In-service experience method, or

(**b**) Accelerated ETOPS method.

(**2**) **ETOPS Beyond 180 Minutes, up to and Including 240 Minutes.** The FAA grants approval for ETOPS beyond 180 minutes only to certificate holders with existing 180-minute ETOPS operating authority for the airplane—engine combination to be operated in the application.

(**3**) **ETOPS Beyond 240 Minutes.** This authority is only granted to operators of two-engine airplanes between specific city pairs. The certificate holder must have been operating at 180 minute or greater ETOPS authority for at least 24 consecutive months, of which at least 12 consecutive months must be at 240-minute ETOPS authority with the airplane—engine combination in the application.

9.4.1.5.2.2. Passenger-Carrying Airplanes with More than Two Engines. There are no minimum in-service experience criteria for certificate holders requesting ETOPS beyond 180 minutes for operations with more than two engines. Those applicants will request approval under the accelerated ETOPS method.

9.4.1.5.3. Validation Flight(s)

Prior to granting ETOPS approval to a certificate holder for operation of a specific airplane—engine combination in an authorized area of operation, the FAA will require actual validation flights on proposed routes that the certificate holder intends to operate within the ETOPS area of operations, designated in the operator's approval request.

Depending on the certificate holder's level of experience in conducting ETOPS and the routes intended to be used in operations, the FAA will determine the number of validation flights required, and the manner in which validation flights may be conducted.

9.4.1.6. FAA ETOPS APPROVAL
9.4.1.6.1. ETOPS OpsSpecs

Following the successful completion of the validation flights, the Flight Standards Service, will authorize the issue of the certificate holder OpSpecs for ETOPS operations providing authorizations and limitations covering at least the following:

(**a**) Approved airplane—engine combinations,

(**b**) Current approved CMP (Configuration, Maintenance, and Procedures) standard required for ETOPS, if appropriate,

(**c**) Authorized geographic area(s) of operation,

(**d**) ETOPS area of operation,

(**e**) Airports authorized for use, including alternates and associated instrument approaches and operating minima,

(f) Approved maintenance and reliability program for ETOPS including those items specified in the type-design-approved CMP standard, if appropriate, and

(g) Identification of the airplanes authorized for ETOPS by make, model, serial, and registration number.

9.4.1.6.2. Processes After Receiving ETOPS Authority

The FAA continuously monitors the world fleet average IFSD rate for two-engine ETOPS authorized airplane—engine combinations to ensure that the levels of reliability achieved in ETOPS remain at the required levels, and it will take the appropriate actions in the event that an acceptable level of reliability is not maintained, or critical deficiencies are detected in the type design or in the conduct of ETOPS operations,

9.4.1.7. POLAR OPERATIONS
9.4.1.7.1. Definition

The North Polar Area is defined as the entire area north of latitude 78 degrees North, and the South Polar Area is defined as the entire area south of latitude 60 degrees South.

9.4.1.7.2. Applicability

Any certificate holder operating an airplane whose route contains any point within the North Polar area or South Polar area as defined above, must comply with the requirements of FAR 121, Appendix P, section III.

9.4.1.7.3. Polar Requirements

The certificate holder applying for authority to fly in the Polar Areas must develop plans in preparation for all polar flights in the North and/or South Polar Areas, as appropriate.

The AC documents the added requirements and identifies equipment and airplane configuration requirements in addition to the requirements discussed for ETOPS Authorization.

9.4.1.8. APPENDICES

Appendix 1. Definitions

Appendix 2. ETOPS Approvals

Appendix P to Part 121 permits certificate holders to seek various levels of ETOPS approvals (75, 90, 120 .... minutes). This Appendix summarizes the details for each approval level and is intended to provide further guidance to the requirements in Appendix P to Part 121.

Appendix 3. ETOPS Approval methods

The two different approval methods available for a certificate holder's use are described in this appendix.

(1) In-Service Experience Method (two-engine ETOPS for up to 180-minute ETOPS).

(2) Accelerated ETOPS Method (up to 180-minute ETOPS for two-engine airplanes and for all ETOPS for passenger-carrying airplanes with more than two engines).

9.4.2. FAR 135 ETOPS

The FAA published the ETOPS final rule on 16 January 2007, with a mandatory compliance date of 13 August 2008.

FAR 135.364 now requires that *"After August 13, 2008, no certificate holder may operate an airplane, other than an all-cargo airplane with more than two engines, on a planned route that **exceeds 180 minutes flying time** (at the one-engine-inoperative cruise speed under standard conditions in still air) from an Adequate Airport outside the continental United States unless the operation is approved by the FAA in accordance with **Appendix G** of this part, **Extended Operations (ETOPS)**."*

The FAA issued the **AC 135-42** on June 2008 *"Extended Operations (ETOPS) and Operations in the North Polar Area"* to provide certificate holders with guidance for obtaining operational approval to conduct ETOPS under FAR 135. The FAA may authorize operations over a route that contains a point farther than 180 minutes flying time from an adequate airport at an approved one-engine inoperative cruise speed under standard conditions in still air. This AC also provides guidance for obtaining authorization to conduct operations under FAR 135 in the North Polar Area.

NOTE: The basic criteria of the AC are similar to those already discussed in Paragraph 9.4.1. Then, we will report only some concepts peculiar to FAR 135 ETOPS.

9.4.2.1. BACKGROUND (CHAPTER 2 OF THE AC)
9.4.2.1.1. ETOPS Regulatory Requirements
To conduct ETOPS, the specified airplane—engine combination must be certificated to the airworthiness standards of transport-category airplanes and be approved for ETOPS. However, Appendix G to FAR 135 allows those **airplanes** manufactured prior to 16 February 2015, to be grandfathered from the ETOPS type-certification requirements. In addition, the **certificate holder** must be approved for ETOPS under FAR 135.

9.4.2.1.2. ETOPS Applicability to FAR 135 Long Range Operations
The FAA and industry analysis of the accidents and incidents involving longer range operations conducted in accordance with FAR 135 indicate that they have been conducted for many years with a high degree of safety without regulatory limitations on range. Before 15 February 2007, no additional regulations had been promulgated. In recent years, several manufacturers have produced new airplanes with range capabilities that could take them well beyond 180 minutes from an airport. As a result, these airplane operations are now compatible with those long-range operations typically associated with large three- and four-engine FAR 121 airplanes. Because of their smaller maximum payload and

seating capacity, despite their range capabilities, these airplanes are authorized to operate according to FAR 135.

9.4.2.1.3. ETOPS Areas of Operations

An ETOPS area of operation is an area within the authorized ETOPS maximum diversion time approved for the operations being conducted. For multiengine airplanes operating under Part 135, it is described as an area beyond 180 minutes from an adequate airport, planned to be no more than 240 minutes from an adequate airport, in still air at normal cruise speed with one-engine inoperative. Because of the impact such distances might have on the diversion time of an airplane, regulatory guidance has been established for the planning, operational, and equipage requirements for such operations. A certificate holder must apply to the Administrator for approval to operate in an ETOPS area using the methodologies in this AC and will be granted ETOPS authority for a specific ETOPS area of operations in their OpSpecs.

The certificate holder will typically request a specific ETOPS area of operation based on an analysis of proposed routings and the availability of airports sufficient to support the operational requirements of the ETOPS regulations.

NOTE: Chapter 2 of the AC also provides information about the following:
ETOPS Risk Management and the Level of Safety
ETOPS Reliability and Systems Suitability Requirements
Preclude and Protect
ETOPS Alternate Airport Requirements
ETOPS In-Service Experience
The basic philosophy is that of the FAR 121 ETOPS, which can be also found in the following chapters:
CHAPTER 3. REQUIREMENTS FOR ETOPS AUTHORIZATION
CHAPTER 4. ETOPS FLIGHT PLANNING.
CHAPTER 5. APPLICATIONS TO CONDUCT ETOPS
CHAPTER 6. FAA APPROVAL

9.4.2.2. APPENDICES

Appendix 1. Definitions
Appendix 2. ETOPS Applications Checklists
Appendix 3. Polar Operations under FAR 135

9.4.3. JAR-OPS ETOPS

We will report an extract of the JAR-OPS 1 requirements.

9.4.3.1. JAR-OPS 1.246 EXTENDED-RANGE OPERATIONS WITH TWO-ENGINED AEROPLANES (ETOPS)

(a) An operator shall not conduct operations beyond the threshold distance determined in accordance with JAR-OPS 1.245 unless approved to do so by the authority.

(b) Prior to conducting an ETOPS flight, an operator shall ensure that a suitable ETOPS en route alternate is available, within either the approved diversion time or a diversion time based on the MEL-generated serviceability status of the aeroplane, whichever is shorter [see also JAR-OPS 1.297(d)].

9.4.3.2. JAR-OPS 1.245 MAXIMUM DISTANCE FROM AN ADEQUATE AERODROME FOR TWO-ENGINED AEROPLANES WITHOUT ETOPS APPROVAL

(a) Unless specifically approved by the authority in accordance with JAR-OPS 1.246(a) (ETOPS approval), an operator shall not operate a two-engined aeroplane over a route that contains a point further from an adequate aerodrome. ...

The paragraph prescribes the maximum distances flown in 60/120/180 minutes with one engine inoperative for aeroplanes of various performance classes,[18] maximum weight and number of passengers, as summarized in IEM-OPS 1.245(a). See Fig. 9.1.

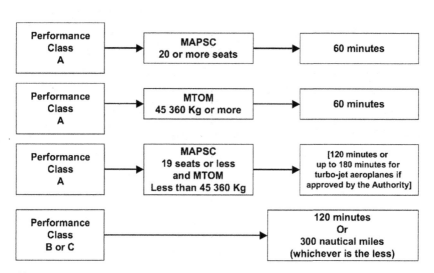

IEM-OPS 1.245(a)
Maximum Distance from an adequate aerodrome
For two-engined aeroplanes without ETOPS Approval
See JAR-OPS 1.245

| Performance Class A | → | MAPSC 20 or more seats | → | 60 minutes |

| Performance Class A | → | MTOM 45 360 Kg or more | → | 60 minutes |

| Performance Class A | → | MAPSC 19 seats or less and MTOM Less than 45 360 Kg | → | [120 minutes or up to 180 minutes for turbo-jet aeroplanes if approved by the Authority] |

| Performance Class B or C | → | | | 120 minutes Or 300 nautical miles (whichever is the less) |

Notes:
MAPSC - Maximum Approved Passenger Seating Configuration
MTOM - Maximum Take-Off Mass

FIGURE 9.1 Summary of IEM-OPS 1.245(a)

---

[18] Performance Class A is defined in JAR-OPS 1 Subpart G; Performance Class B is defined in JAR-OPS 1 Subpart H; Performance Class C is defined in JAR-OPS 1 Subpart I.

9.4.4. EASA ETOPS

OPS 1 embodies JAR-OPS 1; then what reported in Section 9.4.2 is still valid.

9.4.4.1. EASA NPA 2008-01 "EXTENDED RANGE OPERATIONS WITH TWO-ENGINED AEROPLANES ETOPS CERTIFICATION AND OPERATION (AMC 20-6)"

JAA ETOPS/LROPS Ad Hoc Working Group was tasked by the former JAA Regulation Director in 2000 to develop, enhance, and modernize the regulatory material applicable to ETOPS operations. Since then, a considerable amount of work was performed by the JAA ETOPS/LROPS Ad Hoc Working Group. The outcome of this work has been the basis for the current NPA.

The purpose of this NPA, published the 6 March 2008, is to enhance and modernize the airworthiness, continuing airworthiness, and operational considerations for applicants seeking approval for ETOPS of two-engined aeroplanes and in particular it adds additional requirements for applicants seeking approval for diversion time beyond 180 minutes [which is part of extended range operations of two-engined aeroplanes (twins) or ETOPS] at the approved one-engine inoperative speed from an adequate aerodrome. It also introduces new concepts as "early ETOPS" and "accelerated ETOPS."[19]

This NPA does not address the concept of extended range operations for three-engined aeroplanes (tris) and four-engined aeroplanes (quads) (LROPS).

9.5. SAFETY ASSESSMENT OF FOREIGN AIRCRAFT

In this chapter, we have considered a summary of the requirements prescribed by the EASA and FAA for the safety of flight operation, with particular reference to the air-carrier operators.

The rapid growth of civil air traffic all round the world and the proliferation of operators of airlines put the authorities in a position of tightening controls not only on their own operators but also on foreign aircraft operating in their territory.

Under the International Convention on Civil Aviation (Chicago Convention), each country is responsible for the safety oversight of its own air carriers. Other countries can only conduct specific surveillance activities, principally involving inspection of required documents and the physical condition of the aircraft.

Similar initiatives have been taken for long time by the various authorities. To explain this concept, we will report an extract from the "Safety Assessment of Foreign Aircraft (EC SAFA Program)."

9.5.1. Safety Assessment of Foreign Aircraft

This section explains the European Community Safety Assessment of Foreign Aircraft (SAFA) program established by the European Commission (EC) and the role and responsibilities the EASA has in it.

---

[19] "Early ETOPS": when an airliner is delivered with ETOPS on its entry into service. "Accelerated ETOPS": a reduction of in-service experience requirements may be possible when the operator shows to the authority that adequate and validated ETOPS processes are in place.

9.5.1.1. INTERNATIONAL REQUIREMENTS

The international civil aviation is governed by the Convention on International Civil Aviation (commonly known as the Chicago Convention). Under this Convention, the ICAO, a specialized agency of the United Nations, sets the minimum Standards and Recommended Practices for international civil aviation. These standards are contained in 18 Annexes to the Convention. The Individual States remain responsible for regulating their aviation industries but have to take into account the requirements of the Convention and the minimum standards established by the ICAO.

The main standards that apply to airlines are in three of the 18 Annexes.[20]

- Annex 1 deals with personnel licensing including flight crew
- Annex 6 deals with the operation of aircraft
- Annex 8 deals with airworthiness.

The responsibility for implementing Annexes 1 and 8 rests with the State of Registry—that is, the State in which the aircraft is registered. The responsibility for implementing Annex 6 rests with the State of Operator—the State in which the airline is based. Often, the State of Operator and the State of Registry are the same, as airlines tend to operate aircraft registered in the State in which they are based.

Significant increases in the volume of air travel over the last decades or so have made it more of a burden for many States to oversee their airlines in compliance with the Chicago Convention. To maintain confidence in the system, and to protect the interest of the European citizens who may be living in the vicinity of airports or traveling onboard a third-country aircraft, the Community identified the need to effectively enforce international safety standards within the Community. This is done through the execution of ramp inspections on third-country aircraft landing at the airports located in the Member States. The official definition of "third-country aircraft" is an aircraft that is not used or operated under the control of a competent authority of a Community Member State.

The principles of the program are simple: in each EU Member State and those States who have entered into a specific "SAFA" Working Arrangement with EASA,[21] third country aircraft may be inspected. These inspections follow a procedure common to all Member States and are then reported on using a common format. If an inspection identifies significant irregularities, these will be taken up with the airline and the oversight authority. Where irregularities have an immediate effect on safety, inspectors can demand corrective action before they allow the aircraft to leave.

All reported data are stored centrally in a computerized database set up by EASA. The database also holds supplementary information, such as lists of actions carried out following inspections. The information held within this database is reviewed and analyzed by EASA on a regular basis. The EC and

---

[20] See Chapter 3, Section 3.1.1.
[21] There are 40 Member States engaged in the EC SAFA Program.

Member States are informed of any potentially safety hazards identified. On behalf of and in close cooperation with the EC EASA will develop qualitative criteria with the aim to achieve a more focused approach regarding the SAFA inspection priorities. Although there is a legal obligation to perform inspections on third-country aircraft, there is no objection that Member States inspect airlines from other Member States engaged in the EC SAFA Program. It has to be stressed that SAFA inspections are limited to on-the-spot assessments and cannot substitute for proper regulatory oversight. Ramp inspections serve as pointers, but they cannot guarantee the airworthiness of a particular aircraft.

9.5.1.2. AIRCRAFT AND OPERATORS CHECKED

Oversight authorities of the Member States engaged in the EC SAFA Program choose which aircraft to be inspected. Some authorities carry out random inspections while others try to target aircraft or airlines that they suspect may not comply with ICAO standards. In either case, only a very small proportion of foreign aircraft operating into each State are inspected.

Depending on the volume of third-country flights and the availability of inspectors in each Member State, the number of inspections may vary from relatively few to several hundred each year.

Checks may include:

(1) Pilots' licenses
(2) Procedures and manuals that should be carried in the cockpit
(3) Compliance with these procedures by flight and cabin crew safety equipment in cockpit and cabin
(4) Cargo carried in the aircraft
(5) The technical condition of the aircraft.

A checklist of 54 inspection items is used during an SAFA Ramp Check. As the time between arrival and departure (the turnaround time) may not be sufficient to go through the full checklist, only some items may be inspected. It is SAFA policy not to delay an aircraft except for safety reasons.

Since the program began in 1996 as a voluntary ECAC program, the Member States have performed some 37,000 SAFA inspections (status February 2007).

9.5.1.3. RESULTS

Obviously, any major findings will immediately be communicated to all concerned parties. In the case of more serious findings, the oversight authority of the ECAC Member State that performed the ramp check will contact its counterpart in the State responsible for the airline, passing on its findings and asking for any necessary corrective actions. The oversight authority will also inform the aircraft's captain and the headquarters of the airline.

When findings directly affect the safety of the aircraft, its crew and passengers, the Authority of the State of inspection may request immediate corrective action before the aircraft can takeoff. If rectification of the deficiencies requires more time or needs to be performed at another airport, the Authority of the State of inspection may, in coordination with the State responsible for the operation of the aircraft concerned or the State of registration of the aircraft, decide to authorize a positioning flight (a flight to a specific destination without passengers or cargo onboard) and also prescribe the necessary conditions under which the aircraft can be allowed to fly to that specific airport.

In general, all inspection results need to be communicated by the State that performed the inspections to the other EU Member States and to the EC. Whenever an inspection shows the existence of a potential safety threat, or shows that an aircraft does not comply with international safety standards and may pose a potential safety threat, the inspection report will need to be communicated without delay to each EU Member State and the EC. In accordance with Regulation 2111/2005 (establishment of a Community list of air carriers subject to an operating ban within the Community) and based on various other sources of information, the EC may decide on an operating ban in the Community.

9.5.1.4. FURTHER INFORMATION

The Regulation (EC) No. 2111/2005 on "The establishment of a Community list of air carriers subject to an operating ban within the Community and on informing air transport passengers of the identity of the operating air carrier" provides a provision for a decision-making process whereby an airline may be banned from European airspace for safety reasons. Those airlines will then appear on a list, the so-called "Community list."

Having examined the situation in Europe, let us see how this problem is faced on the other side of the Atlantic. For this purpose, we include an extract of the FAA International Aviation Safety Assessment (IASA).

9.5.2. International Aviation Safety Assessment

The US FAA established the IASA program through public policy in August of 1992. The FAA's foreign assessment program focuses *on a country's ability, not the individual air carrier*, to adhere to international standards and recommended practices for aircraft operations and maintenance established by the United Nation's technical agency for aviation, the ICAO.

9.5.2.1. IASA PROGRAM OVERVIEW

In mid-1991, the FAA began to formulate a program to address these concerns. This program included visits to 12 countries with airlines seeking authority to operate to and from the United States. After a trial period, our findings convinced us of the need to formally establish the IASA program. The purpose of the IASA is to ensure that all foreign air carriers that operate to or from the United States are properly licensed and with safety oversight provided by a competent Civil Aviation Authority (CAA) in accordance with ICAO standards.

9.5.2.2. IASA PROCESS OVERVIEW

A foreign air carrier of a sovereign state desiring to conduct foreign air transportation operations into the US files an application with the Department of Transport (DOT) for a foreign air carrier permit under the Federal Aviation Act ....

Consistent with international law, certain safety requirements for operations into the United States are prescribed by FAR 129. FAR 129 specifies that the carrier must meet the safety standards contained in Part 1 (International Commercial Air Transport) of Annex 6 (Operations of Aircraft) to the Convention on International Civil Aviation (Chicago Convention). Before the DOT issues a foreign air carrier permit, it notifies the FAA of the application and requests the FAA's evaluation of the respective CAA's capability for providing safety certification and continuing oversight for its international carriers.

On DOT notification of a pending foreign air carrier application, if the FAA has not made a positive assessment of that country's safety oversight capabilities, the FAA Flight Standards Service will direct its appropriate international field office to schedule an FAA assessment visit to the CAA of the applicant's country.

Once the assessment visits have been completed, the FAA assessment team will return to the United States to compile the findings. Appropriate notifications to the CAA and other US Government officials of the results of the assessments will be made from the Washington, DC, headquarters as soon as possible.

If a CAA is found to be meeting its minimum safety obligations under the Chicago Convention, the FAA will forward a positive recommendation to the DOT. If there is a pending foreign carrier application, the DOT will issue the requested economic authority and the FAA will issue OpSpecs to permit the carrier to begin operations to or from the United States.

When CAAs of countries with existing air carrier service to the United States are found not to meet ICAO standards, the FAA formally requests consultations with the CAA. The purpose of these consultations is to discuss the findings in some detail and explore means to quickly rectify shortcomings found with regard to ICAO annexes, to enable its air carriers to continue service to the United States. During the consultation phase, foreign air carrier operations from that country to the United States will be frozen at existing levels.

The FAA may also intensify its surveillance inspections (ramp checks) on these carriers while they are in the United States. If the deficiencies noted during consultations cannot be successfully corrected within a reasonable period of time, the FAA will notify the DOT that carriers from that country do not have an acceptable level of safety oversight and will recommend that the DOT revoke or suspend their economic operating authority.

After the assessment visit, consultations (if necessary), and notifications are completed, the FAA will publicly release the results of these assessments.

FAA determined that the findings in the IASA program *regarding safety oversight shortcomings must be provided to all US citizens so that they can make informed choices in their international flights.*

The FAA plans to periodically revisit CAAs of countries with air carriers operating in the United States to maintain full familiarity with the methods of those countries' continued compliance with ICAO provisions. The FAA may also find it necessary to reassess a CAA at any time if it has reason to believe that the minimum ICAO standards are not being met.

At present, there are close to 600 foreign air carriers that operate into the United States. There are approximately 103 countries or regional country alliances that have oversight responsibilities for air carriers that either currently operate into the United States, that have air carriers that have applied to operate into the United States, or have a national air carrier that code shares with a US partner air carrier. As of 18 December 2008, the results of 101 completed CAA assessments have been publicly disclosed.

The initial findings have shown that two-thirds of these countries were not fully complying with ICAO standards. Deficiencies found in FAA assessments typically fall into major categories. These categories are almost identical to the deficiencies found by the ICAO in the past. These deficiencies included:

(a) Inadequate and in some cases nonexistent regulatory legislation.
(b) Lack of advisory documentation.
(c) Shortage of experienced airworthiness staff.
(d) Lack of control on important airworthiness-related items such as issuance and enforcement of ADs, minimum equipment lists, investigation of Service Difficulty Reports, and so on.
(e) Lack of adequate technical data.
(f) Absence of AOC systems.
(g) Nonconformance to the requirements of the AOC system.
(h) Lack or shortage of adequately trained flight operations inspectors, including a lack of type ratings.
(i) Lack of updated company manuals for use by airmen.
(j) Inadequate proficiency check procedures.
(k) Inadequately trained cabin attendants.

Some of the same items are also being found on FAA ramp checks of foreign carriers while in this country. This list is long but by no means exhaustive and points out a permanent safety oversight problem that several ICAO Member States need to address within their own CAA. These are also problems that must be corrected before carriers from those CAAs can operate on a regularly scheduled basis to and from the United States.

Desired outcome. The FAA is working to determine that each country meets its obligations under the ICAO and to provide proper oversight to each air carrier operating into the United States. The continued application of this program will result in a lower number of safety-related problems, including accidents, incidents, and an improved level of safety to the public.

9.5.3. General remarks

We have seen two different ways of facing the same problem in Europe and in the United States. The European SAFA is certainly useful, but cannot be the solution to the safety problem. The EASA honestly declares in its explanation that "ramp inspections cannot guarantee the airworthiness of a particular aircraft."

Indeed, try to picture an inspection made at night on an eastern aircraft with documents written in Russian (and tomorrow in Chinese) in a limited space of time because "it is SAFA policy not to delay an aircraft except for safety reasons."

The FAA's approach is (at least philosophically) much more consistent.

At the root of flight, safety is compliance to the "safety minima" known as ICAO Standards and Recommended Practices (SARPs). This compliance has to be demonstrated through the certification and oversight of an airline operator carried out by the state's aviation authority (provided it has the necessary capability and organization).

The capability and organization of this authority is the key to the problem of having air carriers acting according to the rules of the ICAO.

This is why the FAA (also on the basis of the rules of the ICAO) assesses **the authorities' capability** rather than the organization of the air carriers on the basis of a powerful organization that, at least at the time of writing, the EASA/ECAC do not have, although they could rely on a considerable number of national authorities capable of performing something similar to the IASA, if properly coordinated.

The FAA also gives support to countries that are poorly organized from an airworthiness point of view but of course, only for countries with airlines seeking permission to operate to and from the United States.

However, the number of commercial aircraft accidents makes the problem increasingly urgent and concerns not only Europe and the United States but also the entire world.

We must point out that, currently, it is relatively easier—having adequate capital—to set up an airline in any country of the world than to "invent" an effective aviation authority in the same country. Such authorities require several years of grounding and sometimes need external support to reach the necessary expertise; this is something that developing countries in particular are not always capable of achieving. In these cases, it is pointless to blame the authorities of those countries while their aircraft carry on flying around the world.

If the relevant aviation authority is not able to carry out its job, it should delegate its functions to an external competent aviation authority or, according to an ICAO initiative, to a Safety Oversight Group, pooling its resources among groups of nations.

From this perspective, no airline operator should be allowed to operate without certification and oversight made by a competent aviation authority. Of course, that should be coordinated by the ICAO, which would certainly be able to find the appropriate legal enforcement.

9.6. SAFETY MANAGEMENT SYSTEM

NOTE: The following concepts on Safety Management Systems (SMS) are quotes from the (256 pages) ICAO "Safety Management Manual" (Doc. 9859), with the permission of ICAO.

As suggested in other parts of this book, the intent is to provide summary reference and general information on the matter; it is always recommended the good practice of reading the original text in full for practical applications.

9.6.1. General

Everything discussed in this book is based on the concept of "safety," generally defined in Chapter 1 as "absence of danger." We can better explain this concept with the aid of the **ICAO** *Safety Management Manual* (Doc. 9859).

Safety may have different connotations, such as:

zero accidents or serious incidents (view widely held by the traveling public);

freedom from hazards, that is, those factors that cause or are likely to cause harm;

attitudes of employees of aviation organizations toward unsafe acts and conditions;

error avoidance; and

regulatory compliance.

Since the beginning of aviation, the efforts to reduce accident and also incidents[22] have been very effective, but 100 percent safety rate is an unachievable goal. Then, again according to the ICAO manual:

> **Safety** *is the state in which the risk of harm to persons or of property damage is reduced to, and maintained at or below, an acceptable level through a continuing process of hazard identification and* **safety risk management.**

Historically, aviation safety focused on compliance with increasingly complex regulatory requirements. This resulted in the still pervasive notion that safety can be guaranteed as long as rules are followed and that deviation from rules necessarily leads to safety breakdowns.

Without denying the immense importance of regulatory compliance, its limitations as the mainstay of safety have increasingly been recognized, particularly as the complexity of aviation operations has increased.

Accident investigations were to generate safety recommendations aimed at the specific, immediate safety concern identified as causing the safety breakdown, almost exclusively. Little emphasis was placed on the hazardous conditions that, although present, were not *causal* in the occurrence under investigation, even though they held damaging potential for aviation operations under different circumstances.

---

[22] See Note 1 in Chapter 3.

Although this perspective was quite effective in identifying *what* happened, *who* did it, and *when* it happened, it was considerably less effective in disclosing *why* and *how* it happened.

Also the theories on Human Factors could be ineffective without attention to the operational contest in which individuals accomplish their mission. Then, safety must be viewed from a systemic perspective, to encompass organizational, human, and technical factors.

Fundamental is the adoption of a business-like approach to the management of safety, based on the routine collection and analysis of daily operational data. This business-like approach to safety underlies the rationale of **SMS**. In the simplest terms, SMS is the application of business management practices to the management of safety.

Under the ICAO recommendations, Member States should ensure that aircraft operators, aviation maintenance organizations, air traffic services providers, and aerodromes adopt **SMS**.

The purpose of the ICAO manual (Doc. 9859)—a document of 264 pages—is to provide States with:

(a) knowledge of safety management concepts, the ICAO SARPs on safety management contained in Annexes 1, 6, 8, 11, 13, and 14[23] and related GM;

(b) guidance on how to accept and oversee the implementation of the key components of an SMS in compliance with the relevant ICAO SARPs; and

(c) guidance on how to develop and implement a State Safety Program (SSP) in compliance with the relevant ICAO SARPs.

The State, as the signatory to the Chicago Convention, is responsible for implementation of ICAO SARPs affecting flight operations, airspace and navigation services, and aerodromes for which it has responsibility. Generally, these responsibilities include both regulatory functions (licensing, certification, etc.) and safety oversight functions to ensure compliance with regulatory requirements.

Each State must make provisions for the safety of the aviation system within its jurisdiction.

However, each State is one component of the larger global aviation system. In that sense, States also have a responsibility for meeting the requirements of the larger international system.

Annexes 1, 6, 8, 11, 13, and 14 include the requirement for States to establish an SSP, to achieve an acceptable level of safety in civil aviation. An SSP is a management system for the management of safety by the State.

An SSP is defined as an integrated set of regulations and activities aimed at improving safety. It includes specific safety activities that must be performed by the State, and regulations and directives promulgated by the State to support fulfillment of its responsibilities concerning safe and efficient delivery of aviation activities in the State.

---

[23] See Chapter 3 "The International Standards."

A clear understanding of the relationship between an SSP and an SMS is essential for concerted safety management action within States. This relationship can be expressed in the simplest terms as follows:

States are responsible for developing and establishing an SSP; service providers are responsible for developing and establishing an SMS. This is a very important point: States are not expected to develop an SMS; rather, the SSP fulfills the equivalent role. Nevertheless, States are responsible, as part of the activities of their SSP, to accept and oversee the development, implementation, and operational performance of the service provider's SMS.

9.6.2. Understanding safety

As already mentioned, safety is a condition *in which the risk of harm or damage is limited to an acceptable level.* The safety hazards creating risk may become evident after an obvious breach of safety, such as an accident or incident, or they may be *proactively identified* through formal safety management programs *before* an actual safety event occurs. Having identified a safety hazard, the associated risks must be assessed. With a clear understanding of the nature of the risks, a determination can be made as to the *acceptability* of the risks. Those found to be unacceptable must be acted upon.

Safety management is centered on such *a systematic approach to hazard identification and risk management* in the interests of minimizing the loss of human life, property damage, and financial, environmental, and societal losses.

9.6.2.1. HAZARDS

Hazard identification and safety risk management are the core processes involved in the management of safety and, in particular, SMS.

The difference between traditional system safety and present-day safety management is that, because of its engineering roots, system safety focused mostly on the safety implications of technical aspects and components of the system under consideration, somewhat at the expense of the human component. On the other hand, safety management builds on the dogma of system safety (hazard identification and safety risk management), and expands the field of perspective to include Human Factors and human performance as key safety considerations during system design and operation.

The differentiation between hazards and safety risks is oftentimes a source of difficulty and confusion. To develop safety management practices that are relevant and effective, a clear understanding of what is a **hazard** and what is a **safety risk** is essential. A clear understanding of the difference between these two components is also paramount for the practice of safety management.

A hazard is defined as a condition or an object with the **potential** to cause injuries to personnel, damage to equipment or structures, loss of material, or reduction of ability to perform a prescribed function.

Hazards are not necessarily damaging or negative components of a system. It is only **when** hazards interface with the operations of the system aimed at service delivery that their damaging potential may become a safety concern.

Let us make a simple example: wind is a hazard. It is a condition with the potential to cause injuries to personnel, damage to equipment or structures, loss of material, or reduction of ability to perform a prescribed function. A 20-knot wind, by itself, does not necessarily hold potential for damage during aviation operations. In fact, a 20-knot wind blowing directly down the runway will contribute to improving aircraft performance during departure. However, when a 20-knot wind blows in a direction 90 degrees across a runway of intended takeoff or landing, it becomes a crosswind. It is only then, when the hazard interfaces with the operations of the system (takeoff or landing of an aeroplane) that its potential for damage becomes a safety concern.

The damaging potential of a hazard materializes through one or many consequences. In the example of the crosswind above, one consequence of the hazard *crosswind* could be *loss of lateral control.* A further, more serious consequence could be *runway lateral excursion.* An even more serious consequence could be *damage to landing gear.*

Hazards can be grouped into three generic families: natural hazards, technical hazards, and economic hazards.

Natural hazards are a consequence of the habitat or environment within which operations related to the provision of services take place.

Technical hazards are a result of energy sources (electricity, fuel, hydraulic pressure, pneumatic pressure, and so on) or safety-critical functions (potential for hardware failures, software glitches, warnings, and so on) necessary for operations related to the delivery of services.

Economic hazards are the consequence of the sociopolitical environment within which operations related to the provision of services take place.

Hazards may be identified in the aftermath of actual safety events (accidents or incidents), or they may be identified through proactive and predictive processes aimed at identifying hazards before they precipitate safety events. There are a variety of sources of hazard identification. Some sources are internal to the organization while other sources are external to the organization.

Examples of the internal sources of hazard identification available to an organization include flight data analysis; company voluntary reporting system; safety surveys; safety audits; normal operations monitoring schemes; trend analysis; feedback from training; and investigation and follow-up of incidents.

Examples of external sources of hazard identification available to an organization include accident reports; State mandatory occurrence reporting system; State voluntary reporting system; State oversight audits; and information exchange systems.

9.6.2.2. SAFETY RISK

Safety risk is defined as the assessment, expressed in terms of predicted probability and severity, of the consequences of a hazard, taking as reference the worst foreseeable situation.

Using the example of crosswind discussed above, it can be seen that the proposed definition of safety risk allows one to link safety risks with hazards and consequences, thus closing the loop in the hazard-consequence-safety risk trilogy:

(a) a wind of 20 knots blowing directly across the runway is a hazard;

(b) the potential for a runway lateral excursion because a pilot might not be able to control the aircraft during takeoff or landing is one of the consequences of the hazard; and

(c) the assessment of the consequences of a runway lateral excursion, expressed in terms of probability and severity as an alphanumerical convention, is the **safety risk**.

9.6.2.2.1. Safety risk management

Safety risk management is a generic term that encompasses the assessment and mitigation of the safety risks of the consequences of hazards that threaten the capabilities of an organization, to a level **as low as reasonably practicable** (**ALARP**). The objective of safety risk management is to provide the foundation for a balanced allocation of resources between all assessed safety risks and those safety risks the control and mitigation of which are viable. Safety risk management is therefore a key component of the safety management process.

9.6.2.2.2. Safety risk probability

Safety risk probability is defined as the likelihood that an unsafe event or condition might occur. The definition of the likelihood of a probability can be aided by questions such as:

(a) Is there a history of similar occurrences to the one under consideration or is this an isolated occurrence?

(b) What other equipment or components of the same type might have similar defects?

(c) How many personnel are following, or are subject to, the procedures in question?

(d) What percentage of the time is the suspect equipment or the questionable procedure in use?

(e) To what extent are there organizational, management, or regulatory implications that might reflect larger threats to public safety?

We can classify the Safety risk probability as follows

Frequent	Likely to occur many times (has occurred frequently)	5
Occasional	Likely to occur sometimes (has occurred infrequently)	4
Remote	Unlikely to occur, but possible (has occurred rarely)	3
Improbable	Very unlikely to occur (not known to have occurred)	2
Extremely improbable	Almost inconceivable that the event will occur	1

9.6.2.2.3. Safety risk severity

Once the safety risk of an unsafe event or condition has been assessed in terms of probability, the second step in the process of bringing the safety risks of the consequences of hazards under organizational control is the assessment of the severity defined as the possible consequences of an unsafe event or condition, taking as reference the worst foreseeable situation. The assessment of the severity of the consequences of the hazard if its damaging potential materializes during operations aimed at delivery of services can be assisted by questions such as:

(a) How many lives may be lost (employees, passengers, bystanders, and the general public)?
(b) What is the likely extent of property or financial damage (direct property loss to the operator, damage to aviation infrastructure, third-party collateral damage, financial and economic impact for the State)?
(c) What is the likelihood of environmental impact (spillage of fuel or other hazardous product, and physical disruption of the natural habitat)?
(d) What are the likely political implications and/or media interest?

We can classify the Safety risk severity as follows:

Catastrophic **A**	—Equipment destroyed
	—Multiple deaths
Hazardous **B**	—A large reduction in safety margins, physical distress, or a workload such that the operators cannot be relied on to perform their tasks accurately or completely
	—Serious injury
	—Major equipment damage
Major **C**	—A significant reduction in safety margins, a reduction in the operators to cope with adverse operating conditions as a result of increase in workload, or as a result of conditions impairing their efficiency
	—Serious incident
	—Injury to persons
Minor **D**	—Nuisance
	—Operating limitations
	—Use of emergency procedures
	—Minor incident
Negligible **E**	—Little consequences

9.6.2.2.4. Safety risk tolerability

Once the safety risk of the consequences of an unsafe event or condition has been assessed in terms of probability and severity, the third step in the process is the assessment of the tolerability of the consequences of the hazard.

First, it is necessary to obtain an overall assessment of the safety risk. This is achieved by combining the safety risk probability and safety risk severity tables into a safety risk assessment matrix.

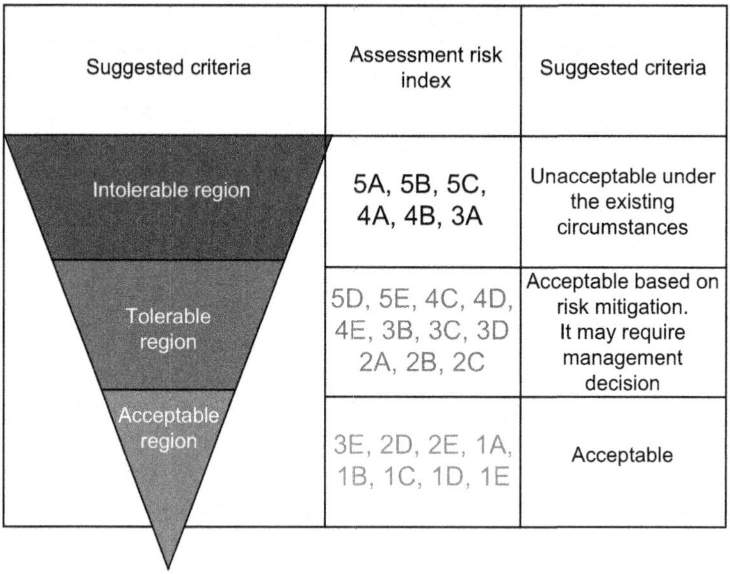

Suggested criteria	Assessment risk index	Suggested criteria
Intolerable region	5A, 5B, 5C, 4A, 4B, 3A	Unacceptable under the existing circumstances
Tolerable region	5D, 5E, 4C, 4D, 4E, 3B, 3C, 3D 2A, 2B, 2C	Acceptable based on risk mitigation. It may require management decision
Acceptable region	3E, 2D, 2E, 1A, 1B, 1C, 1D, 1E	Acceptable

FIGURE 9.2 Safety Risk Tolerability Matrix

For example, a safety risk probability has been assessed as *occasional* **(4)**. The safety risk severity has been assessed as *hazardous* **(B)**. The composite of probability and severity **(4B)** is the safety risk of the consequences of the hazard under consideration.

Second, the safety risk index obtained from the safety risk assessment matrix must then be exported to a *safety risk tolerability matrix* that describes the tolerability criteria. The criterion for a safety risk assessed as 4B is, according to the tolerability table in Figure 9.2, *unacceptable under the existing circumstances*.

In this case, the safety risk falls in the intolerable region of the inverted triangle. The safety risk of the consequences of the hazard is *unacceptable*.

9.6.2.2.5. Safety risk control/mitigation

In the fourth and final step of the process of bringing the safety risks of the consequences of an unsafe event or condition under organizational control, control/mitigation strategies must be deployed. Generally speaking, control and mitigation are terms that can be used interchangeably. Both are meant to designate measures to address the hazard and bring under organizational control the safety risk probability and severity of the consequences of the hazard.

Continuing with the example presented above, the safety risk of the consequences of the hazard under analysis has been assessed as 4B *(unacceptable under the existing circumstances)*. Resources must then be allocated to slide it down the triangle, into the tolerable region, where safety risks are ALARP.

If this cannot be achieved, then the operation aimed at the delivery of services that exposes the organization to the consequences of the hazards in question must be cancelled.

There are three generic strategies for safety risk control/mitigation:

(a) *Avoidance.* The operation or activity is cancelled because safety risks exceed the benefits of continuing the operation or activity. Examples of avoidance strategies include

   (1) operations into an aerodrome surrounded by complex geography and without the necessary aids are cancelled;

(b) *Reduction.* The frequency of the operation or activity is reduced, or action is taken to reduce the magnitude of the consequences of the accepted risks. Examples of reduction strategies include

   (1) operations into an aerodrome surrounded by complex geography and without the necessary aids are limited to daytime, visual conditions;

(c) *Segregation of exposure.* Action is taken to isolate the effects of the consequences of the hazard or build in redundancy to protect against them. Examples of strategies based on segregation of exposure include:

   (1) operations into an aerodrome surrounded by complex geography and without the necessary aids are limited to aircraft with specific performance navigation capabilities.

In evaluating specific alternatives for safety risk mitigation, it must be kept in mind that not all have the same potential for reducing safety risks. The effectiveness of each specific alternative needs to be evaluated before a decision can be taken. Each proposed safety risk mitigation option should be examined from such perspectives as:

(a) *Effectiveness.* Will it reduce or eliminate the safety risks of the consequences of the unsafe event or condition? To what extent do alternatives mitigate such safety risks?

(b) *Cost/benefit.* Do the perceived benefits of the mitigation outweigh the costs? Will the potential gains be proportional to the impact of the change required?

(c) *Practicality.* Is the mitigation practical and appropriate in terms of available technology, financial feasibility administrative feasibility, governing legislation and regulations, political will, and so on?

(d) *Challenge.* Can the mitigation withstand critical scrutiny from all stakeholders (employees, managers, stockholders/State administrations, etc.)?

(e) *Acceptability to each stakeholder.* How much buy-in (or resistance) from stakeholders can be expected? (Discussions with stakeholders during the safety risk assessment phase may indicate their preferred risk mitigation option.)

(f) *Enforceability.* If new rules (Standard Operating Procedures (SOPs), regulations, etc.) are implemented, are they enforceable?

(g) *Durability.* Will the mitigation withstand the test of time? Will it be of temporary benefit or will it have long-term utility?

(h) *Residual safety risks.* After the mitigation has been implemented, what will be the residual safety risks relative to the original hazard? What is the ability to mitigate any residual safety risks?

(i) *New problems.* What new problems or new (perhaps worse) safety risks will be introduced by the proposed mitigation?

Once the mitigation has been accepted, the strategies developed and deployed must, as part of the safety assurance process, be fed back into the organization's defenses, on which the mitigation strategies are based, to ensure integrity, efficiency, and effectiveness of the defenses under the new operational conditions.

9.6.2.2.6. The five fundamentals of safety risk management: summary

The significant concepts regarding safety risk management discussed throughout this chapter can be summarized as follows:

(a) There is no such thing as absolute safety: in aviation, it is not possible to eliminate all safety risks.

(b) Safety risks must be managed to a level ALARP.

(c) Safety risk mitigation must be balanced against:

    (1) time;

    (2) cost; and

    (3) the difficulty of taking measures to reduce or eliminate the safety risk (i.e., managed).

(d) Effective safety risk management seeks to maximize the benefits of accepting a safety risk (most frequently, a reduction in either time and/or cost in the delivery of the service) while minimizing the safety risk itself.

(e) The rationale for safety risk decisions must be communicated to the stakeholders affected by them, to gain their acceptance.

9.6.3. Introduction to Safety Management System

9.6.3.1. SMS AND QMS

Quality management has been established in many segments of the aviation system for a long time. Many aviation organizations have implemented and operated QC and/or QA for a number of years.

A QA program defines and establishes an organization's quality policy and objectives. It ensures that the organization has in place those elements necessary to improve efficiency and reduce service-related risks. If properly implemented, a QA ensures that procedures are carried out consistently and in compliance with applicable requirements, that problems are identified and resolved, and that the organization continuously reviews and improves its procedures, products, and services. QA should identify problems and improve procedures to meet corporate objectives.

The application of QA principles to safety management processes helps ensure that the requisite system-wide safety measures have been taken to support the organization in achieving its safety objectives. However, QA cannot, by itself, as proposed by quality dogma, *assure safety*. It is the integration of QA principles and concepts into an SMS under the safety assurance component that assists an organization in ensuring the necessary standardization of processes to achieve the objective of managing the safety risks of the consequences of the hazards the organization must confront during its activities related to the delivery of services.

QA principles include procedures for monitoring the performance of all aspects of an organization, including such elements as:

(a) design and documentation of procedures (e.g., SOPs);
(b) inspection and testing methods;
(c) monitoring of equipment and operations;
(d) internal and external audits;
(e) monitoring of corrective actions taken; and
(f) use of appropriate statistical analysis, when required.

Aviation organizations have often integrated their QC and QA programs into what is called quality management systems (QMS).

It is accurate to say that SMS and QMS share many commonalities. They both

(a) have to be planned and managed;
(b) depend on measurement and monitoring;
(c) involve every function, process, and person in the organization; and
(d) strive for continuous improvement.

However, in the same way that SMS and QMS share commonalities, there are important differences between both, as well as shortcomings in the effectiveness of QMS to achieve by itself the objective of managing the safety risks of the consequences of the hazards the organization must confront during the activities related to the delivery of services.

Succinctly, then, SMS differs from QMS in that:

(a) SMS focuses on the safety, human, and organizational aspects of an organization (i.e., safety satisfaction); while
(b) QMS focuses on the products and services of an organization (i.e., customer satisfaction).

Once commonalities and differences between SMS and QMS have been established, it is possible to establish a synergistic relationship between both systems. It cannot be stressed strongly enough that the relationship is complementary, never adversarial, and it can be summarized as follows:

(a) SMS builds partly on QMS principles;
(b) SMS should include both safety and quality policies and practices; and
(c) The integration of quality principles, policies, and practices, insofar as SMS is concerned, should be focused toward the support of the management of safety.

The integration of QMS into SMS provides a structured approach to monitor processes and procedures to identify safety hazards and their consequences, and bring the associated safety risks in aviation operations under the control of the organization, function as intended and, when they do not, to improve them.

9.6.4. Conclusions

The ICAO *Safety Management Manual,* from which the synthetic summary of this Section 9.6 has been extracted, provides States with guidance to develop the regulatory framework and the supporting GM for the implementation of safety management systems (SMS) by service providers. It also provides guidance for the development of an SSP, in accordance with the International SARPs contained in Annexes 1, 6, 8, 11, 13, and 14.

It is also worth mentioning the FAA AC No. 120-92 *Introduction to Safety Management Systems for Air Operators.*

This advisory circular introduces the concept of an SMS to aviation service providers (e.g., airlines, air taxi operators, corporate flight departments, and pilot schools) and provides guidance for SMS development.

APPENDIX 9.1.2 EASA CONTINUED AIRWORTHINESS/ MAINTENANCE

Parts	Applicability	NOTES
M	**Continuing airworthiness requirements** 1. The continuing airworthiness of **aircraft** and **components** shall be ensured in accordance with the provisions of **Annex I (Part M)**. 2. Organizations and personnel involved in the continuing airworthiness of aircraft and components, including maintenance, shall comply with the provisions of Annex I and where appropriate those specified in Articles 4 and 5.	**Article 3 of Regulation No. 2042/2003** By derogation from Paragraph 1, the continuing airworthiness of aircraft holding a permit to fly shall be ensured on the basis of the specific continuing airworthiness arrangements as defined in the permit to fly issued in accordance with the Part 21.
145	**Maintenance organization approvals** Organizations involved in the maintenance of **large aircraft** or of **aircraft used for commercial air transport**, and **components** intended for fitment thereto, shall be approved in accordance with the provisions of **Annex II (Part 145)**.	**Article 4 of Regulation No. 2042/2003**

Parts	Applicability	NOTES
66	Certifying staff Certifying staff shall be qualified in accordance with the provisions of Annex III (Part 66).	**Article 5 of Regulation No. 2042/2003** Exception: provision of M.A.607 (b) and M.A.803 of Annex I and in 145.A.30(j) and Appendix IV to Annex II.
147	**Training organization requirements** Organizations involved in the training of personnel referred to in Article 5 shall be approved in accordance with **Annex IV (Part 147)** …	**Article 6 of Regulation No. 2042/2003** Annex IV to be entitled: **(a)** to conduct recognized basic training courses and/or **(b)** to conduct recognized type training courses; **(c)** to conduct examinations; and **(d)** to issue training certificates.

APPENDIX 9.1.5 FAA CONTINUED AIRWORTHINESS/ MAINTENANCE

FARs	Applicability	NOTES
43	**(1)** Aircraft having a US airworthiness certificate **(2)** Foreign-registered civil aircraft used in common carriage or carriage of mail under the provisions of Part 121 or 135 of this chapter, and **(3)** Airframe, aircraft engines, propellers, appliances, and component parts of such aircraft.	**(b)** This part does not apply to any aircraft for which the FAA has issued an Experimental certificate, unless the FAA has previously issued a different kind of airworthiness certificate for that aircraft. **(c)** This part applies to all life-limited parts that are removed from a type-certificated product, segregated, or controlled as provided in Paragraph 43.10. **(d)** This part applies to any aircraft issued a Special airworthiness certificate in the Light-Sport category except: **(1)** The repair or alteration form specified in §§43.5(b) and 43.9(d) is not required to be completed for products not produced under an FAA approval;

FARs	Applicability	NOTES
		(2) Major repairs and major alterations for products not produced under an FAA approval are not required to be recorded in accordance with Appendix B of this part; and **(3)** The listing of major alterations and major repairs specified in Paragraphs (a) and (b) of Appendix A of this part is not applicable to products not produced under an FAA approval.
145	This part describes how to obtain a repair station certificate. This part also contains the rules a certificated repair station must follow related to its performance of maintenance, preventive maintenance, or alterations of an aircraft, airframe, aircraft engine, propeller, appliance, or component part to which Part 43 applies. It also applies to any person who holds, or is required to hold, a repair station certificate issued under this part.	
65	This part prescribes the requirements for issuing the following certificates and associated ratings and the general operating rules for the holders of those certificates and ratings: **(a)** Air traffic control-tower operators. **(b)** Aircraft dispatchers. **(c)** Mechanics. **(d)** Repairmen. **(e)** Parachute riggers.	
147	This part prescribes the requirements for issuing **aviation maintenance technician school certificates** and associated ratings and the general operating rules for the holders of those certificates and ratings.	

FARs	Applicability	NOTES
91	**(a) The Subpart E** prescribes rules governing the maintenance, preventive maintenance, and alterations of US-registered civil aircraft operating within or outside the United States. **(b)** Sections 91.405, 91.409, 91.411, 91.417, and 91.419 of this subpart do not apply to an aircraft maintained in accordance with a continuous airworthiness maintenance program as provided in FAR 121, 129, or Paragraphs 91.1411 or 135.411(a)(2). **(c)** Sections 91.405 and 91.409 of this part do not apply to an airplane inspected in accordance with FAR 125.	
121	**(a)** Except as provided in Paragraph (b) of this section, the **Subpart L** prescribes requirements for maintenance, preventive maintenance, and alterations for all certificate holders.	(b) The Administrator may amend a certificate holder's operations specifications to permit deviation from those provisions of this subpart that would prevent the return to service and use of airframe components, power plants, appliances, and spare parts thereof because those items have been maintained, altered, or inspected by persons employed outside the United States who do not hold US airman certificates. Each certificate holder who uses parts under this deviation must provide for surveillance of facilities and practices to assure that all work performed on these parts is accomplished in accordance with the certificate holder's manual
125	The **Subpart G** prescribes rules, in addition to those prescribed in other parts of this chapter, for the maintenance of airplanes, airframes, aircraft engines, propellers, appliances, each item of survival and emergency equipment, and their component parts operated under this part.	

FARs	Applicability	NOTES
129	The **Subpart B** requires a foreign person or **foreign air carrier** operating a US-registered airplane in **common carriage** to support the continued airworthiness of each airplane. These requirements may include, but are not limited to, revising the maintenance program, incorporating design changes, and incorporating revisions to Instructions for Continued Airworthiness.	
135	The **Subpart J** prescribes rules in addition to those in other parts of this chapter for the maintenance, preventive maintenance, and alterations for each **certificate holder** as follows: **(a)** (1) Aircraft that are type-certificated for a passenger seating configuration, excluding any pilot seat, of nine seats or less, shall be maintained under FAR 91 and 43, and FAR 135.415, 135.416, 135.417, 135.421, and 135.422. An approved aircraft inspection program may be used under FAR 135.419. **(a)** (2) Aircraft that are type-certificated for a passenger seating configuration, excluding any pilot seat, of 10 seats or more, shall be maintained under a maintenance program in FAR 135.415, 135.416, 135.417, and 135.423 to 135.443.	**(b)** A certificate holder who is not otherwise required, may elect to maintain its aircraft under Paragraph (a)(2) of this section. **(c)** Single-engine aircraft used in passenger-carrying IFR operations shall also be maintained in accordance with §135.421 (c), (d), and (e). **(d)** A certificate holder who elects to operate in accordance with §135.364 must maintain its aircraft under Paragraph (a)(2) of this section and the additional requirements of Appendix G of this part.

APPENDIX 9.1.7 FAA CERTIFICATION OF AIR OPERATORS

TYPE OF CERTIFICATE OPERATIONS		OPERATING FAR	KIND OF OPERATION
AIR CARRIER CERTIFICATE **Common Carriage:** • Interstate • Foreign or • Overseas, or • Carriage of mail	Scheduled Operations		
	Example: Multi-engine airplanes with 10 or more passenger seats, OR more than 7500 lb payload capacity	121	Domestic, Flag,
	Example: Airplanes with 9 or fewer passengers AND 7500 lb or less payload capacity, or nay rotorcraft	135	Commuter, On-demand
	Non scheduled Operations		
	Example: Multi-engine aeroplanes with more than 30 passengers OR more than 7500 pounds payload capacity	121	Supplemental
	Example: All cargo operation with aeroplanes with 7500 lb or less payload capacity, or with rotorcraft	135	On-demand
OPERATING CERTIFICATE **Common carriage:** • Intrastate operations	Scheduled Operations		
	Example: Turbojets	121	Domestic,
	Example: Airplanes with 9 or fewer passengers AND 7500 lb or less payload or any rotorcraft	135	Commuter,

TYPE OF CERTIFICATE OPERATIONS		OPERATING FAR	KIND OF OPERATION
	Non Scheduled Operations		
	Example: Multi-engine aeroplanes with more than 30 passengers OR more than 7500 lb payload capacity	121	Supplemental
	Example: Airplanes with 9 or fewer passengers AND 7500 lb or less payload or any rotorcraft	135	On-demand
OPERATING CERTIFICATE **Non common carriage**	Airplanes with 20 or more passengers and 6000 lb or more payload capacity	125	N/A
	Airplanes with less than 20 passengers and less than 6000 lb payload and any rotorcraft	135	On-demand
OPERATOR CERTIFICATE **Agricultural aircraft operations**		137	
Rotorcraft external load operations		133	

Chapter ten

From Airworthiness to "Spaceworthiness"?

We were fascinated by the scene of that shuttle docking to an Earth-orbital space station accompanied by the soft and reassuring music of the "The Blue Danube." It was not the scene of a heroic or extraordinary mission, but of a simple routine trip that would have then continued with a lunar landing craft heading toward a base on the Moon.

Perhaps Stanley Kubrich at the end of the 1960s was a bit optimistic in titling his movie "2001: A Space Odissey." Nevertheless, today the same movie with a different title, "2030: A Space Odyssey" for example, would still be a source of inspiration.

What is happening with space traveling is very similar to what happened at the beginning of the flight era.

In the 1920, the "barnstormers," as they were called, used to entertain people, flying into small towns across the United States, showing their skills and passion, also offering a ride to paying passengers.

People, most of whom had never seen an aeroplane up close, used to pay a small fee in dollars (or sometimes in poultry!) to enjoy the new thrilling experience of flight.

Barnstorming represented the first form of civil aviation in the history of flight and the Federal Government had to create new laws to regulate this new civil aviation business.

Today, there are people willing to pay six-figure number sums in dollars to be able to enjoy a "ride" into space and the new business of "suborbital spaceflights" is becoming a reality.

Another similarity with the beginning of the flight era is represented by the 10 million dollars Ansari X Prize for completing two suborbital spaceflights with at least two passengers in a 2-week period.

This prize was modulated on the 25,000 dollars Orteig Prize that propelled Charles Lindbergh across the Atlantic onboard of his "Spirit of St. Louis" aircraft. The X Prize was won by Burt Rutan Sealed composite.

In suborbital spaceflights, a spacecraft is launched at an altitude of 80−110 km, where the engines are cutoff and then, after a fall wherein the passengers experience a few minutes of absence of weight, it makes a soft **321**

Airworthiness: An Introduction to Aircraft Certification.

landing. A relatively short flight, an amazing view of Planet Earth, an experience to remember forever.

The launches are made directly from the ground or from the air, having been carried at the maximum possible altitude by a specially designed aircraft.

But this is only the first step because there will be increased demand to visit the space station (the current ISS and the future ones) and in the future the bases on the Moon and then......

Furthermore, based on the same principle of suborbital flight, intercontinental transport liners could be realized one day, drastically reducing the traveling time.

Tourism or business? Today on commercial airlines we travel for both reasons. It will be the same for space traveling although somebody prefers to define this as "space exploration."[1]

What we can certainly say is that the civil use of space begins to be a reality and, as it happened in the case of aeronautics, it needs to be regulated.

10.1. THE NEW RULES

In the United States, following the normal rulemaking procedures, the FAA issued a set of requirements on 15 December 2006 titled "**Human Space Flight Requirements for Crew and Space Flight Participants**."

The new rules maintain FAAs commitment to protect the safety of the uninvolved public and call for measures that enable passengers to make informed decisions about their personal safety.

Here is a summary provided by FAA.

The regulations require launch vehicle operators to provide certain safety-related information and identify what an operator must do to conduct a licensed launch with a human on board. In addition, launch operators are required to inform passengers of the risks of space travel generally and the risks of space travel in the operator's vehicle in particular. These regulations also include training and general security requirements for space flight participants.

The regulations also establish requirements for crew notification, medical qualifications and training, as well as requirements governing environmental control and life support systems. They also require a launch vehicle operator to verify the integrated performance of a vehicle's hardware and any software in an operational environment. An operator must successfully verify the integrated performance of a vehicle's hardware and any software in an operational flight environment before allowing any space flight participant on board. Verification must include flight testing.

NOTE: Appendix 10.1 reports the Index of the new requirements.

This new set of requirements, as shown in the Index, represents a significant document of which an outline is given below.

---

[1] Suborbital vehicles could also be used for astronaut training and scientific researches at the limit of the atmosphere.

Subchapter A—General

PART 401—ORGANIZATION AND DEFINITIONS deals with the competent FAA Office of Commercial Space Transportation and with definitions used in the new rules. Here are some useful examples:

Expendable launch vehicle means a launch vehicle whose propulsive stages are flown only once.

Experimental permit or *permit* means an authorization by the FAA to a person to launch or re-enter a reusable suborbital rocket.

Launch means to place or try to place a launch vehicle or re-entry vehicle (RV) and any payload from Earth in a suborbital trajectory, in Earth orbit in outer space, or otherwise in outer space, and includes activities involved in the preparation of a launch vehicle for flight […]

Launch operator means a person who conducts or who will conduct the launch of a launch vehicle and any payload.

Launch vehicle means a vehicle built to operate in, or place a payload in, outer space or a suborbital rocket.

Launch site means the location on Earth from which a launch takes place (as defined in a license the Secretary issues or transfers under this chapter) and necessary facilities at that location.

Re-entry site means the location on Earth where an RV is intended to return. It includes the area within […]

Re-entry vehicle means a vehicle designed to return from Earth orbit or outer space to Earth substantially intact. A reusable launch vehicle (RLV) that is designed to return from Earth orbit or outer space to Earth substantially intact is an RV.

RLV (Reusable Launch Vehicle) means a launch vehicle that is designed to return to Earth substantially intact and therefore may be launched more than one time or that contains vehicle stages that may be recovered by a launch operator for future use in the operation of a substantially similar launch vehicle.

Subchapter B—Procedure

PART 404—REGULATIONS AND LICENSING REQUIREMENT establishes procedures for issuing regulations and for eliminating or waiving requirements for licensing or permitting of commercial space transportation activities.

PART 405—INVESTIGATIONS AND ENFORCEMENT: FAA's monitoring of licensee's facilities and activities; modification, suspension, and revocation of licenses.

PART 406—INVESTIGATIONS, ENFORCEMENT, AND ADMINISTRATIVE REVIEW: a set of rules for hearings and legal actions in case of contentious jurisdiction with the authority.

Subchapter C—Licensing

PART 413—LICENSE APPLICATION PROCEDURES explains how to apply for a license or experimental permit. These procedures apply to all applications for issuing a license or permit, transferring a license, and renewing a license or permit. The following cases are considered:

(**1**) Obtaining a Launch License;

(**2**) License to Operate a Launch Site;

(**3**) Launch and Re-entry of an RLV;

(**4**) License to Operate a Re-entry Site;

(**5**) Re-entry of an RV other than an RLV; and

(**6**) Experimental Permits.

PART 414—SAFETY APPROVALS establishes procedures for obtaining a safety approval and renewing and transferring an existing safety approval.

Safety approval. For purposes of this part, a safety approval is an FAA document containing the FAA determination that one or more of the safety elements listed in Paragraphs (1) and (2) of this definition, when used or employed within a defined envelope, parameter, or situation, will not jeopardize public health and safety or safety of property. A safety approval may be issued independent of a license, [...]

(**1**) Launch vehicle, RV, safety system, process, service, or any identified component thereof; or

(**2**) Qualified and trained personnel, performing a process or function related to licensed launch activities or vehicles.

Safety Element. For purposes of this part, a safety element is any one of the items or persons (personnel) listed in Paragraphs (1) and (2) of the definition of "safety approval" in this section.

Apart from the many details, the application must contain the following technical information:

(**1**) A Statement of Conformance letter, describing the specific criteria the applicant used to show the adequacy of the safety element for which a safety approval is sought, and showing how the safety element complies with the specific criteria.

(**2**) The specific operating limits for which the safety approval is sought.

(**3**) The following as applicable:

    (**i**) Information and analyses required under this chapter that may be applicable to demonstrating safe performance of the safety element for which the safety approval is sought.

    (**ii**) Engineering design and analyses that show the adequacy of the proposed safety element for its intended use, such that the use in a licensed launch or re-entry will not jeopardize public health or safety or the safety of property.

    (**iii**) Relevant manufacturing processes.

    (**iv**) Test and evaluation procedures.

    (**v**) Test results.

    (**vi**) Maintenance procedures.

    (**vii**) Personnel qualifications and training procedures.

PART 415—LAUNCH LICENSE prescribes requirements for obtaining a license to launch a launch vehicle, other than an RLV, and postlicensing requirements with which a licensee must comply to remain licensed.

PART 417—LAUNCH SAFETY
Subpart A—General and License Terms and Conditions
This part sets forth:
(1) The responsibilities of a launch operator conducting a licensed launch of an **expendable** launch vehicle and
(2) The requirements for maintaining a launch license obtained under Part 415 of this chapter.
Subpart B—Launch Safety Responsibilities
This subpart contains public safety requirements that apply to the launch of an orbital or suborbital expendable launch vehicle......
Subpart C—Flight Safety Analysis
This subpart contains requirements for performing the flight safety analysis required by §417.107(f).
Subpart D—Flight Safety System[2]
This subpart applies to any flight safety system that a launch operator uses. The requirements of §417.107(a) define when a launch operator must use a flight safety system. A launch operator must ensure that its flight safety system satisfies all the requirements of this subpart, including the referenced appendices.
Subpart E—Ground Safety
This subpart contains public safety requirements that apply to launch processing and postlaunch operations at a launch site in the United States. Ground safety requirements in this subpart apply to activities performed by, or on behalf of, a launch operator at a launch site in the United States. A licensed launch site operator must satisfy the requirements of Part 420 of this chapter.
NOTE: Part 417 also contains several technical Appendices (from Appendix A to J) with requirements and methodologies to satisfy this part.
PART 420—LICENSE TO OPERATE A LAUNCH SITE: This part prescribes the information and demonstrations that must be provided to the FAA as part of a license application, the bases for license approval, license terms and conditions, and postlicensing requirements with which a licensee shall comply to remain licensed.
PART 431—LAUNCH AND RE-ENTRY OF A REUSABLE LAUNCH VEHICLE (RLV): This part prescribes requirements for obtaining an RLV mission license and postlicensing requirements with which a licensee must comply to remain licensed.
There are two types of RLV mission licenses.
(a) *Mission-specific license.* A mission-specific license authorizing an RLV mission authorizes a licensee to launch and re-enter, or otherwise land, one model or type of RLV from a launch site approved for the mission to a re-entry site or other location approved for the mission. A mission-specific

---

[2] *Flight safety system* means the system that provides a means of control during flight for preventing a hazard from a launch vehicle, including any payload hazard, from reaching any populated or other protected area in the event of a launch vehicle failure.

license authorizing an RLV mission may authorize more than one RLV mission and identifies each flight of an RLV authorized under the license.[...]

(b) *Operator license.* An operator license for RLV missions authorizes a licensee to launch and re-enter, or otherwise land, any of a designated family of RLVs within authorized parameters, including launch sites and trajectories, transporting specified classes of payloads to any re-entry site or other location designated in the license. An operator license for RLV missions is valid for a 2-year renewable term.

We mention in particular:

Subpart C—Safety Review and Approval for Launch and Re-entry of a Reusable Launch Vehicle

The FAA conducts a safety review to determine whether an applicant is capable of launching an RLV and payload, if any, from a designated launch site, and re-entering the RLV and payload, if any, to a designated re-entry site or location, or otherwise landing it on Earth, without jeopardizing public health and safety and the safety of property.

The **AC No. 431.35-2A** provides guidance concerning applying a systematic and logical system safety process for identification, analysis, and control of public safety hazards and risks associated with the operation of RLV and RV systems.

PART 433—LICENSE TO OPERATE A RE-ENTRY SITE: The FAA evaluates on an individual basis an applicant's proposal to operate a re-entry site.

A license to operate a re-entry site authorizes a licensee to operate a re-entry site in accordance with the representations contained in the licensee's application, subject to the licensee's compliance with terms and conditions contained in any license order accompanying the license.

PART 435—RE-ENTRY OF A RE-ENTRY VEHICLE OTHER THAN A REUSABLE LAUNCH VEHICLE (RLV): This part prescribes requirements for obtaining a license to re-enter an RV other than an RLV, and postlicensing requirements with which a licensee must comply to remain licensed.[...]

PART 437—EXPERIMENTAL PERMITS: This part prescribes requirements for obtaining an experimental permit. It also prescribes postpermitting requirements with which a permittee must comply to maintain its permit. Part 413 of this subchapter contains procedures for applying for an experimental permit.

PART 440—FINANCIAL RESPONSIBILITY: This part establishes financial responsibility and allocation of risk requirements for any launch or re-entry authorized by a license or permit issued under this subchapter.

This part, very complex from the legal point of view, contains a *key* paragraph, the Paragraph 440.17, *Reciprocal waiver of claims requirements.*

Trying to make simple the rigorous juridical expression of this complex paragraph, a **reciprocal waiver** of claims for bodily injury, including death, or property damage, regardless of fault, resulting from licensed/permitted activities, must be executed between

- Crew members against United States and its respective Contractors or Subcontractors.[3]
- Spaceflight Participants against United States and its respective Contractors or Subcontractors.
- Licencee/Permittee against Customer[4] and United States and their respective Contractors or Subcontractors (as a three-party reciprocal waiver).

With a broad simplification, each Party shall be responsible for its own damages resulting from licensed/permitted activities.

PART 460—HUMAN SPACEFLIGHT REQUIREMENTS: This is an amendment published on 15 December 2006 inclusive of two Subparts A and B as follow. In synthesis:

Subpart A—Launch and re-entry with crew establishes requirements for crew of a vehicle whose operator is licensed or permitted under this chapter.

- This subpart establishes the qualification and training of the crew members (onboard or remote operator on the ground).
- Paragraph 460.9 states that an operator must inform in writing any individual serving as crew that the United States Government has not certified the launch vehicle and any RV as safe for carrying flight crew or spaceflight participants.
- Paragraphs 460.11 and 13 state the atmospheric conditions that must be maintained within the vehicle and the ability to detect and suppress cabin fire.
- Paragraph 460.17 requires a verification program to ensure the integrated performance of a vehicle's hardware and any software in an operational flight environment before allowing any spaceflight participant onboard during a flight. Verification must include flight testing.

Subpart B—Launch and re-entry with a spaceflight participant establishes requirements for spaceflight participants onboard a vehicle whose operator is licensed or permitted under this chapter.

- Paragraph 460.45 requires that an operator must inform each spaceflight participant in writing about the risks of the launch and re-entry, including the safety record of the launch or RV type, and how this information must be provided.
- Furthermore, an operator must inform each spaceflight participant that the US Government **has not certified the launch vehicle and any RV as safe for carrying crew or spaceflight participants.**

---

[3] *Contractors and subcontractors* means those entities that are involved at any level, directly or indirectly, in licensed or permitted activities, and includes suppliers of property and services, and the component manufacturers of a launch vehicle, RV, or payload.

[4] *Customer* means any person who procures launch or re-entry services from a licensee or permittee; with rights in the payload (or any part of the payload) to be launched or re-entered by the licensee or permittee, including a conditional sale, lease, assignment, or transfer of rights; who has placed property onboard the payload for launch, re-entry, or payload services; or to whom the customer has transferred its rights to the launch or re-entry services.

- Paragraph 460.51 requires an operator to train each spaceflight participant before flight on how to respond to emergency situations, including smoke, fire, loss of cabin pressure, and emergency exit.

10.1.1. The experimental permit

Part 431 prescribes requirements for obtaining an RLV mission license and postlicensing requirements with which a licensee must comply to remain licensed. Part 437 prescribes requirements for obtaining an experimental permit.

Besides a license, an experimental permit is an authorization issued by the FAA to allow an experimental reusable suborbital rocket to launch or re-enter. A permit is an alternative to licensing that is valid for a 1-year renewable term and allows a permittee to conduct an unlimited number of launches and re-entries for a particular suborbital rocket design during that time. The FAA can grant experimental permits more quickly and with fewer requirements than licenses, making it easier for the industry to test new types of reusable suborbital rockets. The scope is to expedite research and development on the vehicles intended to carry passengers on suborbital flights.

Of course, carrying any property or human being for compensation or hire is prohibited under an experimental permit.

The FAA will issue an experimental permit only for research and development to test new reusable suborbital rocket design concepts, new equipment, or new operating techniques, showing compliance with requirements to obtain a license, or crew training before obtaining a license.

As a part of the requirements for obtaining an experimental permit, Paragraph 437.55 requires an operator to perform a hazard analysis and provide the results to the FAA.

The AC No. 437.55-1 provides guidance for applying a systematic and logical hazard analysis to the identification, analysis, and control of public safety hazard and risk associated with the launch and re-entry of a reusable rocket under an experimental permit.

10.1.2. General remarks

The FAA issued regulations establishing requirements for crew and spaceflight participants (passengers) involved in private human spaceflight. The new rules maintain FAA's commitment to protect the safety of the uninvolved public and call for measures that enable passengers to make informed decisions about their personal safety.

These human space requirements face a new type of passenger flight for which there is no consolidated experience. For this reason, the FAA published some regulations that we could define as "preliminary" while we are waiting to define them technically on the basis of future operational experience. We will therefore consider for the time being the bureaucratic aspect of these regulations, considering the licensing process, the allocation of responsibilities, and so on.

It is actually worth remembering, as provided by Part 460, that FAA, on the basis of the "commercial Space Launch Amendments Act" of 2004, **does not certificate** the launch vehicles and any RV as safe for carrying crew and spaceflight participants. Furthermore, an operator must inform each spaceflight participant about the possible hazard and risks[5] in all the flights. These regulations also include training and general security requirements for spaceflight participants.

These hazards have to be identified through an analysis conducted by the operator.[6]

It is then a matter of an "informed consent" regime ("fly at your own risk") requiring the spaceflight participants and other parties to take a personal responsibility for the risks they face, as specified in Part 440.

This is obviously a pragmatic approach. If the several private initiatives in this sector would have had to wait for a complete set of regulations similar to the commercial aviation before starting operations, these new initiatives would have simply been suffocated.

We therefore find here the confirmation of what we quoted in Chapter 2: *"Generally, a standard does not precede aeronautical progress, it follows it and sometimes accompanies it."* In fact, recognizing that this is a new industry, the US law required a phased approach in regulating commercial human spaceflight, with regulatory standards evolving as the industry matures.

We have so far considered what happens in the United States where a new industry is flourishing, strongly supported by the FAA leading the way on the regulations front for space exploration and tourism.

In Europe, many initiatives for space tourism are in progress; US companies would be willing to operate in Europe, but a relevant legislation still does not exist.

It is possible that some national authority will engage in this sort of initiative. However, it has to be seen if Europe will want to follow the United States and give EASA the responsibility for licensing space planes, space tourism businesses, and spaceports, and thus controlling safety issues related to European space tourism operations.

Being at the core of Europe's space activities, the European Space Agency (ESA)[7] issued a position document on Space Tourism, claiming the needs to

---

[5] Part 401 defines *Risk* as a measure that accounts for both the probability of occurrence of a hazardous event and the consequence of that event to persons or property. Generally speaking, *hazard* may be defined as the potential to cause harm, whereas *risk* is the probability and consequence of harm.

[6] A "Guide to Reusable Launch and Reentry Vehicle Reliability Analysis" has been already produced by the FAA (April 2005), a "Guide to Reusable Launch and Reentry Vehicle Software and Computing System Safety" (July 2006), and a "Guide to Reusable Launch and Reentry Vehicle Reliability Analysis Federal (January 2010)."

[7] ESA is an international organization with 17 Member States. ESA's job is to draw up the European space program and carry it through. ESA's programs are designed to find out more about Earth, its immediate space environment, our Solar System, and the Universe, as well as to develop satellite-based technologies and services, and to promote European industries. ESA also works closely with space organizations outside Europe.

have a coordinated and corporate approach with respect to these activities. The document outlines the different aspects linked to Space Tourism, which may have an impact on ESA, and proposes the major features of such a position. In particular, *ESA should contribute in the development of a regulatory frame for Space Tourism in Europe, involving both civil aviation regulatory authorities and competent bodies from the EC, aiming also at a more level playing field for all worldwide players, and supporting the interests of European industry.*

However, the challenge offered by space tourism has global validity. Currently, in fact there are activities of note in Australia, Canada, China, France, Germany, India, Israel, Japan, Russia, Singapore, the United Arab Emirates, the United Kingdom, and the United States and certainly others will follow suit.

This means that an ICAO's commitment on this issue is demanded, as it currently happens for the international civil aviation.

In this sense, the International Association for the Advancement of Space Safety (IAASS)[8] has undertaken a study about the regulation and license at the international level of commercial or private space initiatives and private space stations. This study is being carried out in cooperation with the ICAO, because the legal and regulatory practices and approaches to safety that have been developed for global aviation might be productively applied to private space initiatives as well.

The *IAASS Study: " An ICAO for Space?"* white paper was developed by the "IAASS ICAO for Space?" IIAA Working Group (WG). The contents of this white paper were developed with the intent to generate international discussions on the topic.

A new era has commenced. New regulations will follow its development. There will certainly be an evolution toward civil spacecraft type certification and we will probably read the expression: "*Spaceworthiness* **type certificates.**"

---

[8] The IAASS was formed in 2005, sponsored by the ESA, the National Aviation and Space Administration, the Japanese Aerospace Exploration Agency, the Federal Space Agency of Russia (Roscosmos), the Canadian Space Agency, the French National Center for Space Studies (CNES), the German Space Agency (DLR), and the Italian Space Agency among others and it is dedicated to the idea of creating international standards for space safety.

APPENDIX 10.1 INDEX OF 14 CFR CHAPTER III

14 CFR Chapter III—Commercial Space transportation, Federal Aviation administration, Department of Transportation

List of Acronyms

- **ACARE** Advisory Council for Aeronautical Research in Europe
- **ACO** Aircraft Certification Office
- **AC** Advisory Circular
- **ADOAP** Alternative Procedures to Design Organization Approval
- **AD** Airworthiness Directive
- **AEC** Airframe-Engine Combination
- **AEG** Aircraft Evaluation Group
- **AI** Action Item
- **AFM** Aircraft Flight Manual
- **ALARP** As Low As Reasonably Practicable
- **AMC** Acceptable Means of Compliance
- **A-NPA** Advanced-Notice of Proposed Amendment
- **AOC** Air Operator Certificate
- **APIS** Approved Production Inspection System
- **APU** Auxiliary Power Unit
- **ARC** Airworthiness Review Certificate
- **ARC** Aviation Rulemaking Committee
- **ASTRAEA** Autonomus System Technology Related Airborne Evaluation and Assessment
- **ATC** Air Traffic Control
- **ATCs** Additional Technical Conditions
- **ATM** Air Traffic Management
- **AVS** Air Vehicle Station
- **AWO** All Weather Operation

- **BASA** Bilateral Aviation Safety Agreement

- **CAA** Civil Aeronautics Administration
- **CAA** Civil Aviation Authority
- **CAB** Civil Aeronautics Board
- **CAMO** Continuing Airworthiness Management Organization
- **CAMP** Continuing Airworthiness Maintenance Program
- **CAP** Civil Aviation Publication
- **CAR** Civil Aviation Regulations
- **CASA** Civil Aviation Safety Authority
- **CCL** Compliance Checklist
- **CFR** Code of Federal Regulations
- **CM** Certification Manager

- **C of A**   Certificate of Airworthiness
- **CPI**   Certification Process Improvement
- **CPP**   Certification Program Plan
- **CRD**   Comment Response Document
- **CRI**   Certification Review Item
- **CRSs**   Compliance Record Sheets
- **CRS**   Certificate of Release to Service
- **CS**   Certification Standard
- **CSTA**   Chief Scientist and Technical Adviser
- **CVE**   Certification Verification Engineer

- **DA**   Designated Authority
- **DAR**   Designated Airworthiness Representative
- **DAS**   Design Assurance System
- **DDP**   Declaration of Design and Performance
- **DER**   Designated Engineering Representative
- **DMIR**   Designated Manufacturing Inspection Representative
- **DO**   Design Organization
- **DOA**   Design Organization Approval
- **DOT**   Department of Transportation

- **EAA**   Experimental Aircraft Association
- **EAAWG**   European Aging Aircraft Working Group
- **EASA**   European Aviation Safety Agency
- **ECAC**   European Civil Aviation Conference
- **EC**   European Commission
- **ECO**   Engine Certification Office
- **ELA**   European Light Aircraft
- **ELT**   Emergency Locator Transmitter
- **ENAC**   Ente Nazionale Aviazione Civile
- **EPA**   European Part Approval
- **ESA**   European Space Agency
- **ETOPS**   Extended Range Twin Engine Operations
- **ETSO**   European Technical Standard Order
- **EU**   European Union
- **EUROCAE**   European Organization for Civil Aviation Equipment

- **FAA**   Federal Aviation Administration
- **FAR**   Federal Aviation Regulations
- **FOEB**   Flight Operations Evaluation Board
- **FSB**   Flight Standardization Board
- **FTS**   Flight Termination System
- **FUJA**   Future of JAA

- **GA**   General Aviations
- **GM**   Guidance Material

- **HB**   Hot Balloon

- **IASA** International Aviation Safety Assessment
- **IAASS** International Association for Advancement of Space Safety
- **ICA** Instructions for Continued Airworthiness
- **ICAO** International Civil Aviation Organization
- **IFR** Instrumental Flight Rules
- **IFSD** In-Flight Shutdowns
- **IPA** Implementation Procedures of Airworthiness
- **IRs** Implementing Rules

- **JAAB** JAA Board
- **JAA C** JAA Committee
- **JAA EB** JAA Executive Board
- **JAA FB** JAA Foundation Board
- **JAA** Joint Aviation Authority
- **JAA DOA** JAA Design Organization Approval
- **JAA LO** JAA Liaison Office
- **JAA TO** JAA Training Office
- **JAR** Joint Aviation Requirements
- **JPA** Joint Parts Approval

- **LSA** Light Sport Aircraft
- **LUAS** Light UAS

- **MAC** Manager of Applications Certification
- **MAPSC** Maximum Approved Passenger Seating Configuration
- **MASPS** Minimum Aviation System Performance Standard
- **MEL** Minimum Equipment List
- **MIDO** Manufacturing Inspection District Office
- **MIL- HDBH** Military Handbook
- **MIO** Manufacturing Inspection Office
- **MMEL** Master Minimum Equipment List
- **MOA** Maintenance Organization Approval
- **MoC** Means of Compliance
- **MOE** Maintenance Organization Exposition
- **MRB** Maintenance Review Board
- **MS** Military Standard
- **MSL** Mean See Level
- **MTOA** Maintenance Training Organization Approval
- **MTOM** Maximum Take Off Mass
- **MTOW** Maximum Take Off Weight

- **NAA** National Aviation Authority
- **NAS** National Airspace System
- **NASA** National Aeronautics and Space Administration
- **NATO** North Atlantic Treaty Organization
- **NPA** Notice of Proposed Amendment
- **NPRM** Notice of Proposed Rulemaking

- **NRS** National Resources Specialists
- **NTSB** National Transport Safety Boar

- **ODAR** Organizational Designated Airworthiness Representative
- **OPS** Operations
- **OSTIV** Organisation Scientifique et Technique International du Vol à Voile

- **PC** Production Certificate
- **PCA** Primary Certification Authority
- **PCM** Project Certification Manager
- **PMA** Parts Manufacturer Approval
- **PM** Project Manager
- **POA** Production Organization Approval
- **POC** Production Oversight Coordinator
- **POE** Production Organization Exposition
- **POM** Production Organization Manager
- **PPC** Powered Parachute
- **PSCP** Project −Specific Certification Plan
- **PSP** Partnership for Safety Plan
- **PTVP** Post Type Validation Principles

- **QA** Quality Assurance
- **QC** Quality Control
- **QMS** Quality Management System
- **QSST** Quiet Supersonic Transport

- **RAI** Registro Aeronautico Italiano
- **RLV** Reusable Launch Vehicle
- **ROA** Remotely Operated Aircraft
- **RPV** Remotely Piloted Vehicle
- **RTCA** Radio Technical Commission for Aeronautics
- **RV** Reentry Vehicle

- **SAFA** Safety Assessment of Foreign Aircraft
- **SARPs** Standards and Recommended Practices
- **SID** Supplemental Inspection Document
- **SMS** Safety Management System
- **SOF** Safety of
- **SOPs** Standard Operating Procedures
- **SSBJ** Supersonic Business Jet
- **S-LSA** Special Light Sport Aircraft
- **SSIP** Supplemental Structural Inspection Program
- **SSP** State Safety Program
- **SST** Supersonic Transport
- **STC** Supplemental Type Certificate

- **TCH** Type Certificaate Holder
- **TCB** Type-Certification Board

- **TC**   Type Certificate
- **TCDS**   Type Certification Data Sheets
- **TPV**   Type Validation Principles
- **ToRs**   Terms of reference
- **TSO**   Technical Standard Order

- **UASSG**   UAS Study Group
- **UAS**   Unmanned Aircraft System
- **UAV**   Uninhabitated Aerial Vehicle
- **UAV**   Unmanned Aerial Vehicle

- **VFR**   Fisual Flight Rules
- **VLA**   Very Light Aeroplane
- **VLR**   Very Light Rotorcraft
- **Vso**   Stalling speed (landing configuration)

- **WSC**   Weight Shift Contro

Bibliography

Cardi, A. (1998) Gli allegati Tecnici ICAO in Italia. *Volabilita*, No. 38.
Cardi, A. (2002) *Annessi ICAO*. Universita degli Studi di Modena e Reggio Emilia, 3 May.
De Florio, F. (1980) La regolamentazione per la sicurezza degli aeromobili. *Volabilita*, No. 7.
De Florio, F. (1996) Le nuove frontiere ed il RAI. *Volabilita*, No. 31.
De Florio, F. (1999) Airworthiness for UAV. *UAVS International Conference Proceedings*, Paris.
De Florio, F. (2003) AERONAVIGABILITÀ IBN Editore
Falessi, C. (1997) I sessanta anni del RAI. *Volabilita*, No. 37.
Flight International, Flight Group, Reed Business Information Ltd.
Lloyd, E. and Tye, W. (1982) *Systematic Safety*. Civil Aviation Authority, London, July.
Marasà, B. (1998) JAR-OPS: benvenute in Italia! *Volabilita*, No. 38.
RAI-ENAC, Regolamento Tecnico (Technical Regulations).
Registro Aeronautico Italiano (RAI) Linea Guida MAV, Doc. 25 (1994); RAI-ENAC, 'Circolari' No. 44 (25 October 1996); No. 30B (18 December 1998).
Rich, B.R. and Janos, L. (1994) *Skunk Works*. Warner Books.
ICAO (2009) Safety Management Manual (SMM) (Doc 9859)
European Commission (2009) Hearing on LUAS-Conclusions.
EASA (2009) Type Certification procedures (C.P008-02).
UVS International The '*Global Perspective 2008/2009 and 2009/2010*'

Internet resources

EASA: www.easa.eu.int
FAA: www.faa.gov
ICAO: www.icao.int
JAA: www.jaa.nl

Index